Soil Carbon Storage

Soil Carbon Storage
Modulators, Mechanisms and Modeling

Edited by

Brajesh K. Singh
Western Sydney University, Penrith South, NSW, Australia

Academic Press is an imprint of Elsevier
125 London Wall, London EC2Y 5AS, United Kingdom
525 B Street, Suite 1800, San Diego, CA 92101-4495, United States
50 Hampshire Street, 5th Floor, Cambridge, MA 02139, United States
The Boulevard, Langford Lane, Kidlington, Oxford OX5 1GB, United Kingdom

Notices
Knowledge and best practice in this field are constantly changing. As new research and experience broaden our
understanding, changes in research methods, professional practices, or medical treatment may become necessary.

Practitioners and researchers must always rely on their own experience and knowledge in evaluating and using any
information, methods, compounds, or experiments described herein. In using such information or methods they
should be mindful of their own safety and the safety of others, including parties for whom they have a professional
responsibility.

To the fullest extent of the law, neither the Publisher nor the authors, contributors, or editors, assume any liability for
any injury and/or damage to persons or property as a matter of products liability, negligence or otherwise, or from
any use or operation of any methods, products, instructions, or ideas contained in the material herein.

British Library Cataloguing-in-Publication Data
A catalogue record for this book is available from the British Library

Library of Congress Cataloging-in-Publication Data
A catalog record for this book is available from the Library of Congress

ISBN: 978-0-12-812766-7

For Information on all Academic Press publications
visit our website at https://www.elsevier.com/books-and-journals

Working together
to grow libraries in
developing countries

www.elsevier.com • www.bookaid.org

Publisher: Candice Janco
Acquisition Editor: Candice Janco
Editorial Project Manager: Tasha Frank
Production Project Manager: Vijayaraj Purushothaman
Designer: Mark Rogers

Cover image prepared by Vanesa Gonzalez-Quiñones and Ghost Media Australia

Typeset by MPS Limited, Chennai, India

Contents

CHAPTER 6 **Soil Nutrients and Soil Carbon Storage: Modulators and Mechanisms** **167**

Catriona A. Macdonald, Manuel Delgado-Baquerizo, David S. Reay, Lettice C. Hicks and Brajesh K. Singh

*Marta Dondini, Mohamed Abdalla, Fitri K. Aini, Fabrizio Albanito,
Marvin R. Beckert, Khadiza Begum, Alison Brand, Kun Cheng,
Louis-Pierre Comeau, Edward O. Jones, Jennifer A. Farmer,
Diana M.S. Feliciano, Nuala Fitton, Astley Hastings, Dagmar N. Henner,
Matthias Kuhnert, Dali R. Nayak, Joseph Oyesikublakemore, Laura Phillips,
Mark I.A. Richards, Vianney Tumwesige, William F.A. van Dijk,
Sylvia H. Vetter, Kevin Coleman, Joanne Smith and Pete Smith*

List of Contributors

Mohamed Abdalla
University of Aberdeen, Aberdeen, United Kingdom

Fitri K. Aini
CIFOR—Center for International Forestry Research Bogor, Indonesia

Fabrizio Albanito
University of Aberdeen, Aberdeen, United Kingdom

Richard D. Bardgett
The University of Manchester, Manchester, United Kingdom

Marvin R. Beckert
University of Aberdeen, Aberdeen, United Kingdom

Khadiza Begum
University of Aberdeen, Aberdeen, United Kingdom

Mark A. Bradford
Yale University, New Haven, CT, United States

Alison Brand
University of Aberdeen, Aberdeen, United Kingdom

Kun Cheng
Nanjing Agricultural University, China

Kevin Coleman
Rothamsted Research, Harpenden, United Kingdom

Louis-Pierre Comeau
Fredericton Research and Development Centre, Fredericton, Canada

Jonathan R. De Long
The University of Manchester, Manchester, United Kingdom

Manuel Delgado-Baquerizo
University of Colorado, Boulder, CO, United States

Marta Dondini
University of Aberdeen, Aberdeen, United Kingdom

Jennifer A. Farmer
University of Aberdeen, Aberdeen, United Kingdom

Diana M.S. Feliciano
University of Aberdeen, Aberdeen, United Kingdom

Nuala Fitton
University of Aberdeen, Aberdeen, United Kingdom

Ellen L. Fry
The University of Manchester, Manchester, United Kingdom

Iain P. Hartley
University of Exeter, Exeter, United Kingdom

Astley Hastings
University of Aberdeen, Aberdeen, United Kingdom

Dagmar N. Henner
University of Aberdeen, Aberdeen, United Kingdom

Lettice C. Hicks
University of Edinburgh, Edinburgh, United Kingdom

Edward O. Jones
University of Aberdeen, Aberdeen, United Kingdom

Senani B. Karunaratne
Western Sydney University, Penrith, NSW, Australia; Department of the Environment and Energy, Parkes, ACT, Australia

Matthias Kuhnert
University of Aberdeen, Aberdeen, United Kingdom

Anitha Kunhikrishnan
Elizabeth Macarthur Agricultural Institute, Menangle, NSW, Australia

Catriona A. Macdonald
Western Sydney University, Penrith, NSW, Australia

Dali R. Nayak
University of Aberdeen, Aberdeen, United Kingdom

Joseph Oyesikublakemore
University of Aberdeen, Aberdeen, United Kingdom

Laura Phillips
World Food Programme, Dhaka, Bangladesh

David S. Reay
University of Edinburgh, Edinburgh, United Kingdom

Mark I.A. Richards
University of Aberdeen, Aberdeen, United Kingdom

Raj Setia
Punjab Remote Sensing Centre, Ludhiana, Punjab, India

Bhupinder P. Singh
Elizabeth Macarthur Agricultural Institute, Menangle, NSW, Australia

Brajesh K. Singh
Western Sydney University, Penrith, NSW, Australia

Joanne Smith
University of Aberdeen, Aberdeen, United Kingdom

Pete Smith
University of Aberdeen, Aberdeen, United Kingdom

Pankaj Trivedi
Colorado State University, Fort Collins, CO, United States

Vianney Tumwesige
African Centre for Clean Air, Kampala, Uganda

William F.A. van Dijk
Province of Noord-Holland, Haarlem, The Netherlands

Sylvia H. Vetter
University of Aberdeen, Aberdeen, United Kingdom

Matthew D. Wallenstein
Colorado State University, Fort Collins, CO, United States

Martin Wiesmeier
Technical University of Munich, Freising, Germany; Bavarian State Research Center for Agriculture, Freising, Germany

Stephen A. Wood
Yale University, New Haven, CT, United States; The Nature Conservancy, Arlington, VA, United States

Preface

In a letter to all state governors on a Uniform Soil Conservation Law, in 1937, The President of United State of America Franklin D. Roosevelt stated:

> "A nation that destroys its soils destroys itself."

I believe Roosevelt's statement is the most appropriate political statement made in 20th Century because soil is vital for all terrestrial life, including human survival. There is ample historical evidence to suggest that the rise and fall of human civilizations are linked to soil health. Soil consists of minerals, organic matter, biodiversity, water and air, while underpinning economic, social, and environmental services. The ability of soil to provide these services is largely determined by organic matter—of which soil organic carbon is the main component. Soil organic matter critically contributes to store and supply water and nutrients for primary production, purifies water, and regulates climate via locking carbon in soil and minimizing emissions. There is no surprise, then, that since prehistoric times soil organic carbon has been managed by growers and other stakeholders through multiple approaches including the application of manure.

In recent years, researchers have made enormous progress toward advancing the fundamental science which facilitated a number of policy frameworks (both at national and global levels) to increase soil carbon storage. However, large uncertainties remain regarding the magnitude, mechanisms, and predictions based on modeling which constrains effective implementation of policy and management tools. This book aims to provide an effective snapshot of current knowledge and gaps, as well as to contribute ideas and options to fill some of these knowledge gaps with the goal to provide a focussed message. This by no means includes comprehensive details of the topic as, in reality, a separate book could be written on each of the eight topics included here. My aim was to assemble information in a way which can provide enough details for scientists, but also be accessible to nonscientific stakeholders—e.g., policy advisers, farming consultants, and intergovernmental agencies. I have tried to achieve this by encouraging authors to include real data to support their arguments and provide a take home message and concluding remarks for nonspecialists. The first chapter is written mainly to introduce the subject and highlights the importance of soil carbon for human civilization and survival while addressing global challenges, including food security, environmental sustainability, and climate change mitigation and adaptation. Chapters 2–4 describe the role of biotic modulators and Chapter 5, Climate, Geography, and Soil Abiotic Properties as Modulators of Soil C Storage, and Chapter 6, Soil Nutrients and Soil Carbon Storage: Modulators and Mechanisms, deal with abiotic modulators of soil carbon storage. Human activities have a significant impact on soil carbon storage and Chapter 7, Agricultural Management Practices and Soil Organic Carbon Storage, and Chapter 8, Impact of Global Changes on Soil C Storage—Possible Mechanisms and Modeling Approaches, describe how management practices and climate change can impact soil carbon storage as well as suggesting approaches to minimize these impacts. The final chapter describes different modeling approaches which can be used to predict soil carbon stock under future climate and management practice scenarios.

I am fully aware that a number of key issues are not covered in this book, for example (but not limited to) soil inorganic carbon, methane flux, and physical storage of carbon, which all are important components of the global budget. However, I have kept them out for two reasons: (1) These topics are covered by other literature and books, and (2) to keep our discussion focussed on the matters-on-hand. There are some significant overlaps between different chapters which were deliberately promoted to demonstrate that all these variables interact with each other to influence the overall carbon storage in soils.

In writing and editing this book, I am grateful to all the authors for providing authoritative insight on the current understanding, knowledge gaps, and suggestions on how can we effectively apply potential solutions to advance science, but also implement policy and management frameworks, to improve soil carbon storage. Each chapter of the book was peer reviewed by at least two external reviewers who not only provided critical scientific comments, but also suggested approaches to clarify the message for both scientists and nonacademic stakeholders. I would like to specially thank reviewers for their hard work on improving the quality of the book.

I am confident that the book provides an effective snapshot and will serve both academic and nonacademic audiences and will lead to further scientific debate and studies, improved policy framework, and the adoption of management practices around the world for increased soil carbon storage.

Brajesh K. Singh

SOIL CARBON: INTRODUCTION, IMPORTANCE, STATUS, THREAT, AND MITIGATION

Pankaj Trivedi[1], Bhupinder P. Singh[2] and Brajesh K. Singh[3]

[1]Colorado State University, Fort Collins, CO, United States [2]Elizabeth Macarthur Agricultural Institute, Menangle, NSW, Australia [3]Western Sydney University, Penrith, NSW, Australia

1.1 INTRODUCTION

The world is facing multiple challenges including food security, environmental sustainability, soil protection, and climate change. These challenges need to be tackled as a priority to avoid human-made catastrophe in coming decades (Banwart et al., 2014a,b) and to double farm productivity by 2050 to feed ~ 9.6 billion people (Lal, 2006; Alexandratos and Bruinsma, 2012). This increase in farm productivity needs to come from shrinking arable lands (Singh and Trivedi, 2017), where one-third of food production has been lost due to reduction in soil organic carbon (SOC) and land degradation, hence requiring management improvements to restore C-poor degraded lands (Hazell and Wood, 2008; Brevik, 2010; Pimentel and Burgess, 2013). In addition to advancing food security, soil resources need to be managed in sustainable ways to meet multiple global needs, including mitigating and adapting to climate change, improving the quality and quantity of water resources, promoting biodiversity, preserving human heritage, preventing desertification, alleviating poverty, and being an engine of new industries and economic growth (Lal, 2007).

SOC is a critical natural resource and contributes significantly to achieving all these goals, such as by directly storing more C in soils, while improving soil health and ecosystem functions. SOC is considered a key indicator for soil health because of its contributions to food production, mitigation and adaptation to climate change, and role in water storage and purification. The SOC content is almost $50\% - 58\%$ of soil organic matter (SOM) (Pribyl, 2010), which stores nutrients for plants, improves soil structural stability to enhance soil fertility, and ultimately provides food. With an optimal amount of SOC, the filtration capacity of soils further supports the supply of clean water. Turnover of SOC in terrestrial ecosystems is dynamic and human impacts can turn SOC into either a net sink or a net source of greenhouse gases (GHG) to the atmosphere. Although the overall impact of climate change on SOC stocks is highly variable according to the region and soil type, rising temperatures and increased frequency of extreme drought events are likely to lead to increased loss of SOC. Significant scientific progress has been achieved in understanding and explaining SOC dynamics. The dynamics of these processes highlight the importance of quantifying global C fluctuations to ensure maximum benefits of SOC to human well-being, food production, and water and climate regulation. However, protection and monitoring of SOC stocks at national

Soil Carbon Storage. DOI: https://doi.org/10.1016/B978-0-12-812766-7.00001-9

and global levels still face complicated challenges impeding effective on-the-ground policy designs and regionally adapted implementation. This chapter highlights the importance of SOC for ecosystem services and intergovernmental policies. Further, this chapter describes the potential and challenges of soil C sequestration and how knowledge of SOC dynamics can be applied to address multiple global challenges and deliver ecosystem benefits within increasingly complex policy frameworks (see Box 1.1 for definitions).

1.1.1 CARBON CYCLE

The carbon (C) cycle consists of the transfer of C in different forms between the atmosphere, living organisms (biosphere), and soil (pedosphere) (Orgiazzi et al., 2016). The current level of atmospheric CO_2 concentration is a balance between C fixation *via* photosynthesis and C loss via respiration. In terrestrial systems, about 123 Gigaton (Gt) of C is assimilated by primary production and about 120 Gt are respired back roughly half by autotrophic respiration and another half by microbial respiration (Högberg et al., 2001; Nordgren et al., 2003; Singh et al., 2010) (Fig. 1.1). Human activities have significantly modified the global C cycling by enhancing the release of significant CO_2 through fossil fuel burning and industrial activities. About 9 Gt are added in the atmosphere by

---

BOX 1.1

Definition of soil: Soil is a natural body comprised of solids (minerals and organic matter), liquid, and gases that occurs on the land surface, occupies space, and is characterized by one or both of the following: horizons, or layers, that are distinguishable from the initial material as a result of additions, losses, transfers, and transformations of energy and matter or the ability to support rooted plants in a natural environment (Soil Science Society of America, 2017).

Definition of soil health: Soil health is the capacity of soil to function as a living system, within ecosystem and land use boundaries, to sustain plant and animal productivity, maintain or enhance water and air quality, and promote plant and animal health. Healthy soils maintain a diverse community of soil organisms that help to control plant disease, insect and weed pests, form beneficial symbiotic associations with plant roots; recycle essential plant nutrients; improve soil structure with positive repercussions for soil water and nutrient holding capacity, and ultimately improve crop production. A healthy soil prevents pollution of environment and contributes to mitigating climate change by maintaining or increasing its carbon content (Doran and Zeiss, 2000; FAO and ITPS, 2015).

Definition of soil quality: Soil quality can be defined as the fitness of a specific kind of soil, to function within its capacity and within natural or managed ecosystem boundaries, to sustain plant and animal productivity, maintain or enhance water and air quality, and support human health and habitation (Karlen et al., 1997; Arshad and Martin, 2002). Many times, soil quality is used as synonymous to soil health.

Definition of ecosystem functions: Ecosystem functions (sometimes called as ecological processes) relate to the structural components of an ecosystem (e.g., vegetation, water, soil, atmosphere, and biota) and how they interact with each other, within an ecosystem and across ecosystems. Primarily, these are exchanges of energy and nutrients in the food chain which are vital to the sustenance of plant and animal life on the planet as well as the decomposition of organic matter and production of biomass made possible by photosynthesis (Tilman et al., 2014).

Definition of ecosystem services: Ecosystem services is the technical term given to the goods and services provided by the ecosystem that directly benefits, sustain and support the well-being of people (Guerry et al., 2015).

Definition of soil organic matter (SOM): As per a new paradigm, SOM is a continuum of progressively decomposing organic compounds of plant, animal, and microbial origin that are stabilized with clay minerals and soil aggregates through microbial, physical, and chemical processes (Lehmann and Kleber, 2015).

Definition of soil organic carbon (SOC): SOC is a measure of carbon contained within SOM. SOM usually contains approximately 58% C; therefore, a factor of 1.72 can be used to convert OC to SOM.

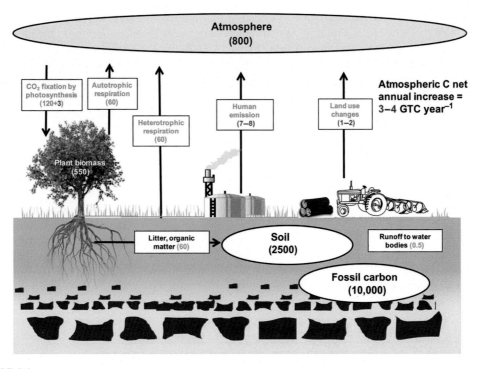

FIGURE 1.1

Simplified terrestrial carbon cycle. The values in bracket of square boxes represent the exchange of C between land and atmosphere in Gigatons (Gt) of C per year. Green numbers are natural fluxes; red numbers are human contributions (Gt of C per year). The values in bracket in oval indicate the amount of C in different pools (Gt).

human activities consisting of 7−8 Gt from the fossil fuel burning and 1−2 Gt via land-use change primarily via deforestation (Fig. 1.1). The current global C budget is thus unbalanced due to higher CO_2 emission into the atmosphere than C sink in soil and vegetation, mainly driven by human activities. House et al. (2003) reported that terrestrial C uptake was in the range of 0.3−4.0 Gt C year^{-1} and 1.6−4.8 Gt C year^{-1} for the 1980s and 1990s, respectively. Le Quéré et al. (2015) presented a new estimate of the present-day global carbon budget and pointed out that on average 45% of the total CO_2 emitted from anthropogenic activities (mainly fossil fuel burning and land use change) stayed in the atmosphere in the past decades. They also suggested that the efficiency of C sinks could have decreased in the past decades. This creates an unbalanced global budget and has led to a sustained increase of atmospheric CO_2 concentration, the main driver of global warming. The further consequences of this imbalance in C budget are unpredictable feedback responses of climate change (detailed in Chapter 8: Impact of Global Changes on Soil C Storage—Possible Mechanisms and Modeling Approaches). Understanding the global C cycle and how it interacts with climate change is a key research challenge that is crucial for the future of our planet.

The process of C sequestration includes CO_2 fixation, transfer of fixed C to plant biomass and soils, and stabilization of organic C in soil where it is stored as SOC following interactions with soil microbes, minerals, and aggregates. Terrestrial C sequestration, with application of improved management methods (see Chapter 7: Agricultural Management Practices and Soil Organic Carbon Storage), could sequester more than 0.5 Gt C year^{-1} by 2040 mitigating from 6% to 23% of the emissions by mid-century and accumulating over 40 Gt of C by 2100 (Thomson et al., 2008; Post et al., 2009). Increased SOC storage occurs via multiple factors that result in increased C inputs and reduced C losses (or both) and is influenced by soil texture, clay mineralogy, depth, bulk density, aeration, and proportion of coarse fragments (see Chapter 5: Climate, Geography and Soil Abiotic Properties as Modulators of Soil C Storage for details). Carbon residence time is a key factor affecting C sequestration potential in different soils (Post et al., 2009; Luo et al., 2015). Interactions with minerals can provide physicochemical protection of C that is determined by processes responsible for creation, turnover, and stabilization of soil aggregates at multiple, often hierarchical scales (Tisdall and Oades, 1982; Jastrow and Miller, 1997; Six et al., 2004). These stabilization mechanisms operate at different time scales (Lützow et al., 2006), and their interactions lead to a continuum of SOC pools with residence times that can range from less than a year to centuries and even millennia (Post et al., 2009). Furthermore, the effectiveness and relative importance of different factors related to C stabilization depends on various factors, including soil type, management practices, historical contingency, and climate conditions (Post et al., 2009; Delgado-Baquerizo et al., 2017b).

1.2 HUMAN CIVILIZATIONS AND SOIL ORGANIC CARBON

Since the dawn of human civilization, SOM and SOC have been recognized as a key to manage soil fertility and improving agronomic productivity. This topic is well covered in previous publications (Lal, 2007, 2014, 2016). Briefly, the words "human" and "humus" (mixture of dark, colloidal poly-dispersed organic compounds with high molecular weights and relatively resistant to decomposition) are intricately linked and SOM has been considered by ancient civilizations to be a "concoction" of sustainability and productivity (Lal, 2016). Archeologists have identified examples of human societies that have been brought to the limit of sustainability by SOM/SOC depletion, even resulting, in some cases, in the decline and fall of their civilization (Olson, 1981; Fig. 1.2). The experiences of past societies provide ample historical basis for linking soil quality to SOM/SOC and long-term prospects of managing SOM/SOC for multiple societal benefits (Lal, 2016). For example, Asian farmers were able to cultivate the same field for as long as 40,000 years by maintaining soil fertility and managing SOM/SOC via manuring and recycling (Lal, 2014). A Sanskrit text written in about 1500 BC quoted: "Upon this handful of soil our survival depends. Husband it and it will grow our food, our fuel and our shelter and surround us with beauty. Abuse it and the soil will collapse and die, taking humanity with it (Shiva et al., 2016)." Agronomic functions of SOM/SOC and their connection to cropping and farming sustainability has been recognized and preached by the philosophers and religious leaders from ancient times. In 1400 BC, after their arrival in Canaan, Moses asked his followers to "bring back some fruits from lands with fertile soils as those represent healthy people" (Lal, 2016). In the 4th century BC, an Indian Scholar, Chanukya/Kautilya in his manual *Artha Sathra* encouraged land managers to improve soil functions by applying manure and proposed mechanisms to manage soil fertility and water conservation. In

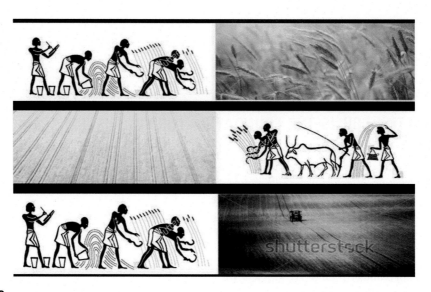

FIGURE 1.2

Drawing illustrates linkage between soil and human civilizations (particularly farming). All through human civilization, soil organic matters and soil organic carbon have been recognized as key manage soil health and farm productivity.

Source: LilKar/Shutterstock, Valentin Valkov/Shutterstock, Thumbelina/Shutterstock, Svend77/Shutterstock.

middle history around 12th century, a Moorish philosopher, Ibn-Al-Awwam stated in his book *Kitab-Al-Felha* that the first step of agriculture is recognition of soils and how to distinguish that which is of a good quality and that which is of an inferior quality (Lal, 2016).

In modern civilization, the role of SOM/SOC in controlling the capacity of soil resources to deliver agricultural and environmental services and sustain human societies at both local (e.g., fertility maintenance) and global (e.g., mitigation of atmospheric C emissions) scales is well established (Tiessen et al., 1994; Wolf and Snyder, 2003). Emphasizing the importance of agronomic practices that increase SOM, MK Gandhi wrote "to forget how to dig the earth and to tend soil is to forget ourselves." U.S. President Franklin D. Roosevelt, who in 1935 signed legislation aimed at combating soil erosion and preserve natural resources, made a powerful statement on the importance of maintaining soil quality by stating: "A nation that destroys its soils destroys itself." Increasingly, farmers, scientists, landowners, environmental experts, and policymakers understand that SOM/SOC management is critical in order to improve soil health and that addressing its complex challenges is urgently required to meet global concerns including food security, climate change mitigation, and water conservation (Doran et al., 1994; Lal, 1997, 2016). Innovative initiatives by government and intergovernmental agencies such as United States Department of Agriculture (USDA), Food and Agriculture Organization (FAO), USDA, FAO, Global Soil Partnerships (GSP), United Nations Convention to combat Desertification (UNCCD), and United Nations Environmental Program (UNEP), as well as newer public–private entities such as the Soil Health Partnership, the Global Soil Biodiversity Initiative, and the Soil Health Institute (in United States), are already making important progress in coordinating, innovating, and investing to monitor the status, threats, and propose mitigation/agronomic approaches to build SOC. A notable change is underway, and momentum is increasing around the opportunity to manage SOC to address important social and environmental challenges.

1.3 HEALTHY SOIL IS THE FOUNDATION FOR ECOSYSTEM SERVICES AND TERRESTRIAL LIFE

Healthy soil is the cornerstone of terrestrial life on Earth, providing habitat for biodiversity, facilitating food production, effective water filtration and storage, and regulating the climate. It is fundamental to soil fertility by retaining and releasing nutrients for plant growth. Plants being the primary producers, capture C from the atmosphere, energy from sunlight, and transfer the photosynthesized C to soil for growth and activities of other organisms (heterotrophs) via root exudation and litter deposition. These forms of SOM are then recycled by heterotrophs (microbes and fauna) into different forms of soil C, which interact with soil minerals and microbes, and release nutrients (e.g., nitrogen, phosphorus, and sulfur) from SOM turnover to support plant growth. In healthy "functional" soils, SOC is a basis of critical physicochemical and biological processes (Lal, 2014) and it drives almost all key services that underpin the role of healthy soils in maintaining terrestrial life (Fig. 1.3).

All aspects of humanity, including peace and war, rural and urban setting, poverty and affluence, can be affected by soil health (Keesstra et al., 2016). Although advancement in agriculture technology throughout the past century has increased food production, their wide-scale application has significantly reduced soil quality with a loss of as much as 60% in SOC content (Lal, 2004). The degradation of soils has undermined the productivity and resilience of croplands while causing significant environmental impacts resulting in increased GHG emissions, nutrient loss, and soil erosion. In 2015, a report estimated that in the United States alone, the societal and environmental cost of soil degradation is up to $85.1 billion annually through unintended effects on human health, property, energy, endangered species, loss of biodiversity, eutrophication, contamination, agriculture productivity, and resilience. As highlighted in the first principle established by the revised World Soil Charter (FAO, 2015, p. 2): "Soils are a key enabling resource, central to the creation of a host of goods and services integral to ecosystems and human well-being. The maintenance or enhancement of global soil resources is essential if humanity's overarching need for food, water, and energy security is to be met. In particular, the projected increases in food, fiber, and fuel production required to achieve food and energy security will place increased pressure on the soil." The 17 Sustainable Development Goals (SDGs) of the 2030 Agenda for Sustainable Development which were adopted by world leaders in September 2015 explicitly identified the need to restore degraded soils and improve soil health.

Maintaining SOC stock at an equilibrium or increasing SOC content toward an optimal level for the local environment can contribute to meeting several global challenges including: (1) Provision of food and energy for soil biodiversity to further improve nutrient availability for plant uptake, and support productivity of aboveground plants and animals; (2) Ensuring chemical reaction, transformation and exchange for ensuring optimal nutrient and water availability; (3) Improving soil physical structure for optimal provision of habitats and exchange of air, water, and gases; and (4) Mitigating climate change effects by C sequestration and offsetting GHG emissions.

1.4 SOIL ORGANIC CARBON IS THE MOST IMPORTANT COMPONENT OF SOIL QUALITY AND HEALTH

Soil organic C is critical to soil health and a threshold/critical level of 1.5%−2.0% of SOC in agro-ecosystems is proposed as essential to maintain: (1) Appropriate soil structure and aggregates; (2) Water retention, release and use efficiency and resilience to abiotic stresses—e.g., drought, heat wave; and (3) Nutrient retention, release and use efficiency and gas exchanges to regulate climate change (Banwart et al., 2014a,b). Soil organic C is typically segregated into distinct "pools" or "fractions" with unique characteristics and specific turnover times (Parton et al., 1994). These pools have been used to indicate the sustainability of farming practices from a soil quality or C sequestration perspective and include plant residues, particulate organic matter (POM), humus C, and recalcitrant organic C. The POM fraction (Cambardella and Elliott, 1992) is a fraction of SOM that is composed of particulate (>0.05 mm), partially decomposed plant and animal residues, fungal hyphae, spores, root fragments, and seeds (Causarano et al., 2008). The POM fraction provides an important energy source for soil microbes, and the proportion of organic matter in this fraction can be directly correlated with soil fertility (Lehman et al., 2015). Soil microbial biomass C and potentially mineralized C are considered as active fractions of SOM and are important for supplying plant nutrients, decomposing organic residues, and developing soil structure (Franzluebbers and Stuedemann, 2008). These fractions together with POM and soil aggregation are important indicators of dynamic soil quality because they are responsive to change in management practices (Franzluebbers, 2002; Causarano et al., 2008). The recalcitrant pool is of utmost interest for C sequestration as an option of mitigating climate change given its ability to resist rapid decomposition, but is also important for the chemical health of soil by its influence on cation exchange capacity (CEC). Different soil management practices are known to alter not only total SOC, but also the relative proportion of C residing in these different pools. For example, management practices that produce high amounts of organic materials such as pastures and native vegetation will have higher proportions of residue and POM. On the other hand, agricultural systems characterized by continuous cropping, long fallow periods, tillage, and stubble burning or grazing typically have low proportions of these, relatively labile, C fractions. Knowledge of how C pools/fractions change in response to management can provide valuable information on likely soil functioning and health (See Chapter 2: Plant Communities as Modulators of Soil Carbon Storage, Chapter 7: Agricultural Management Practices and Soil Organic Carbon Storage, and Chapter 8: Impact of Global Changes on Soil C Storage—Possible Mechanisms and Modeling Approaches for more details). It must also be noted that in recent years however the long-standing theory that suggests that SOM is composed of inherently stable and chemically unique compounds has been challenged and a "continuum model" that suggests that SOM is a continuum of progressively decomposing organic compounds has been proposed (Lehmann and Kleber, 2015).

1.5 SOIL ORGANIC CARBON IS A MAJOR DRIVER OF ECOSYSTEM SERVICES

Ecosystem functions include the physicochemical and biological processes that occur within the ecosystem to maintain terrestrial life. Ecosystem services are the set of ecosystem functions that are directly linked to benefit human well-being (Kremen, 2005). Soil organic C regulates most ecosystem services including provisioning, supporting, regulating, and cultural services (Fig. 1.3). These ecosystem services can be broadly categorized as: (1) Provisioning services—SOM serves as the basis for food and fiber production via influencing soil structure, nutrient and water availability; (2) Regulatory services—SOM reduces soil erosion, water run-off, and attenuation of toxic pollutants, and regulates climate change via offsetting greenhouse gas emissions; (3) Supporting services—the formation and breakdown of SOM, which can influence soil characteristics such as soil fertility and soil biodiversity; and (4) Cultural services—SOM influences the soil to retain diverse cultures of the past, the nature of the landscape and preserve archeological remains (Banwart et al., 2014a,b).

1.6 SOIL ORGANIC CARBON AND PROVISIONING SERVICES

1.6.1 PROVISION OF FOOD, FIBER, AND TIMBER

Fertile soils and sustainable practices in managed agro-ecosystems generally maintain high levels of crop, pasture, and animal production to ensure sufficient supply of food, fiber, timber, and

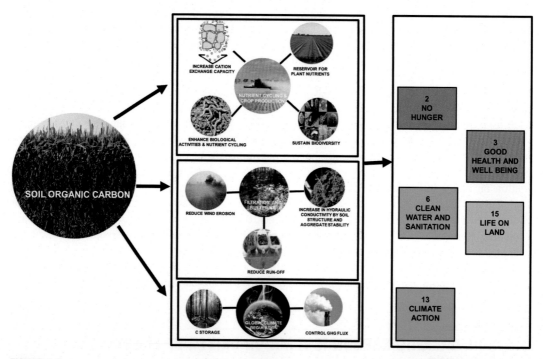

FIGURE 1.3

Soil organic carbon provides multiple functions and contributes significantly to multiple ecosystem services and is critical for the delivery of the sustainable development goals.

woodchips, thus meeting the needs of the growing human population through supporting these provisioning services (Banwart et al., 2014a,b). At the same time, fertile soils support the growth of native vegetation (forests, woodland, and grasslands) and thus contribute to the ecosystem services that native vegetation provides, including the provision of food for humans, feed for livestock, and other resources (firewood, building timbers, paper products, etc.). These managed and natural systems provide a range of other ecosystem services, including regulating and supporting services for a range of key provisioning services, while also playing a vital role in the provision of fresh water and medicines (FAO, 2017). Maximizing provisioning services from managed and native systems would result in tradeoffs with other ecosystem services they provide, but ecologically sustainable management of native vegetation and improved agricultural practices may substantially benefit supporting services and alleviate tradeoffs among the ecosystem services (Power, 2010; Smith et al., 2015).

1.6.2 PROVISION OF CLEAN WATER

Soil organic C plays an important role in water filtration and purification and soil water holding capacity by improving soil structural stability and microbial activity. Most studies show a positive relationship between SOC and water holding capacity. For example, several studies have demonstrated an increase in water content with increasing SOC content wherein an increase of 1% SOM can add 1.5% additional moisture by volume at field conditions (Haynes and Naidu, 1998; Wolf and Snyder, 2003). Soils with high levels of SOC confer resilience and promote stress resistance against drought compared to soils with low levels of SOC (Yuste et al., 2011). Porosity and connectivity of pores are two critical components of water retaining capacity of soils. SOM improves the physical conditions of soil by enhancing organo-mineral interactions and soil structure development thus providing greater pore space for water and air holding (Srinivasarao et al., 2015). Increasing the water holding capacity is essential during low rainfall periods and drought, as soil retains moisture and also slowly releases water for plant uptake (Jung et al., 2007). The water stored in soil serves as the source for 90% of the world's agricultural production and represents about 65% of global fresh water (Amundson et al., 2015). While plant growth and surface mulches can help protect the soil surface, a stable, well-aggregated soil structure that resists surface sealing and facilitates water infiltration during intense rainfall events will decrease the potential for downstream flooding. Soil chemical properties and microbial activity, on other hand, play important roles in removing contaminants from the water to minimize the pollution of ground and surface water.

1.7 REGULATING SERVICES

1.7.1 CLIMATE REGULATION

Soils represent a massive stock of C and act both as a buffer against atmospheric CO_2 increase and as a potential sink for additional C depending on the balance between photosynthesis, the respiration of decomposer organisms and stabilization of C in soils (Dungait et al., 2012). The soil C storage at 3 m depth is 3.3 times the size of atmospheric pool (760 Gt) and 4.5 times the size of the biotic pools (560 Gt) (Lal, 2004; Trivedi et al., 2013). Further, soil C sequestration can meaningfully contribute to a portfolio of mitigation approaches and potentially offset a significant fraction of diffuse CO_2 emissions for which direct capture is not yet feasible (King, 2011). In particular, it has been estimated that through judicious management, the world agricultural and degraded soils could sequester equivalent to 5%−15% of global fossil fuel emissions (Lal, 2004). While these rates will offset only a fraction of emissions from fossil fuels, results from integrated assessment analyses (Edmonds et al., 1999; Rosenberg et al., 1999) indicate that soil C sequestration may have an important strategic role due to its potential for early deployment and low costs options to mitigate climate change. However, the influence of climate change on the soil C storage and sink remains a major area of uncertainty considering the feedback from various tightly linked biotic and abiotic factors on soil C cycles (see Chapter 8: Impact of Global Changes on Soil C Storage—Possible Mechanisms and Modeling Approaches for details). Various studies have suggested that conversion of native land for agriculture practices have resulted in 40%−60% reduction in SOC from preclearing levels that has resulted in the emission of approximately 150 Gt of CO_2 into the atmosphere (Guo and Gifford, 2002). The estimates of SOC loss varied, however, depending on factors such as annual precipitation, plant species, and the length of study periods (Trivedi et al., 2016a). It has been suggested that recapturing even a small fraction of these legacy emissions through improved and innovative land management practices that help building SOC in agro-ecosystems will not only lead to lowering the rate of GHG emissions, but also increase various ecosystem services including primary productivity and soil functionality.

1.7.2 REGULATION OF SOIL STRUCTURE

Soil organic C binds mineral particles to form aggregates (Oades, 1993), whose distribution and stability influence the soil's physical properties, including pore size distribution, bulk density, soil strength, and soil erodibility (Tatarko, 2001; Chapter 3: Microbial Modulators and Mechanisms of Soil Carbon Storage). Loss of organic matter affects soil structure and aggregate formation, with negative impacts on biological diversity, soil fertility, crop production potential, erosion control, water retention, matter exchange between soil, atmosphere, groundwater, and the filtering, buffering, and transforming capacity of terrestrial ecosystems (Huber et al., 2001; Kirchmann and Andersson, 2001). Soil management practices that improve formation of soil aggregates through an increase in SOC have demonstrated minimizing soil and nutrient losses through erosion mitigation and eutrophication of surface waters (Srinivasarao et al., 2015). The success of these soil management systems, however, relies strongly on site-specific characteristics, farmers' awareness and knowledge, and often on weather events.

1.8 SUPPORTING SERVICES

These services underpin the delivery of all other services and benefits that are obtained by the natural environment, including biomass production, soil formation, nutrient cycling, and provision of habitat for biodiversity. Therefore, understanding their responses to key drivers, such as climate change, land use change, and nutrient enrichment, is vital for the sustainable management of global land and water resources. SOM is a key attribute with influence on soil's capacity to sustain and provide supporting services. Thus, SOC is not only a component of soil capital, but a relative change in SOC is potentially a practical indicator of changes in flows of supporting ecosystem services and associated soil natural capital. Relative changes in SOC concentrations can be correlated with changes in soil biodiversity, which further impacts on other provisional and supporting ecosystem services. For example, a decline in SOC concentrations has been shown to have a negative effect on the supporting services which in turn reduce both maximum yield and fertilizer use efficiency in agro-ecosystems (Brady et al., 2015). Supporting services are all strongly interrelated and, in many cases, underpinned by a vast array of physical, chemical, and biological interactions. Our understanding of the connectivity among and within different variables (namely, flow of nutrients, energy, and information) and the relative contribution of interconnected variables in driving supporting services is generally limited.

1.8.1 NUTRIENT CYCLING

1.8.1.1 Reservoir of nutrients

SOM is an important source of nutrients for plants. With the exception of fertilizers, SOM provides the largest pool of macronutrients with >95% of N and S and 20%−75% of P found in SOM (Duxbury et al., 1989; Baldock and Nelson, 2000). Heterotrophic organisms mineralize SOM wherein some soil nutrients are used in the synthesis of new biomass, wherein a portion is immobilized and another portion is released as plant-available forms. In relation to N supply from plant residues, 1% SOC is considered as the threshold value below which an effective N supply is reduced (Loveland and Webb, 2003). Soil aggregates may provide an important transient storage capacity for macronutrients and the size of aggregate classes can influence nutrient availability. Studies have shown greater mineralization of both C and N in macro-aggregate associated SOM compared with microaggregates (Nimmo and Perkins, 2002; Trivedi et al., 2015, 2017). Land management practices can affect the nutrient status and release from SOM. For example, conservational agricultural practices that increase the SOM content may reduce the availability of nutrients to crops with a gradual accumulation of nutrients in SOM over time (Duxbury et al., 1989).

There has been an ongoing debate on "using" SOM to release nutrients for plant growth in shorter terms (Janzen, 2006) versus "saving" SOM to provide long-term benefits with respect to C sequestration and enhancing long-term soil quality (Lal, 2004). These two processes may occur in distinct components of SOM wherein the relatively labile fast-cycling POM serves as a source of nutrients while the recalcitrant slow cycling mineral associated C serves as a long-term storehouse of SOC (Wood et al., 2016; Lehmann and Kleber, 2015; Trivedi et al., 2017). Different sized aggregates show an uneven distribution of labile *versus* recalcitrant C wherein the amount of labile C decreased from macro- to microaggregates and vice versa (Trivedi et al., 2015, 2017). The current paradigm suggests that accumulation of different SOM fractions is controlled by different mechanisms, potentially leading to different relationships with management outcomes (Janzen, 2006; Schmidt et al., 2011; Wood et al., 2016; Trivedi et al., 2017). Thus, understanding the controls on each of these pools and quantifying their impacts on different services provided by SOC will help in designing effective soil management practices for sustainable agriculture.

1.8.1.2 *Cation exchange capacity*

A high soil CEC is regarded as favorable as it contributes to the capacity of soils to retain plant nutrient cations, thus reducing the potential for leaching in soil. The contribution of SOM/SOC to CEC varies between 25% and 90% (Stevenson, 1994) and there is generally a very good correlation between SOC and CEC (McGrath et al., 1988). Humus, being highly negatively charged, has the potential to hold enormous cations, which enable soils to retain plant-available nutrients for a longer period of time such as via cation bridging. Soils with reduced CEC lead to low pH resulting in soil acidification, which further lowers crop productivity (Cork et al., 2012). Soil organic C influences soil pH and buffering capacity (Cayley et al., 2002). Soil organic C has a buffering capacity of over 300 times than that of clay minerals such as illite and kaolinite. As a result, soils with high SOC are less susceptible to acidification compared to highly weathered soils that are low in SOC.

1.8.2 PROVISION OF GENETIC RESOURCES

Soil biodiversity (including organisms such as bacteria, fungi, protozoa, insects, worms, other invertebrates, and mammals), supported by SOM pools, enhances the metabolic capacity of soils and plays a crucial role in soil health and ecosystem functioning. The highly diverse underground communities are immersed in a framework of networks that: (1) Determine the net flux of C between the atmosphere and soils, and (2) Cycle SOM thus influencing nutrient availability. The revised World Soil Charter states that soils are a key reservoir of global biodiversity which ranged from microorganisms to flora and fauna. The biodiversity has a fundamental role in supporting soil functions and, therefore, ecosystem goods and services associated with soils. It is thus necessary to maintain soil biodiversity to safeguard these functions.

The quality and quantity of SOC/SOM directly influences soil biodiversity and activities, as it is the main source of energy for their survival and growth. Indirectly, it regulates the soil biodiversity by influencing the habitat properties such as soil aggregates, pore size, and connectivity. The quality and quality of SOM/SOC determines the abundance, diversity, and activities of soil communities, but these are circular interactions, where soil biodiversity also determines the quality and quantity of SOC, and there are multiple interactions which influence the rate and aspects of C cycling and ecosystem functions (Delgado-Baquerizo et al., 2016a,b; Trivedi et al., 2016b; Delgado-Baquerizo et al., 2017a,b). Recent studies provide evidence that soil diversity and functional community diversity are strongly linked to enzymes that degrade SOC, suggesting microbial community regulation of SOC storage (Trivedi et al., 2016a). Soil microbial respiration accounts for 50% ($\sim$60 billion tonnes year^{-1}) of the net terrestrial flux while microbes are the main regulators of rate of decomposition (Karhu et al., 2014). However, soil fauna also plays a significant role by manipulating and controlling microbial communities even though they have minor direct contribution via litter fragmentation, partial digestion of litters, and promoting direct contacts between litter and microbial communities (Orgiazzi et al., 2016).

There is increasing evidence that SOC is one of the main drivers of microbial diversity, community structure, and abundance at global and regional scales (Delgado-Baquerizo et al., 2016a; Louis et al., 2016). Studies also suggest that SOC quality and soil aggregation create distinct niches for different microbial communities. For example, the SOC associated with macroaggregates (relatively labile SOC) promotes microbial diversity and community structure which are distinct to microbial communities associated with SOC of microaggregates (Trivedi et al., 2015, 2017; Rillig et al., 2017). Higher diversity provides a higher reservoir of gene pool for harnessing their functional capabilities for the betterment of the ecosystem and human societies. Loss of belowground diversity linked to soil C loss has significant consequences, hence understanding relationships between soil biodiversity and C cycling is critical for projecting how the loss of diversity under continued environmental alteration by humans will impact global C cycling processes (De Graaff et al., 2015) and other ecosystem functions (Delgado-Baquerizo et al., 2016a,b; 2017a). Current research indicates that soil biodiversity can be maintained and partially restored if managed sustainably. Promoting the ecological complexity and robustness of soil biodiversity through improved management practices represents an underutilized resource with the ability to ultimately improve human health (Nielsen et al., 2015). Other than being drivers of ecosystem functions, soil organisms have also been a source of many industrial products including medicines, genes for genetically modified crops, and chemicals/enzymes for food and chemical industries worth trillion of dollars (Singh, 2010). Going forward, harnessing natural resources (microbes, fauna, flora) together with SOM, is considered as the most effective approach for sustainable increase in farm productivity, mitigating climate change, and restoring degraded environments. Further evidence of the relationships between soil biodiversity and functioning with regard to SOC dynamics and primary productivity at farm scales can help in bridging the knowledge gaps in the biotic regulation of SOC turnover and plant productivity. This will represent a major advancement, not only in ecology, but also in agriculture in the context of global climate change and food security (Lemanceau et al., 2015).

1.9 CULTURAL SERVICES

1.9.1 HUMAN AND PLANETARY HERITAGE

Soil is one of the main sources of information on the prehistoric culture of humankind. For example, archeological remains preserved in water-saturated peatlands for about 2500 years have provided important information on the origin of human civilization. Similarly, ancient cities sat beneath the soil for thousands of years and provided preserved relics of ancient civilizations (Banwart et al., 2014a,b). Some portions of SOC survive millennia and therefore can provide critical information on human and planetary heritage. Stable isotope composition of SOC can provide evidence for pedogenic and climatic conditions for past millennia (Kovda et al., 2016). Similarly, measurements of ^{14}C in SOC have provided evidence on soil processes since the first nuclear explosion in 1950s (Lal, 2007). In addition, land use as a sense-of-place has been embodied within the cultural and spiritual beliefs of indigenous peoples throughout the world.

1.9.2 RECREATIONAL AND ESTHETIC EXPERIENCES

Soils provide esthetic and recreational values through landscape, particularly in Globally Important Agricultural Heritage Systems (Altieri and Toledo, 2011). They have also been used as an esthetic approach to raise soil awareness in contemporary art (Feller et al., 2015). Because SOM provides regulating services associated with water retention and purification, it is key for meeting recreational and esthetic values provided by water bodies that include swimming, water sports, and fishing (van den Belt and Blake, 2014). Certain soil types are more valuable than others in providing recreational benefits to the communities. Retisols that typically carry a temperate needle-leaf evergreen forest/woodland on often steeply sloping land are highly valued for forestry, recreation, and watershed protection. Gardening is one of the most important recreational activities across the globe (Comerford et al., 2013) and is recognized as a viable treatment to cure a wide range of mental and emotional conditions (Rice and Remy, 1998). Patients suffering from physical trauma and surgery recover rapidly just by viewing gardens while "healing gardens" have become an important component of hospital designs (Cooper Marcus and Barnes, 1999).

1.10 STATUS, THREAT, AND OPPORTUNITY

Status: Many estimates of global SOC stocks have been published during the past 70 years to support the calculation of potential CO_2 emissions from the soils under land use and land cover change and/or climate change scenarios (Don et al., 2011). More recent studies have reported a global soil estimate of roughly 1417−1500 Gt of C stored in the first meter of soil and about 716 Gt organic C in the top 30 cm (Köchy et al., 2015). There are fewer estimates of global SOC stocks below 1 m. Global SOC stocks to a depth of 3 m are estimated at 2344−3000 Gt (Guo and Gifford, 2002; Jobbágy and Jackson, 2000). Large variations in the SOC stock estimates may be attributed to various factors including the lack of large-scale dataset, analysis methods applied, extrapolation, and the uncertainties concerning certain soil types such as Arctic and peatlands in South Asia.

Both SOC stocks and their contributions to total C stock vary with latitude and among climatic regions (Fig. 1.4; more details in Chapter 5: Climate, Geography and Soil Abiotic Properties as Modulators of Soil C Storage). Most of the SOC is stored at northern latitudes, particularly in the "Boreal Moist" (356.7 Pg C) and "Cool Temperature Moist" regions (201.3 Pg C). With 117.6 and 88.2 Mg C ha^{-1}, these climatic regions also have the highest SOC densities. Regions near the equator in wet and moist tropical forests (including Amazonia's rainforest, the Congo basin and South-Eastern Asia) also known as the "green lungs" of the planet, store vast amount of C either in vegetation or soils (Deng et al., 2016). In fact, for the terrestrial pool of SOC, biomass is the most important pool only in "Tropical wet" and "Tropical Moist" climate regions while soil stores more SOC than the biomass in all the other climatic regions (Scharlemann et al., 2014; Fig. 1.5). As a general propensity, soil dominates the terrestrial C pool in cooler climates while vegetation dominates terrestrial C pools in tropical regions. A large fraction of boreal forest and tundra regions has additional C stored beneath the permafrost layer. In response to climate change scenarios, thawing of permafrost could release large amounts of C as CO_2, or in swamps and bogs, as CH_4, which could further amplify climate change (Schuur and Abbott, 2011). Within different vegetation classes, broadleaf forests (509.4 Gt C) represent the largest stock for terrestrial SOC (Lefèvre et al., 2017). These forest types contain approximately one quarter of all terrestrial SOC in either soils or the biomass. All the vegetation classes except for "Forest/Cropland Mosaic" store more C in soil compared to biomass. Historic trends in the fluctuations of SOC have shown that conversion of natural to agro-ecosystems in the past has led to a decline in the SOC stocks. However, the magnitude of the historic loss differs among soils, climate, and the adopted management practices. For example, a 2017 study indicated that climate legacies help to predict global soil C stocks in natural

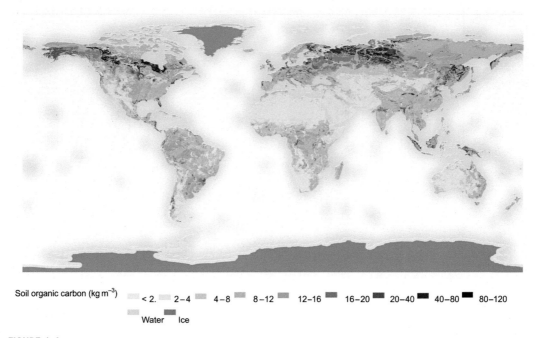

Soil organic carbon (kg m^{-3})   < 2.   2–4   4–8   8–12   12–16   16–20   20–40   40–80   80–120

Water   Ice

FIGURE 1.4

Global map of soil organic carbon based on FAO-UNESCO. The map shows global distribution of soil organic carbon to a depth of one meter.

Reproduced with permission from Global Biodiversity atlas.

terrestrial ecosystems, whereas SOC in arable land is strongly correlated with current climate conditions (Delgado-Baquerizo et al., 2017a,b). These findings emphasize the importance of considering how climate legacies influence soil C content, thus allowing improvements in quantitative predictions of global C stocks under different climatic scenarios.

Different soils have different capacity to store SOC. At the global level, the SOC is concentrated in five major soil orders: histosols (357 Gt), inceptisols (352 Gt), entisols (148 Gt), alfisols (127 Gt), and oxisols (119 Gt) (Eswaran et al., 1993). SOM tends to increase as the clay content increases. Under similar climate conditions, the organic matter content in fine textured (clayey) soils is two to four times that of coarse textured (sandy) soils (Prasad and Power, 1997). This increase depends on two mechanisms. First, bonds on the surface of clay particles with organic matter are less vulnerable to the decomposition process. Second, soils with higher clay content increase the potential for aggregate formation.

Threat: Global stocks of SOC are under threat from multiple activities including dramatic changes in land use and climate change, with consequences for the loss of ecosystem services, increase in GHG emissions, and acceleration of global warming (Lal, 2010a,b). The current rate of SOC loss due to land use change (deforestation) and related land-use activities (tillage, biomass burning, residue removal, excessive fertilizers, erosion, and drainage of peatlands) is between 0.7

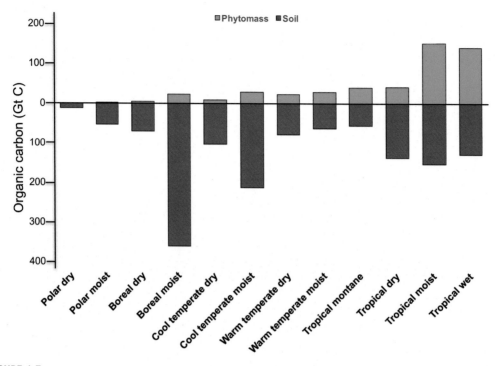

FIGURE 1.5

Distribution of terrestrial (soil and vegetation) organic carbon by IPCC climate region in soil.

The figure was adapted from Scharlemann, J.P., Tanner, E.V., Hiederer, R. and Kapos, V., 2014. Global soil carbon: understanding and managing the largest terrestrial carbon pool. Carbon Manag. 5(1), 81–91 with permission which was based on data from Hiederer, R., Kőchy, M., 2011. Global Soil Organic Carbon Estimates and the Harmonized World Soil Database. EUR 25225 EN. Publications Office of the EU, Luxembourg. (Hiederer and Kőchy, 2011) and Ruesch, A., Gibbs, H., 2008. New Global Biomass Carbon Map for the Year 2000 Based on IPCC Tier-1 Methodology. Carbon Dioxide Information Analysis Center. Oak Ridge National Laboratory, Oak Ridge, TN (Ruesch and Gibbs, 2008).

and 2.1 Gt C year^{-1}. Conversion of natural vegetation for agriculture land-use systems has led to a decrease of 60% and 75% of SOC stocks in temperate and tropical regions, respectively (FAO and ITPS, 2015). Today, 33% of land is moderately to highly degraded due to erosion, salinization, compaction, acidification, and chemical pollution of soils. The Intergovernmental Technical Panel on Soils (ITPS) concluded on the basis of the Seven Regional Assessments completed for the 2015 state of the Worlds' Soil resource that currently the SOC status in fair only in North America and poor in other regions, including Asia, Europe and Eurasia, Latin America, and the Caribbean Islands, Southwest Pacific, and Southwest pacific, Africa South of the Sahara, Near East and North Africa (FAO and ITPS, 2015). Soil erosion is the major land degradation process that accounts for up to 1.2 Gt of C emitted into the atmosphere each year. The annual soil losses in Africa, South America, and Asia are estimated at 39–74 Gt, which corresponds to C emissions of 0.16–0.44 Gt per year. Further, loss of productive soils severely damage food production and food security, amplify food-price volatility, and potentially plunge millions of people into hunger and poverty. But, the report also provides some evidence that the loss of soil resources and functions can be avoided.

Global SOC stock is sensitive to multiple climate change drivers. Drier, warmer conditions are expected to coincide with greater potential for the loss of SOC and associated soil functions. It is estimated that drylands in Central Asia lost approximately 0.46 Gt of soil C during the decade of drought between 1998 and 2008, which was possibly related to protracted La Niña episodes (Li et al., 2015). Climate change is positively correlated with the increasing rates of soil respiration which is the second largest terrestrial C flux (Bond-Lamberty and Thomson, 2010). A recent study predicts that for 1°C of warming, about 30 Gt of soil C will be released into the atmosphere, or about twice as much as could be emitted annually due to human-related activities (Crowther et al., 2016). This is particularly concerning because previous climate studies predicted that the planet is likely to warm by 2°C by mid-century (Gornall et al., 2010). Warming will result in the loss of permafrost that will expose accumulated C in cold regions to much greater rates of microbial decomposition (Schuur and Abbott, 2011). Although most of the studies have predicted SOC losses in response to warming, it should be noted that there are several other biological processes—such as accelerated plant growth as a result of CO_2 increases and warming—that could dampen or enhance the effect of SOC loss feedback (Nie et al., 2015; Crowther et al., 2016). Understanding these interacting processes at a global scale is critical to better predict the effect of climate change on SOC stocks.

Opportunity: Beneficial management of SOC offers opportunities not only to avoid negative consequences, but also to enhance a wide range of soil functions and ecosystem services. The most cost-effective mitigation options include afforestation, sustainable forest management, and reducing deforestation, with large differences in their relative importance across regions. In agriculture, best cropland management options are many for supporting and enhancing soil functions, including reduced tillage, crop rotation, integrated nutrient management, adding cover crops and particularly legumes during fallow periods, incorporating perennial vegetation, optimal grazing, and soil restorations with organic amendments (Machmuller et al., 2015). Thompson et al. (2008) estimated that by using current best land management practices, terrestrial C sequestration can reach a peak rate of 0.5−0.7 Gt C year^{-1} by mid-century with contributions from agricultural soils (0.21 Gt C year^{-1}), reforestation (0.31 Gt C year^{-1}), and pasture (0.31 Gt C year^{-1}). Climate smart agriculture that includes farm biodiversity, carbon farming, farmland conservation, integrated livestock and crop systems, renewable energy, and water conservation has been proposed to reduce agriculture's contribution to climate change by soil C sequestration and offsetting greenhouse gas emissions (Chapter 7: Agricultural Management Practices and Soil Organic Carbon Storage and Chapter 9: Projecting Soil C Under Future Climate and Land-Use Scenarios (Modeling)). Land managers, farmers, and producers can abate emissions and enhance soil C sequestration using several methods, but these stakeholders must be supported by education, incentives, and better decision support tools for the most appropriate approach on case-based situations.

A more unconventional root focused intervention has been proposed to breed plants for larger, deeper root systems, hence increasing plant C inputs and soil C sinks (Lynch and Wojciechowski, 2015). Paustian et al. (2016) have estimated that a sustained increase in root C inputs might add about 1 Gt CO_2 (cq) year^{-1} or more if applied over a large portion of global cropland area. In fact, it is well recognized that perennial grasses that incorporate greater root C can maintain higher SOC stocks compared to annual crops (Conant et al., 2001). In the United States, Advanced Research Projects agency—Energy (ARPA) has initiated a new program, Rhizosphere Observations Optimizing Terrestrial Sequestration (ROOTS), that aims to improve crop breeding for enhancing root traits and soil functions, allowing for greater C storage in both plants and soils. Addition of plant-derived C following biomass carbonization (biochar) can also increase soil C stocks (Singh et al., 2012, 2015). The carbonized biomass decomposes slower than fresh plant residues and can be retained in the soil over several decades or longer depending on the amendment type, soil conditions, and nutrient content (Schmidt et al., 2011; Singh et al., 2012; Singh and Cowie, 2014; Lehmann and Joseph, 2015). By considering the scenario of a 30-year start period for the adoption of biomass carbonization to the level of 1% of global NPP, Post et al. (2009) estimated that a characteristic storage time of 80 years will yield a net sequestration of 19 Gt C in the next century.

1.11 VALUE OF ECOSYSTEM SERVICES PROVIDED BY SOC

Soil organic C is an asset that provides multiple ecological and societal benefits and, therefore, demonstrating the economic values of these benefits can promote SOC storage by practitioners and land owners/managers (Pascual et al., 2014), while providing valuable information for policy-makers. The different components of the value of SOC differ both conceptually and with respect to how they can be measured or manifested. There are various methods for quantifying SOC values that differ with respect to the types of values they are suitable for, or able to assess. Policies have been developed that have placed an economic value to SOC storage. Many C-valuation mechanisms are in place, and despite challenges over the past decade, the valuation of C is expanding rapidly both at local and national scales (Newell et al., 2013). A number of regulated and voluntary markets have valued C accumulation in terrestrial ecosystems including forests and grasslands, and protocols have been developed with a clear view of the economic importance of SOC storage (Kelly and Schmitz, 2016). Cost-effective SOC storage, whereby C stocks in agricultural lands are enhanced, can create a value of up to US$480 billion, while increasing food and water security (http://www.greengrowthknowledge.org/resource/value-land-prosperous-lands-and-positive-rewards-through-sustainable-land-management).

Economic valuation of SOC provides information to help assess how efficiently a particular land management can reallocate goods and services from soil to different (and often competing) uses. In the United States, Wander and Nissen (2004) found the total value of SOC to be US $3.15 tons^{-1} C ha^{-1} year^{-1}(t^{-1} C ha^{-1} year^{-1}) from the top 30 cm, the majority being from productivity enhancement ($2.73 t^{-1} C ha^{-1} year^{-1}) and smaller benefits from fertilizer replacement ($0.40 t^{-1}C ha^{-1} year^{-1}) and water quality enhancement ($0.02 t^{-1} C ha^{-1} year^{-1}). Belcher et al. (2003) estimated the value of SOC changes in the Canadian Prairies to be between a loss of $0.03 t^{-1} C ha^{-1} year^{-1} and a gain of $0.74 t^{-1} C ha^{-1} year^{-1}, depending on soil type, crop rotation, and soil management methods. Petersen and Hoyle (2016) estimated the marginal value of SOC (the value of a soil with more SOC, by 1 t^{-1} C ha^{-1}, than a standard soil) to be AU $7.1−8.7 t^{-1} C ha^{-1} year^{-1}, depending on rainfall zone and crop type. Approximately 75% of this value was the estimated C sequestration, 20% was the N replacement, and 5% was the estimated productivity improvement value. These estimated benefits equated to the value of $130−160\,t^{-1}$ $C\,ha^{-1}$ over 50 years (Petersen and Hoyle 2016). It must be noted that most of these estimates focus exclusively on the on-site benefits of SOC in production systems and do not take into account the off-site benefits, including improved surface and groundwater quality. Farmers value environmental assets differently, and may adopt practices that increase SOC due to the environmental and societal values they place on SOC (Lal, 2014).

We argue that putting an economic value of SOC that takes into consideration all the ecosystem and societal services it provides, will signal the scarcity of the resource from a social viewpoint, and also the extent to which investment in enhancing SOC should be prioritized, relative to other investments (Pascual et al., 2014). Valuation of SOC will help policymakers to determine what type of economic instruments or incentives are necessary to align privately and socially optimal soil conservation decisions. Mapping the ecosystem service values of SOC needs to account for who appropriates the different values (private vs social values), whether the values are direct or indirect, and how to best emphasize the natural insurance value of SOC (Pascual et al., 2014). Combining SOC storage with other conservational efforts (i.e., biodiversity and wildlife) may well enhance the economic arguments for increasing SOC and conservation even further. Such assessments would illustrate the immense economic importance of SOC, and enable scientists, policymakers, and stakeholders to reverse the trends of SOC degradation which is bound to impact on the sustainability and productivity of ecosystems and the future well-being of the planet. While the demonstration of the economic value of SOC is important from a policy perspective, the exact value of SOC to humankind and terrestrial life is immeasurable—given the central role of SOC.

1.12 POLICIES ON SOIL CARBON

Because SOM is a major contributor of soil health and can be built up, or lost by, the type of land management, it is recognized as one of the most relevant targets for human well-being and conservation interventions. The 2015 Status of the World Soil Resources report (FAO and ITPS, 2015) highlighted that SOC is critical in the global C balance, and national governments must set specific targets to maintain or ideally increase SOC storage. These reports emphasize that the most significant threats to soil function at the global scale are soil erosion, loss of SOC, and nutrient imbalance. The reports also point that warming-induced changes in soil temperature and moisture regimes may increase the SOM decomposition rate and intensify the risks of erosion and desertification.

The introduction of the United Nations SDGs offers a unique and welcome opportunity to direct joint activities toward ending poverty, protecting the planet, and ensuring prosperity for all. These SDGs build on the success of the Millennium Development Goals, while including new areas such as climate change, economic inequality, innovation, sustainable consumption, peace and justice, among other priorities. These goals are interconnected—often the key to success of one will involve tackling issues more commonly associated with others. Soil science plays an important role in realizing a number of SDGs focusing on food, water, climate, health, biodiversity, and sustainable land use (Fig. 1.3). As such, SOC is included in the monitoring of SDG indicator 15.3.1, under which belowground and aboveground C stocks are one of the three indicators (along with land cover and land cover change, and land productivity) to determine the proportion of land that is degraded over the total land area. Addressing the increasing trends in soil and land degradation primarily due to the loss of SOC is a critical challenge for sustainable development of agroecosystems since soil degradation processes can have adverse effects on nearly all the ecosystem services provided by SOC, including food security, water quality and availability, human health, and social and economic well-being of human society. Consistent increment in the SOC stocks will increase the amount and quality of land resources necessary to support ecosystem functions and services and enhance food security, particularly in developing countries that are worst affected by desertification and land degradation.

The SDG 15.3.1 advocates the establishment of a baseline for C stocks, aboveground and belowground, with an emphasis on SOC. The Intergovernmental Panel on Climate Change (IPCC) has conceptualized guidelines for the assessment of SOC and its stock changes in the context of offsetting GHG emissions. However, despite widespread recognition of the importance of SOM, there are no reliable, quantitative targets for the amount of SOM required to achieve SDG relevant impacts, such as soil health, C sequestration, and nutrient reductions in waterways, biodiversity conservation, and sustainable land-use. The immediate challenge for soil scientists is to develop multi-disciplinary, system-level, harmonized methodologies for standardized sampling protocols, data collection and sharing, robust modeling, as well as adaptation and implementation by stakeholders. The Global Soil Partnership (GSP) and members of Food and Agriculture Organization (FAO) of the United Nations are currently working on the establishment of the Global Soil Information System to improve science-based soil management by quantifying the relationships between: SOM and crop yield, and C storage, biodiversity outcomes and nutrient retention. The Global Soil Organic Carbon (GSOC) map is proposed to be released by December 2017 which will contribute to develop SOC stocks as an indicator of land degradation (as proposed in SDG 15.3.1). Such an approach that aims to improve information on SOC stocks can constitute a unique option to reinforce the current IPCC assessment and for reporting to the United Nations Framework Convention on Climate Change (UNFCCC), the UNCCD, and SDG 15.3. These initiatives could further provide valuable information to support the IPCC sixth assessment report (AR6) and its products by contributing to the methodology report(s) to refine the 2006 IPCC Guidelines for National Greenhouse Gas Inventories.

At the 21st Climate Change Conference, the French Ministry of Agriculture launched the "4 per 1000" initiative. This initiative calls on nations to increase the C content in the top 40cm of their soils by 0.4% per year. This equates to annually removing approximately 3.5 Gt of C from the atmosphere. This initiative has been signed off by 32 nations, including Germany, France, United Kingdom, and Australia, as well as dozens of agricultural and civil society agencies. The "4 per 1000" initiative also aims to strengthen existing synergies between the three Rio Conventions—UNFCCC, UNCCD and the Convention on Biological Diversity (CBD)—and the Committee for Food Security (CFS), GSP, and SDGs. The initiative aims to reach its goals by: (1) Implementing training programs for farmers and agricultural advisors which aim to enhance organic matter in soils; (2) Contributing financially to C sequestration development projects; and (3) Developing policies that promote sustainable management of soils. In the United States, the National Science and Technology Council's Soil Science Interagency Working Group developed a framework for the federal strategic plan for soil science development. This framework identifies "climate and environmental change" as one of the three overarching "Challenge and Opportunities" categories, and recognizes the need for "terrestrial C sequestration" in soils and forests. In the United States, several states have framed and adopted policies to build adequate SOM that can sequester soil C with the primary objective to reduce overall GHG emission and increase soil health including "Healthy Soils Initiative" in California, "Carbon Sequestration Enhancement Act" in Oklahoma, "Concurrent Restoration on Carbon Sequestration on Rangelands" in Utah, and "Regenerative Soils Program legislation" in Vermont.

In spite of the fact that SOC plays a key role in soil quality and environmental health, its importance has mostly not been translated into international actions. This disconnect between the science and policy arenas at local, national, and global levels can only be resolved with an innovative framework that provides simple and clear messages to all stakeholders, equating SOC with societal priorities like growth, income, jobs, and social welfare (van Wesemael et al., 2011). The starting point should be the key cross-cutting role of SOC toward high profile topics such as food security, environmental sustainability, climate change scenarios, societal development, human well-being, and bioeconomy.

1.13 CONCLUSION

Land use change, such as conversion of native vegetation to cropping systems, and unsustainable agricultural management practices caused significant loss of SOC and land degradation over several decades. As SOC is critical for the maintenance of multiple ecosystem services related to soil health and functioning, it is considered as a strong determinant of global food and nutritional security. Improved management strategies to enhance SOC storage can support provision of essential ecosystem services while offering part of the solution to a warming climate (Box 1.2). Sustainable soil management practices that increase SOC using scientific evidence-based and local knowledge, and proven approaches and technologies, can increase nutritious food supply, provide a valuable means for climate regulation, and safeguard multiple ecosystem services. A new focus on SOC storage at all levels of governance for soil management would better enable the full potential of

---

BOX 1.2 TAKE HOME MESSAGE

- Soil stores more carbon (C) than vegetation and the atmosphere together
- Soil C is an important resource and provides key ecosystem services including provision of food and fiber, habitats of biodiversity, climate regulation, water filtration and purification, human heritage
- A significant portion of SOC has been lost due to land-use change and is under threat of further loss due to climate change, soil erosion, and inappropriate land management practices.
- An appropriate management based on scientific evidence can not only minimize SOC loss but can restore additional SOC and can contribute directly to address key global challenges including food security, environmental sustainability, and climate change mitigation and adaption.
- Increasing SOC storage can significantly improve our ability to achieve aims of multiple global and national policies including Sustainable Development Goals, and aims of IPCC, FAO, UNCCD.
- An effective integrated approach is needed which can consolidate current national and intergovernmental policies, in order to achieve stated goals of increment in SOC stocks.

ecosystem services to be realized. Therefore, several suggested soil conservation practices need to be implemented to increase SOC storage and reach the maximum potential of climate change mitigation and adaptation, as well as food productivity. It is also an essential step toward developing a framework for soil management, not only to avoid negative consequences of climate change, but also to enhance the wide range of available soil functions. However, financial, technical, logistic, institutional, knowledge, and resource and socio-cultural barriers along with physical factors and their interactions, all influence the global adaptation of practices and policies aimed at building and preserving SOC in terrestrial ecosystems. Despite some recognized solutions to overcome human induced barriers, global adoption rates of sustainable soil management practices remain below the level necessary to achieve SDGs. Going forward, greater specificity and accuracy, as well as improved methods are required to measure, account for, monitor and report SOC pools. We urgently require effective integration of national and global initiatives that engage all stakeholders including land managers, scientists, policy advisors, and science advocates to draft and implement policies to manage SOC to address the key challenges of food security, mitigation, and adaptation to climate change. Such initiatives will provide a transition toward a productive, resilient agriculture based on sustainable soil management and generating incomes, hence ensuring the sustainable development of land-based resources that last to meet the needs of future generations.

REFERENCES

Alexandratos, N., Bruinsma, J., 2012. World Agriculture Towards 2030/2050: The 2012 Revision, No. 12-03. ESA Working Paper, FAO, Rome, p. 4.

Altieri, M.A., Toledo, V.M., 2011. The agroecological revolution in Latin America: rescuing nature, ensuring food sovereignty and empowering peasants. J. Peasant Stud. 38 (3), 587–612.

Amundson, R., Berhe, A.A., Hopmans, J.W., Olson, C., Sztein, A.E., Sparks, D.L., 2015. Soil and human security in the 21st century. Science 348 (6235), 1261071.

Arshad, M.A., Martin, S., 2002. Identifying critical limits for soil quality indicators in agro-ecosystems. Agric. Ecosyst. Environ. 88 (2), 153–160.

Baldock, J.A., Nelson, P.N., 2000. Soil organic matter. In: Sumner, M.E. (Ed.), Handbook of Soil Science. CRC Press, Boca Raton, pp. 25–84.

Banwart, S., Black, H., Cai, Z., Gicheru, P., Joosten, H., Victoria, R., et al., 2014a. Benefits of soil carbon: report on the outcomes of an international scientific committee on problems of the environment rapid assessment workshop. Carbon Manag. 5 (2), 185–192.

Banwart, S.A., Noellemeyer, E., Milne, E. (Eds.), 2014b. Soil Carbon: Science, Management and Policy for Multiple Benefits, vol. 71. CABI, Wallingford.

Belcher, K.W., Boehm, M.M., Zentner, R.P., 2003. The economic value of soil quality under alternative management in the Canadian prairies. Can. J. Agric. Econ./Rev. can. d'agroecon. 51 (2), 175–196.

van den Belt, M., Blake, D., 2014. Ecosystem services in New Zealand agro-ecosystems: a literature review. Ecosyst. Serv. 9, 115–132.

Bond-Lamberty, B., Thomson, A., 2010. Temperature-associated increases in the global soil respiration record. Nature 464 (7288), 579.

Brady, M.V., Hedlund, K., Cong, R.G., Hemerik, L., Hotes, S., Machado, S., et al., 2015. Valuing supporting soil ecosystem services in agriculture: a natural capital approach. Agron. J. 107 (5), 1809–1821.

Brevik, E.C., 2010. Soil health and productivity, Soils, Plant Growth and Crop Production., vol. I. p. 106.

Cambardella, C.A., Elliott, E.T., 1992. Particulate soil organic-matter changes across a grassland cultivation sequence. Soil Sci. Soc. Am. J. 56 (3), 777–783.

Causarano, H.J., Franzluebbers, A.J., Shaw, J.N., Reeves, D.W., Raper, R.L., Wood, C., 2008. Soil organic carbon fractions and aggregation in the Southern Piedmont and Coastal Plain. Soil Sci. Soc. Am. J. 72 (1), 221–230.

Cayley, J.W.D., McCaskill, M.R., Kearney, G.A., 2002. Changes in pH and organic carbon were minimal in a long-term field study in the Western District of Victoria. Austr. J. Agric. Res. 53 (2), 115–126.

Comerford, N.B., Franzluebbers, A.J., Stromberger, M.E., Morris, L., Markewitz, D., Moore, R., 2013. Assessment and evaluation of soil ecosystem services. Soil Horiz. 54, 3.

Conant, R.T., Paustian, K., Elliott, E.T., 2001. Grassland management and conversion into grassland: effects on soil carbon. Ecol. Appl. 11 (2), 343–355.

Cooper Marcus, C., Barnes, M., 1999. Healing Gardens. John Wiley & Sons, New York.

Cork, S., Eadie, L., Mele, P., Price, R., Yule, D., 2012. The Relationships Between Land Management Practices and Soil Condition and the Quality of Ecosystem Services Delivered From Agricultural land in Australia. Kiri-ganai Research Pty Ltd., Canberra, Australia.

Crowther, T.W., Todd-Brown, K.E.O., Rowe, C.W., Wieder, W.R., Carey, J.C., Machmuller, M.B., et al., 2016. Quantifying global soil carbon losses in response to warming. Nature 540 (7631), 104–108.

De Graaff, M.A., Adkins, J., Kardol, P., Throop, H.L., 2015. A meta-analysis of soil biodiversity impacts on the carbon cycle. Soil 1 (1), 257.

Delgado-Baquerizo, M., Maestre, F.T., Reich, P.B., Jeffries, T.C., Gaitan, J.J., Encinar, D., et al., 2016a. Microbial diversity drives multifunctionality in terrestrial ecosystems. Nat. Commun. 7, 10541.

Delgado-Baquerizo, M., Maestre, F.T., Reich, P.B., Trivedi, P., Osanai, Y., Liu, Y.R., et al., 2016b. Carbon content and climate variability drive global soil bacterial diversity patterns. Ecol. Monogr. 86 (3), 373–390.

Delgado-Baquerizo, M., Powell, J.R., Hamonts, K., Reith, F., Mele, P., Brown, M.V., et al., 2017a. Circular linkage between soil biodiversity, fertility and plant productivity are limited to topsoil at the continental scale. New Phytol. 215 (3), 1186–1196.

Delgado-Baquerizo, M., Eldridge, D.J., Maestre, F.T., Karunaratne, S.B., Trivedi, P., Reich, P.B., et al., 2017b. Climate legacies drive global soil carbon stocks in terrestrial ecosystems. Sci. Adv. 3 (4), e1602008.

Deng, L., Zhu, G.Y., Tang, Z.S., Shangguan, Z.P., 2016. Global patterns of the effects of land-use changes on soil carbon stocks. Glob. Ecol. Conserv. 5, 127–138.

Don, A., Schumacher, J., Freibauer, A., 2011. Impact of tropical land-use change on soil organic carbon stocks- a meta-analysis. Glob. Change Biol. 17, 1658–1660.

Doran, J.W., Zeiss, M.R., 2000. Soil health and sustainability: managing the biotic component of soil quality. Appl. Soil Ecol. 15 (1), 3–11.

Doran, J.W., Parkin, T.B., Doran, J.W., Coleman, D.C., Bezdicek, D.F., Stewart, B.A., 1994. Defining and assessing soil quality. In: Doran, J.W., Coleman, D.C., Bezdicek, D.F., Stewart, B.A. (Eds.), Defining Soil Quality for a Sustainable Environment. Soil Science Society of America Special Publication, 35. Soil Science Society of America, Madison, Wisconsin, pp. 3–21.

Dungait, J.A., Hopkins, D.W., Gregory, A.S., Whitmore, A.P., 2012. Soil organic matter turnover is governed by accessibility not recalcitrance. Glob. Change Biol. 18 (6), 1781–1796.

Duxbury, J.M., Smith, M.S., Doran, J.W., 1989. Soil organic matter as a source and sink of plant nutrients. In: Coleman, D.C., Oades, J.M., Uehara, G. (Eds.), Dynamics of Soil Organic Matter in Tropical Ecosystems. University of Hawaii Press, Honolulu.

Edmonds, J.A., Dooley, J.J., Kim, S.H., 1999. Long-term energy technology needs and opportunities for stabilizing atmospheric CO_2 concentrations. In: Climate Change Policy: Practical Strategies to Promote Economic Growth and Environmental Quality. Monograph Series on Tax, Trade, and Environmental

Policies and Economic Growth from the American Council for Capital Formation, Washington, DC, pp. 81−107.

Eswaran, H., Van Den Berg, E., Reich, P., 1993. Organic carbon in soils of the world. Soil Sci. Soc. Am. J. 57 (1), 192−194.

FAO, 2015. The state of food and agriculture. Social Protection and Agriculture: Breaking the Cycle of Rural Poverty. FAO, Rome, Italy, 129 pp.

FAO, 2017. Ecosystem Services & Biodiversity (ESB). <http://www.fao.org/ecosystem-services-biodiversity/background/provisioning-services/en/> (accessed 08.09.17.).

FAO and ITPS, 2015. Status of the World's Soil Resources (SWSR)—Main Report. Food and Agriculture Organization of the United Nations and Intergovernmental Technical Panel on Soils, Rome, Italy.

Feller, C., Landa, E.R., Toland, A., Wessolek, G., 2015. Case studies of soil in art. Soil 1 (2), 543−559.

Franzluebbers, A.J., 2002. Soil organic matter stratification ratio as an indicator of soil quality. Soil Tillage Res. 66 (2), 95−106.

Franzluebbers, A.J., Stuedemann, J.A., 2008. Early response of soil organic fractions to tillage and integrated crop−livestock production. Soil Sci. Soc. Am. J. 72 (3), 613−625.

Gornall, J., Betts, R., Burke, E., Clark, R., Camp, J., Willett, K., et al., 2010. Implications of climate change for agricultural productivity in the early twenty-first century. Philos. Trans. R. Soc. Lond. B: Biol. Sci. 365 (1554), 2973−2989.

Guerry, A.D., Polasky, S., Lubchenco, J., Chaplin-Kramer, R., Daily, G.C., Griffin, R., et al., 2015. Natural capital and ecosystem services informing decisions: From promise to practice. Proc. Natl. Acad. Sci. 112 (24), 7348−7355.

Guo, L.B., Gifford, R.M., 2002. Soil carbon stocks and land use change: a meta analysis. Glob. Change Biol. 8 (4), 345−360.

Haynes, R.J., Naidu, R., 1998. Influence of lime, fertilizer and manure applications on soil organic matter content and soil physical conditions: a review. Nutr. Cycl. Agroecosyst. 51 (2), 123−137.

Hazell, P., Wood, S., 2008. Drivers of change in global agriculture. Philos. Trans. R. Soc. Lond. B: Biol. Sci. 363 (1491), 495−515.

Hiederer, R., Köchy, M., 2011. Global Soil Organic Carbon Estimates and the Harmonized World Soil Database. EUR 25225 EN. Publications Office of the EU, Luxembourg.

Högberg, P., Nordgren, A., Buchmann, N., Taylor, A.F., 2001. Large-scale forest girdling shows that current photosynthesis drives soil respiration. Nature 411 (6839), 789.

House, J.I., Prentice, I.C., Ramankutty, N., Houghton, R.A., Heimann, M., 2003. Reconciling apparent inconsistencies in estimates of terrestrial CO_2 sources and sinks. Tellus 55B, 345−363.

Huber, S., Syed, B., Freudenschuß, A., Ernstsen, V., Loveland, P., 2001. Proposal for a European Soil Monitoring and Assessment Framework. Technical Report No. 61. European Environment Agency, Copenhagen, 58 pp.

Janzen, H.H., 2006. The soil carbon dilemma: shall we hoard it or use it? Soil Biol. Biochem. 38 (3), 419−424.

Jastrow, J.D., Miller, R.M., 1997. Soil aggregate stabilization and carbon sequestration: feedbacks through organomineral associations. Soil Process. Carbon Cycle 207−223.

Jobbágy, E.G., Jackson, R.B., 2000. The vertical distribution of soil organic carbon and its relation to climate and vegetation. Ecol. Appl. 10 (2), 423−436.

Jung, M., Le Maire, G., Zaehle, S., Luyssaert, S., Vetter, M., Churkina, G., et al., 2007. Assessing the ability of three land ecosystem models to simulate gross carbon uptake of forests from boreal to Mediterranean climate in Europe. Biogeosciences 4, 647−656.

Karhu, K., Auffret, M.D., Dungait, J.A., Hopkins, D.W., Prosser, J.I., Singh, B.K., et al., 2014. Temperature sensitivity of soil respiration rates enhanced by microbial community response. Nature 513 (7516), 81−84.

Karlen, D.L., Mausbach, M.J., Doran, J.W., Cline, R.G., Harris, R.F., Schuman, G.E., 1997. Soil quality: a concept, definition, and framework for evaluation (a guest editorial). Soil Sci. Soc. Am. J. 61 (1), 4–10.

Keesstra, S.D., Quinton, J.N., van der Putten, W.H., Bardgett, R.D., Fresco, L.O., 2016. The significance of soils and soil science towards realization of the United Nations Sustainable Development Goals. Soil 2 (2), 111.

Kelly, E.C., Schmitz, M.B., 2016. Forest offsets and the California compliance market: Bringing an abstract ecosystem good to market. Geoforum 75, 99–109.

King, G.M., 2011. Enhancing soil carbon storage for carbon remediation: potential contributions and constraints by microbes. Trends Microbiol. 19 (2), 75–84.

Kirchmann, H., Andersson, R., 2001. The Swedish system for quality assessment of agricultural soils. Environ. Monit. Assess. 72 (2), 129–139.

Köchy, M., Hiederer, R., Freibauer, A., 2015. Global distribution of soil organic carbon–Part 1: Masses and frequency distributions of SOC stocks for the tropics, permafrost regions, wetlands, and the world. Soil 1 (1), 351–365.

Kovda, I.V., Morgun, E.G., Lebedeva, M.P., Oleinik, S.A., Shishkov, V.A., 2016. Identification of carbonate pedofeatures of different ages in modern chernozems. Eurasian Soil Sci. 49 (7), 807–823.

Kremen, C., 2005. Managing ecosystem services: what do we need to know about their ecology? Ecology Lett. 8 (5), 468–479.

Lal, R., 1997. Degradation and resilience of soils. Philos. Trans. R. Soc. Lond. B: Biol. Sci. 352 (1356), 997–1010.

Lal, R., 2004. Soil carbon sequestration impacts on global climate change and food security. Science 304 (5677), 1623–1627.

Lal, R., 2006. Managing soils for feeding a global population of 10 billion. J. Sci. Food Agric. 86 (14), 2273–2284.

Lal, R., 2007. Soil science and the carbon civilization. Soil Sci. Soc. Am. J. 71 (5), 1425–1437.

Lal, R., 2010a. Enhancing eco-efficiency in agro-ecosystems through soil carbon sequestration. Crop Sci. 50 (Supplement_1), S-120.

Lal, R., 2010b. Managing soils and ecosystems for mitigating anthropogenic carbon emissions and advancing global food security. Bioscience 60 (9), 708–721.

Lal, R., 2014. Societal value of soil carbon. J. Soil Water Conserv. 69 (6), 186A–192A.

Lal, R., 2016. Soil health and carbon management. Food Energy Secur. 5 (4), 212–222.

Le Quéré, C., Moriarty, R., Andrew, R.M., Peters, G.P., Ciais, P., Friedlingstein, P., et al., 2015. Global carbon budget 2014. Earth Syst. Sci. Data 7 (1), 47–85.

Lefèvre, C., Rekik, F., Alcantara, V., Wiese, L., 2017. Soil Organic Carbon: The Hidden Potential. Food and Agriculture Organization of the United Nations (FAO), Rome, Italy, 77 pp.

Lehmann, J., Joseph, S. (Eds.), 2015. Biochar for Environmental Management: Science, Technology and Implementation. Routledge, London.

Lehmann, J., Kleber, M., 2015. The contentious nature of soil organic matter. Nature 528 (7580), 60–68.

Lehman, R.M., Cambardella, C.A., Stott, D.E., Acosta-Martinez, V., Manter, D.K., Buyer, J.S., et al., 2015. Understanding and enhancing soil biological health: the solution for reversing soil degradation. Sustainability 7 (1), 988–1027.

Lemanceau, P., Maron, P.A., Mazurier, S., Mougel, C., Pivato, B., Plassart, P., et al., 2015. Understanding and managing soil biodiversity: a major challenge in agroecology. Agron. Sustain. Dev. 35 (1), 67–81.

Li, C., Zhang, C., Luo, G., Chen, X., Maisupova, B., Madaminov, A.A., et al., 2015. Carbon stock and its responses to climate change in Central Asia. Glob. Change Biol. 21 (5), 1951–1967.

Louis, B.P., Maron, P.A., Viaud, V., Leterme, P., Menasseri-Aubry, S., 2016. Soil C and N models that integrate microbial diversity. Environ. Chem. Lett. 14 (3), 331–344.

Loveland, P., Webb, J., 2003. Is there a critical level of organic matter in the agricultural soils of temperate regions: a review. Soil Tillage Res. 70 (1), 1–18.

Luo, Z., Wang, E., Zheng, H., Baldock, J.A., Sun, O.J., Shao, Q., 2015. Convergent modelling of past soil organic carbon stocks but divergent projections. Biogeosciences 12 (14), 4373–4383.

Lynch, J.P., Wojciechowski, T., 2015. Opportunities and challenges in the subsoil: pathways to deeper rooted crops. J. Exp. Bot. 66 (8), 2199–2210.

Lützow, M.V., Kögel-Knabner, I., Ekschmitt, K., Matzner, E., Guggenberger, G., Marschner, B., et al., 2006. Stabilization of organic matter in temperate soils: mechanisms and their relevance under different soil conditions—a review. Eur. J. Soil Sci. 57 (4), 426–445.

Machmuller, M.B., Kramer, M.G., Cyle, T.K., Hill, N., Hancock, D., Thompson, A., 2015. Emerging land use practices rapidly increase soil organic matter. Nature Commun. 6, 6995.

McGrath, S.P., Sanders, J.R., Shalaby, M.H., 1988. The effects of soil organic matter levels on soil solution concentrations and extractibilities of manganese, zinc and copper. Geoderma 42, 177–188.

Newell, R.G., Pizer, W.A., Raimi, D., 2013. Carbon markets 15 years after Kyoto: Lessons learned, new challenges. J. Econ. Perspect. 27 (1), 123–146.

Nie, M., Bell, C., Wallenstein, M.D., Pendall, E., 2015. Increased plant productivity and decreased microbial respiratory C loss by plant growth-promoting rhizobacteria under elevated CO_2. Sci. Rep. 1–5.

Nielsen, U.N., Wall, D.H., Six, J., 2015. Soil biodiversity and the environment. Annu. Rev. Environ. Resour. 40, 63–90.

Nimmo, J.R., Perkins, K.S., 2002. 2.6 Aggregate stability and size distribution. Methods of Soil Analysis, Part, 4. Soil Society of America, Madison, Wisconsin, pp. 317–328.

Nordgren, A., Ottosson Löfvenius, M., Högberg, M.N., Mellander, P.E., Högberg, P., 2003. Tree root and soil heterotrophic respiration as revealed by girdling of boreal Scots pine forest: extending observations beyond the first year. Plant Cell Environ. 26 (8), 1287–1296.

Oades, J.M., 1993. The role of biology in the formation, stabilization and degradation of soil structure. Geoderma 56 (1–4), 377–400.

Olson, G.W., 1981. Archaeology: lessons on future soil use. J. Soil Water Conserv. 261, 261–264.

Orgiazzi, A., Panagos, P., Yigini, Y., Dunbar, M.B., Gardi, C., Montanarella, L., et al., 2016. A knowledge-based approach to estimating the magnitude and spatial patterns of potential threats to soil biodiversity. Sci. Tot. Environ. 545, 11–20.

Parton, W.J., Schimel, D.S., Ojima, D.S., Cole, C.V., 1994. A general model for soil organic matter dynamics: sensitivity to litter chemistry, texture and management. In: Bryant, R.B., Arnold, R.W. (Eds.), Quantitative Modelling of Soil Forming Processes. pp. 147–167. ASA, CSSA and SSA, Madison, Wisconsin. SSSA Special Publication 39.

Pascual, U., Phelps, J., Garmendia, E., Brown, K., Corbera, E., Martin, A., et al., 2014. Social equity matters in payments for ecosystem services. BioScience 64 (11), 1027–1036.

Paustian, K., Lehmann, J., Ogle, S., Reay, D., Robertson, G.P., Smith, P., 2016. Climate-smart soils. Nature 532 (7597), 49–57.

Petersen, E.H., Hoyle, F.C., 2016. Estimating the economic value of soil organic carbon for grains cropping systems in Western Australia. Soil Res. 54 (4), 383–396.

Pimentel, D., Burgess, M., 2013. Soil erosion threatens food production. Agriculture 3 (3), 443–463.

Post, W.M., Amonette, J.E., Birdsey, R., Garten, C.T., Izaurralde, R.C., Jardine, P.M., et al., 2009. Terrestrial biological carbon sequestration: science for enhancement and implementation. Carbon Sequestration and Its Role in the Global Carbon cycle. American Geophysical Union, Washington, DC, pp. 73–88.

Power, A.G., 2010. Ecosystem services and agriculture: tradeoffs and synergies. Philos. Trans. R. Soc. Lond. B: Biol. Sci. 365 (1554), 2959–2971.

Prasad, R., Power, J.F., 1997. Soil Fertility Management for Sustainable Agriculture. Lewis Publishers, New York, NY, 356 pp.

Pribyl, D.W., 2010. A critical review of the conventional SOC to SOM conversion factor. Geoderma 156 (3), 75−83.

Rice, J.S., Remy, L.L., 1998. Impact of horticultural therapy on psychosocial functioning among urban jail inmates. J. Offender Rehabil. 26 (3−4), 169−191.

Rosenberg, N.J., Izaurralde, R.C., Malone, E.L. (Eds.), 1999. Carbon Sequestration in Soils: Science, Monitoring and Beyond. Battelle Press, Columbus, OH, 201 pp.

Rillig, M.C., Muller, L.A., Lehmann, A., 2017. Soil aggregates as massively concurrent evolutionary incubators. ISME J. 11 (9), 1943−1948.

Ruesch, A., Gibbs, H., 2008. New Global Biomass Carbon Map for the Year 2000 Based on IPCC Tier-1 Methodology. Carbon Dioxide Information Analysis Center, Oak Ridge National Laboratory, Oak Ridge, TN.

Scharlemann, J.P., Tanner, E.V., Hiederer, R., Kapos, V., 2014. Global soil carbon: understanding and managing the largest terrestrial carbon pool. Carbon Manag. 5 (1), 81−91.

Schmidt, M.W., Torn, M.S., Abiven, S., Dittmar, T., Guggenberger, G., Janssens, I.A., et al., 2011. Persistence of soil organic matter as an ecosystem property. Nature 478 (7367), 49.

Schuur, E.A., Abbott, B., 2011. Climate change: high risk of permafrost thaw. Nature 480 (7375), 32−33.

Shiva, V., Deutscher, Y., Hurtado, L., Pembamoyo, E., Lemons, M., Adhikari, J., et al., 2016. Religion and Sustainable Agriculture: World Spiritual Traditions and Food Ethics. University Press of Kentucky, Kentucky.

Singh, B.K., 2010. Exploring microbial diversity for biotechnology: the way forward. Trends Biotechnol. 28 (3), 111−116.

Singh, B.K., Trivedi, P., 2017. Microbiome and the future for food and nutrient security. Microb. Biotechnol. 10 (1), 50−53.

Singh, B.K., Bardgett, R.D., Smith, P., Reay, D.S., 2010. Microorganisms and climate change: terrestrial feedbacks and mitigation options. Nat. Rev. Microbiol. 8 (11), 779.

Singh, B.P., Cowie, A.L., 2014. Long-term influence of biochar on native organic carbon mineralisation in a low-carbon clayey soil. Sci. Rep. 4, 3687.

Singh, B.P., Cowie, A.L., Smernik, R.J., 2012. Biochar carbon stability in a clayey soil as a function of feedstock and pyrolysis temperature. Environ. Sci. Technol. 46 (21), 11770−11778.

Singh, B.P., Fang, Y., Boersma, M., Collins, D., Van Zwieten, L., Macdonald, L.M., 2015. In situ persistence and migration of biochar carbon and its impact on native carbon emission in contrasting soils under managed temperate pastures. PLoS One 10 (10), e0141560.

Six, J., Bossuyt, H., Degryze, S., Denef, K., 2004. A history of research on the link between (micro) aggregates, soil biota, and soil organic matter dynamics. Soil Tillage Res. 79 (1), 7−31.

Smith, P., Cotrufo, M.F., Rumpel, C., Paustian, K., Kuikman, P.J., Elliott, J.A., et al., 2015. Biogeochemical cycles and biodiversity as key drivers of ecosystem services provided by soils. Soil Discuss. 2 (1), 537−586.

Soil Science Society of America, 2017. Glossary of Soil Science Terms <https://www.soils.org/publications/soils-glossary/>.

Srinivasarao, C., Lal, R., Kundu, S., Thakur, P.B., 2015. Conservation agriculture and soil carbon sequestration. Conservation Agriculture. Springer International Publishing, Switzerland, pp. 479−524.

Stevenson, F.J., 1994. Humus Chemistry. Genesis, Composition, Reactions. Wiley and Sons, New York.

Tatarko, J., 2001. Soil aggregation and wind erosion: processes and measurements. Ann. Arid Zone 40 (3), 251−264.

Thompson, D.W., Kennedy, J.J., Wallace, J.M., Jones, P.D., 2008. A large discontinuity in the mid-twentieth century in observed global-mean surface temperature. Nature 453 (7195), 646.

Thomson, A.M., Izaurralde, R.C., Smith, S.J., Clarke, L.E., 2008. Integrated estimates of global terrestrial carbon sequestration. Glob. Environ. Change 18 (1), 192−203.

Tiessen, H., Cuevas, E., Chacon, P., 1994. The role of soil organic matter in sustaining soil fertility. Nature 371 (6500), 783−785.

Tilman, D., Isbell, F., Cowles, J.M., 2014. Biodiversity and ecosystem functioning. Annu. Rev. Ecol. Evol. Syst. 45, 471−493.

Tisdall, J.M., Oades, J., 1982. Organic matter and water-stable aggregates in soils. Eur. J. Soil Sci. 33 (2), 141−163.

Trivedi, P., Anderson, I.C., Singh, B.K., 2013. Microbial modulators of soil carbon storage: integrating genomic and metabolic knowledge for global prediction. Trends Microbiol. 21 (12), 641−651.

Trivedi, P., Rochester, I.J., Trivedi, C., Van Nostrand, J.D., Zhou, J., Karunaratne, S., et al., 2015. Soil aggregate size mediates the impacts of cropping regimes on soil carbon and microbial communities. Soil Biol. Biochem. 91, 169−181.

Trivedi, P., Delgado-Baquerizo, M., Anderson, I.C., Singh, B.K., 2016a. Response of soil properties and microbial communities to agriculture: implications for primary productivity and soil health indicators. Front. Plant Sci. 7.

Trivedi, P., Delgado-Baquerizo, M., Trivedi, C., Hu, H., Anderson, I.C., Jeffries, T.C., et al., 2016b. Microbial regulation of the soil carbon cycle: evidence from gene−enzyme relationships. ISME J. 10 (11), 2593−2604.

Trivedi, P., Delgado-Baquerizo, M., Jeffries, T.C., Trivedi, C., Anderson, I.C., Lai, K., et al., 2017. Soil aggregation and associated microbial communities modify the impact of agricultural management on carbon content. Environ. Microbiol. 19 (8), 3070−3086.

Wander, M., Nissen, T., 2004. Value of soil organic carbon in agricultural lands. Mitig. Adapt. Strateg. Glob. Change 9 (4), 417−431.

van Wesemael, B., Paustian, K., Andrén, O., Cerri, C.E., Dodd, M., Etchevers, J., et al., 2011. How can soil monitoring networks be used to improve predictions of organic carbon pool dynamics and CO_2 fluxes in agricultural soils? Plant Soil 338 (1−2), 247−259.

Wolf, B., Snyder, G.H., 2003. Sustainable soils. The Place of Organic Matter in Sustaining Soils and Their Productivity. Food Products Press, New York, 352 pp.

Wood, S.A., Sokol, N., Bell, C.W., Bradford, M.A., Naeem, S., Wallenstein, M.D., et al., 2016. Opposing effects of different soil organic matter fractions on crop yields. Ecol. Appl. 26 (7), 2072−2085.

Yuste, J.C., Penuelas, J., Estiarte, M., Garcia-Mas, J., Mattana, S., Ogaya, R., et al., 2011. Drought-resistant fungi control soil organic matter decomposition and its response to temperature. Glob. Change Biol. 17 (3), 1475−1486.

PLANT COMMUNITIES AS MODULATORS OF SOIL CARBON STORAGE

2

Ellen L. Fry, Jonathan R. De Long and Richard D. Bardgett
The University of Manchester, Manchester, United Kingdom

2.1 INTRODUCTION

The carbon (C) storage capacity of an ecosystem's soils is contingent on the production and composition of plant communities, as they are the primary contributors of organic C to terrestrial ecosystems. Stored soil C consists of inorganic C from weathered and eroded rock, and organic forms from plants, animals, and microbes. Plant-derived soil C can be viewed as a continuum of degradation, from intact plant material to low molecular weight C from root exudates to molecules in the last stages of decomposition. Recent calls to increase soil C storage in light of anthropogenic climate change and soil degradation mean that land management and agricultural practises are now being optimized to increase the sequestration and stabilization of C in the soil (Panagos et al., 2015; Keesstra et al., 2016). Soil is the second largest actively cycling global C pool after the ocean (Janzen, 2004). Conservative estimates predict that globally 500–3000 Gt of organic C resides in the top 3 m of soil (Jobbagy and Jackson, 2000; Scharlemann et al., 2014), with the majority occurring in the top 1 m ($\sim$1500 Gt) (Batjes, 1996). The amount of photosynthetically derived C available for assimilation into long-term storage is dependent on plant communities, which can be considered in terms of community or vegetation type (forest, peatland and so on), or within-community species composition (hereafter referred to as community composition), both of which will be discussed in this chapter. Annual C uptake ranges from 0.2 g C m^{-2} per year in polar deserts to 12 g C m^{-2} per year in broadleaf evergreen forests, with a global average of 2.4 g C m^{-2} per year (Schlesinger, 1990). Therefore, optimizing C sequestration potential of terrestrial ecosystems is one of the most effective ways to increase their C sink capacity, thereby mitigating climate change (Schlesinger, 1990; De Deyn et al., 2008; Smith et al., 2008).

Standing biomass of terrestrial plants comprises 560 Gt of C and adds about 60 Gt of C year^{-1} to the soil (Stockmann et al., 2013). However, this plant-derived C is added to soil in a wide range of molecular forms that vary in their complexity. As a result, the subsequent fate of this C is complex. For example, low molecular weight metabolic compounds such as sugars and amino acids can have a residence time of minutes, while larger and more complex structural molecules such as lignin or cellulose can persist in the soil up to 1200 years (Trumbore, 2000). However, there is increasing evidence that "lability" or "recalcitrance" of molecular forms is misleading, and that adsorption of molecules into "protected" organomineral complexes or aggregates can occur at any molecular weight. Therefore, C protection status is actually more indicative of long-term storage (Dungait et al., 2012).

Soil Carbon Storage. DOI: https://doi.org/10.1016/B978-0-12-812766-7.00002-0

Stored plant-derived soil C concentrations correspond with root depth, which varies dramatically across vegetation types (Jobbagy and Jackson, 2000). This is also closely linked with volume of precipitation—typically the greater the volumes of precipitation, the deeper the roots and the higher the C stock (Jobbagy and Jackson, 2000). Additionally, litter holds approximately $80-160$ Gt of C globally (Matthews, 1997), with litter production estimated at $90-100$ Gt C dm^{-1} per year and an estimated pool size of leaf litter of 29.3 Gt C, and this varies with vegetation type and land-use change. Soil organic C (SOC) is deepest in shrublands (which have the deepest roots of all plants; Canadell et al., 1996), followed by forests and grasslands. Nonetheless, there is a huge disparity in soil depth and vegetation types across the globe, and this means that the amount and type of soil C stored varies considerably (De Deyn et al., 2008). A key problem for accurately assessing soil C stocks is that such stocks have traditionally not been measured to depths greater than 1 m across contrasting biomes, and in many cases to even shallower depths, which consistently underestimates the true soil C stock (Batjes, 1996; Ward et al., 2016). This is a particularly pressing problem in nearly every major ecosystem, as nearly all biomes have plants with rooting depths that go far deeper than 1 m (Canadell et al., 1996).

Variation within and between plant communities influences soil C stocks through modulating the amount of C that is taken up, how quickly it is cycled, and how well protected it is from metabolism and loss through the transformation and stabilization of C-based molecules. Carbon storage potential, therefore, depends strongly on the plant community type and plant tissue turnover within the community. This variation is extremely complex because plant communities themselves assemble in response to climate and soil type, and a myriad of other factors. All of these factors directly impact on primary productivity, tissue type and quality, and interactions with soil microbial and faunal communities. Therefore, there is still much to be done to enable C modelers to improve the accuracy of their projections, especially with regard to global change factors such as pollution, extreme weather events, and land-use change.

In this chapter, we describe the general mechanisms by which plants influence soil C cycling across vegetation types, moving from photosynthetic pathways, to microbial interactions with plants, and finally to the transfer of photosynthetically derived C into the soil via root exudation, as well as root and litter inputs. Having explored the main processes, we then take a finer scale approach, considering the chemical composition of different C-skeleton molecules and how this contributes to soil C storage across a range of plant communities. Then, both physical and microbial mechanisms for incorporation of C molecules into the soil organic matter (SOM) are assessed. We then move to the influence of plant-soil feedbacks (PSFs) on soil C, which incorporates both abiotic and biotic effects on plant communities. Having identified the main processes and building blocks of C in plant communities, we then consider the role of changes in plant community composition on soil C sequestration, exploring the role of plant species identity, biomass, species, and functional diversity. Finally, we address large, broad-scale effects of weather and climate, including climate change on plant communities, with implications for C storage.

2.2 PLANT COMMUNITY EFFECTS ON SOIL CARBON CYCLING

2.2.1 PHOTOSYNTHETIC ASSIMILATION OF CARBON

Plants are the primary agents for C addition to terrestrial ecosystems. As autotrophs, they have the ability to harvest sunlight, water, and carbon dioxide (CO_2) to create solid carbohydrates for use as structural components, energy stocks, and so-called "C currency" with microbes (Fig. 2.1). Rates of

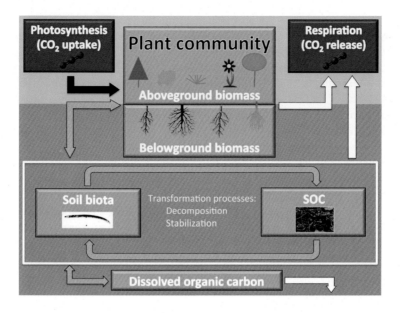

FIGURE 2.1

A simplified conceptual diagram of how the plant community can impact on C assimilation into (*black arrow*), cycling within (*gray arrows*) and loss from (*white arrows*) the ecosystem. Photosynthesis is the process by which CO_2 is assimilated into the system. However, the composition of the plant community (i.e., conservative vs. acquisitive species, contrasting functional traits) can affect how much C is stored in standing aboveground and belowground biomass. The plant community also can control how much C is either shunted to the belowground subsystem (via, e.g., shoot and root litter, root exudates, mycorrhizal inputs) or lost to the atmosphere through autotrophic respiration. Once C finds its way from the plant community into the soil subsystem, it can be transformed into soil biota biomass (i.e., microbial, soil fauna) or soil organic C (SOC) that is protected in aggregates or bound to soil particles. Transformations of C between different pools in the soil can occur through various decomposition (i.e., physical, chemical breakdown) and stabilization (i.e., aggregation via mycorrhizal hyphae, root mucilage, formation of compounds resistant to decay) processes. Further, organic C in the soil is lost through heterotrophic respiration, as dissolved organic C via leaching or it may be taken up again by the plant community.

Photo credits: Jonathan R. De Long, Tyler Logan.

photosynthesis vary according to time of day, season, climate, and plant community type. Photosynthesis is limited by a wide range of variables including water, nutrients, light availability, temperature, and biotic factors such as defoliation and pathogen attack. All of these factors vary with vegetation type. Photosynthetic machinery is extremely sensitive to external stimuli and can respond within seconds to changing light levels, meaning that the supply of photosynthate to the soil and microbial symbionts could be subject to fluctuation (Geiger and Servaites, 1994).

Seasonal dynamics become increasingly important in soil C cycling as latitude increases. Photosynthetic capacity is strongly seasonally dependent, and so, in areas that have a pronounced summer-winter cycle, most of the plant driven C assimilation will take place during the summer months. Summer heat can create difficulty for plants—loss of water can be reduced through increased stomatal control, but this has consequences for C uptake, which also occurs through the stomata. At very high latitudes, plant communities have solved the problem of reduced winter daylight by becoming evergreen—boreal forests and tundra species retain their foliage in order to optimize every photon possible. Deciduous perennial species and summer annuals can cope with photosynthesizing only during the summer months, with an adaptation to either lose leaves and "hibernate" or achieve a full lifecycle within one growing season.

The wide range of climatic and seasonal patterns seen in terrestrial systems has resulted in three photosynthetic pathways that have evolved in higher plants, and these have different implications for C storage and cycling. These pathways are C_3 in temperate plants and forests, C_4 in hotter, drier areas, and Crassulacean Acid Metabolism (CAM), which is found in very arid conditions. The latter two evolved because, in hotter temperatures, plants close their stomata to retain water, which results in O_2 being used in the Calvin cycle rather than CO_2, a process known as photorespiration (Hatch and Slack, 1966; Sage, 2016). C_4 plants separate the initial CO_2 fixation from the Calvin cycle machinery in space, while CAM plants separate them in time (i.e., night and day). This allows photosynthesis and productivity to remain high. More productive species, such as those typically found in crops with a C_4 type photosynthetic pathway (e.g., sorghum or maize), tend to exhibit higher photosynthetic rates, partly because of increased leaf surface areas and because higher growth rates require the rapid genesis of structural materials (Guo et al., 1994; Gitelson et al., 2015). Systems heavily dominated by C_4 plants include maize, sugar cane and sorghum crops, palms, tropical, and subtropical grasslands (e.g., Costa Rica; Powers and Veldkamp, 2005), Hawaiian tropical forests (Pearcy and Calkin, 1983), and saline desert plants such as *Bienertia* spp. (Sage, 2016). The C_4 pathway is not found in Gymnosperms, meaning that needle-leaved species are less common in lower- than in higher latitudes, and tend to be found mostly at higher elevations in subtropical areas (Bond, 1989). This is because in hotter temperatures or at low CO_2, C_3 plants cannot distinguish between CO_2 and O_2 for uptake (Edwards et al., 2010). Therefore, plant communities are dominated by C_3 where it's cool, shady, or elevated, and C_4 where it's hot and dry, saline or marshy—note that tropical rainforest vegetation tends to be C_3 because the high humidity in these systems means that stomatal closure is not often as necessary as in arid locations, although Hawaii is an exception; Powers and Veldkamp, 2005.

Once photosynthetically assimilated, C is partitioned between a number of C pools including respiration, growth, root exudation, chemical or structural defense, and mycorrhizal mutualism. However, allocation of C to specific pools is strongly determined by plant community type. For example, plant communities adapted to harsh environmental conditions such as the subarctic tundra tend to exhibit conservative growth strategies, directing more C to belowground organs in order to better withstand temperature or microbially-induced nutrient stresses (De Long et al., 2015). Plants with more conservative growth strategies invest more C into producing either structural defense (e.g., thorns, spines, stinging hairs; Pérez-Harguindeguy et al., 2000), chemical defense (e.g., tannins, terpene, polyphenols; Iason et al., 2012) or structural and storage organs (e.g., tubers, thick tap roots; Puijalon et al., 2008), because they tend to occur in very resource-poor areas. Plants with more acquisitive growth habits will have highly plastic growth strategies (Grassein et al., 2010) and invest more C into cheaper, more expendable structures in order to maximize photosynthesis and resource foraging. Essentially, plant communities that are dominated by fast-growing acquisitive species allocate C to maximize growth over defense, operating under the premise of compensatory growth (McNaughton, 1983). Nevertheless, it must be noted that conservative versus acquisitive growth strategies among plant communities are strongly dictated by nutrient availability, which subsequently influences C allocation. This highlights the necessity of considering both abiotic and biotic factors that control C allocation within and between plant communities. There is also a well-studied trade-off between stand age and C acquisition in forests, with C uptake decelerating as the stand ages (Ryan et al., 1997) and certain ecosystems may stop accumulating C once they hit a ret-rogressive phase (i.e., after several millennia since last disturbance; Bansal et al., 2012). However, declines in C uptake have typically been observed in ecosystems where there is a lack of diversity of plant species and stand ages (Hardiman et al., 2013; Castro-Izaguirre et al., 2016).

2.2.2 **RHIZODEPOSITION**

Rhizodeposition is a blanket term for a number of ways plants input C to the soil, including muci-lage created by the root, sloughing of root cells, or exudation. Some of this is passive or seasonal, and comprises a wide range of molecular C forms, with varying molecular weights. The plant com-munity type and composition will determine the amount, location, and type of rhizodeposition, with species-specific patterns in phenology and rhizosphere microbial, and faunal communities observ-able as a direct result of contrasting rhizodeposition regimes (Murphy et al., 2016). Exudates are passed from roots to the rhizosphere in various low molecular weight forms. Their passive diffusion principally includes sugars, organic acids, and amino acids. However, hormones, enzymes, and phe-nolics are also released from roots in more targeted ways for purposes such as nutrient release from SOM (e.g., phosphatases; Jones, 1998), inducing nodulation in legumes (phenolics; e.g., flavonoids; Phillips, 1992), and allelopathic suppression of other plants (Inderjit et al., 2011).

In forest ecosystems, estimates suggest that between 1% and 10% of the total photosynthate is lost to exudation (Qiao et al., 2014; Yin et al., 2014), with herbaceous species exuding up to 40% of their total assimilated C (Marschner, 1995). Because these exudates are simple C forms, if they are not immediately adsorbed onto clay surfaces (Cotrufo et al., 2013), they are unlikely to be stored in the soil for very long or make up a significant component of the SOM. This is particularly likely in temperate and tropical forests and grasslands, where soils are warm and aerated (Maier et al., 2010). Instead, these C compounds are likely to be utilized by soil microbial and mesofauna and respired as CO_2, thereby resulting in C loss from the system. Alternatively, the exudates might be absorbed into the soil food web, thereby becoming immobilized in the soil system. Studies have shown that the par-titioning between dissolved organic C(DOC) and CO_2 efflux by microbes is controlled by the lignin/lignin + cellulose ratio in plant litter. Smaller, nonstructural C forms in the material tend to be incor-porated into DOC and later, microbial biomass, while lignin and other large structural molecules form coarse particulate organic matter (POM; Cotrufo et al., 2015; Soong et al., 2015). Therefore, it is logical to suggest that the much lower molecular weight exudates are most likely to be absorbed and mineralized into CO_2. The tradeoff between incorporation and loss is termed substrate use effi-ciency, and it is likely to be context dependent, with the relationship between microbial activity and substrate quality predicting microbial assimilation of C differently across contrasting plant commu-nity types (Xu et al., 2014). For example, exudates can become immobilized in the soil if they are added to a soil layer where microbial activity is very low, e.g., deep in the soil where oxygen and warmth are limited (Maier et al., 2010). Many systems have very low microbial activity because of waterlogged or permafrosted conditions, which may also be conducive to preservation of exudates in the form of DOC. Examples of such systems include mangrove swamps (Adame and Fry, 2016), peatlands (Strack et al., 2015), and tundra (Hicks Pries et al., 2013).

As well as exudates, root tips add C to the soil in the form of mucilage and sloughed root cap and border cells, which aid lubrication and protection for progression of the roots through the soil as they grow (Bengough and McKenzie, 1997). Mucilages comprise a complex mix of sugars, fatty acids and amino acids, and are made up of generally higher molecular weight polysaccharides than root exudates (Dennis et al., 2010). Studies have shown that maize mucilage is highly reactive because of hydroxyl groups, which bind strongly to mineral complexes, thus immobilizing the C (Watt et al., 1993). This finding is particularly interesting in the context of land-use change, plant community composition, and C storage (see Chapter 6: Soil Nutrients and Soil Carbon Storage: Modulators and Mechanisms). As monoculture communities (i.e., maize fields) replace diverse plant communities, C is likely to be lost from the system (Box 2.1; Murty et al., 2002). However, selecting crops such as maize that are capable of producing such hydroxyl groups at their root tips might actually improve soil C sequestration in agricultural systems relative to current intensive practises (Fig. 2.4). Furthermore, younger root tips will be more active than older ones, and lay down more C into the soil (Jones et al., 2009), because of the strong correlation between aging and metabolic activity (Walker et al., 2003). Therefore, a faster growing vegetation type (i.e., grasslands), with high seasonal turnover and many annual plants, is likely to have a higher tissue turnover rate and net C loss through mineralization than slower moving, more mature stands (i.e., boreal or nemoral forests). This type of fast-growing vegetation typifies early successional communities, such as those that colonize newly exposed ground during glacial retreat (Bokhorst et al., 2017) (Fig. 2.2).

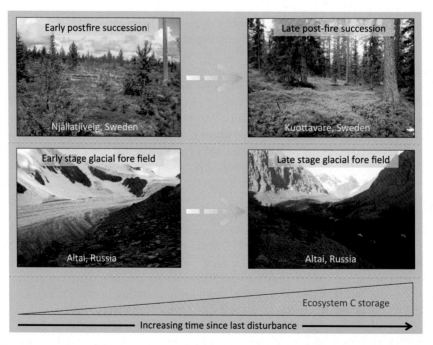

FIGURE 2.2

Temporal dynamics can play a strong role in driving ecosystem-level C storage due to shifts in the plant community. With increasing time since a last major disturbance such as fire or glaciation, both above- and belowground C stocks tend to increase. These increases are due to increasing biomass of the plant community, increasing complexity of the soil food web and a shift from domination of fast-growing, acquisitive plant species in early post-disturbance communities to slow-growing, conservative plant communities (Bonan and Shugart, 1989; Vilmundardóttir et al., 2015). However, it must be noted that if an ecosystem is not disturbed for several millennia, it may reach a so-called "retrogression" stage, where C storage plateaus (Bansal et al., 2012).

Photo credits: Babs Stuiver, Jonathan R. De Long.

BOX 2.1 A COMMUNITY OF ONE: IMPLICATIONS OF MONOCULTURE PLANT COMMUNITY TRAITS FOR C STORAGE

A great deal of ecological research has focused on how mixed plant communities under more natural settings impact on soil C sequestration, while more species depauperate systems, including monocultures, have been somewhat overlooked in research efforts, although this is being addressed. Such monocultures range from high input, deep tillage systems, to natural near monoculture ectomycorrhizal forests to forestry plantations and permaculture cropping, such as palm oil plantations (see Fig. 2.3). However, evidence is mounting that the adoption of practices which maximize plant trait diversity within cropping systems, such as intercropping, crop rotation, or through the selection of cultivars, or cultivar mixtures, that have traits associated with soil C sequestration can also help. For example, intercropping between maize, faba bean, and wheat increased soil C storage by 4%, due (at least in part) to functional trait complementarity (Cong et al., 2015) and studies of grasslands show that high-diversity mixtures can increase soil C storage (De Deyn et al., 2009; Fornara and Tilman, 2008; Lange et al., 2015). A metaanalysis by West and Post (2002) showed that enhancing crop rotational complexity can increase soil C storage, likely as the result of changes to crop residue input and composition. Further, Six et al. (2006) found that using legumes as a cover crop between rotations enhances microbial biomass and thereby could enhance soil C sequestration. This could be the result of trait differences between legumes (i.e., lower C:N ratios, ability to fix N) and common crop species such as oats, wheat, and corn (i.e., all grasses with high relative growth rates). Furthermore, fast-growing plantation tree species grown for biofuel production such as *Populus* spp. or *Salix* spp. have a great potential to enhance soil C sequestration and offset greenhouse gas emissions (Rytter, 2012). However, there are extreme levels of intra- and interspecific trait variation that have been observed for these tree genera (Brunner et al., 2004; Hanley and Karp, 2014), which can equate to markedly different amounts of C being stored in the soil. Finally, a review by Kho and Jepsen (2015) showed that palm oil plantations sequestered less C than old-growth tropical forests at the ecosystem level. However, such plantations do have the capacity to act as C sinks when compared to fallow forest agro-systems (i.e., 9% increase). This is likely due, in part, to fallow forests being dominated by pioneer plants with highly acquisitive traits and labile litter, as compared to palm oil trees (Kho and Jepsen, 2015). Therefore, in addition to changing management practices, greater soil stocks of C could be sequestered in agricultural and planation systems if crops, or mixtures of crops, with appropriate traits are selected.

FIGURE 2.3

Plantations that result from changing land-use represent plant communities that typically consist of one plant species (i.e., monoculture communities). Palm oil plantations in Costa Rica that have taken the place of diverse tropical rainforests (and only support a fraction of the plant species of intact rainforests) result in decreased soil C storage potential (Kho and Jepsen, 2015).

Photo credit: Andrea Vincent.

FIGURE 2.4

Maize (left hand side) and soybean (right hand side) crops in Ames, Iowa, United States, represent examples of monoculture plant communities and intensive agricultural practices. The repeated, annual tilling such systems experience results in large losses of SOM (soil organic matter) due to increased erosion, thereby limiting C storage more than permaculture systems. Crop rotations have been proposed as a potential mechanism to help improve soil C sequestration, but this is not the case when comparing continuous maize cultivation versus maize-soybean rotations (West and Post, 2002). The mucilage exuded by maize is highly reactive, leading to the formation of mineral complexes and thereby promoting the formation of SOM (Watt et al., 1993). As a result, maize might be a better choice for promoting soil C sequestration compared to certain other agricultural crops such as soybeans.

Photo credit: Guillaume Bay.

2.2.2.1 Priming effects

Priming effects are microbially-mediated processes that occur when the breakdown of fresh C inputs, in the form of litter or exudates, enables the microbial community to incidentally breakdown older, stored C (Fontaine et al., 2011). These inputs tend to be more energy-rich than the previously stored C, which means that when decomposer communities exude cellulolytic enzymes, or other enzymes designed to depolymerize C-based polymers in order to mobilize N, SOM is also mineralized by proxy. Research indicates that the main drivers of these priming effects are fungi (bacteria target more soluble litter fractions; Poll et al., 2008). In experimental (mostly crop plant) systems, the rhizosphere priming effect observed was up to 380% of added substrate, so the net result is a substantial decrease in stored, stable C stock (Cheng et al., 2014). Mycorrhizal fungi (particularly arbuscular-mycorrhizal fungi (AMF)) are key priming agents, partly because they have heightened requirements for N in the SOM, so the C is broken down incidentally, and partly to make labile C available to saprotrophs, which in turn decompose older stored C (Cheng et al., 2014). However, most of the studies that identify this priming effect have been conducted under conditions of elevated CO_2, which alters stoichiometric balances. Under ambient CO_2, the priming effect of mycorrhiza is far weaker, so more work is needed to determine their role in natural conditions (Phillips et al., 2012; Cheng et al., 2014). Therefore, SOM is likely to be vulnerable under future scenarios of elevated CO_2 and potentially other climate change drivers such as drought (Talbot et al., 2008). Suppression of mineralization (i.e., reverse priming) has also been observed, but it is less common. It has been hypothesized that this is as a result of preference for simpler substrates, and indeed, priming does seem to be directly stimulated by inputs of labile substrates (Paterson and Sim, 2013). Further, plant phenology seems to have a clear effect on priming, with peak priming occurring during flowering and much lower rates occurring at other life stages (Cheng et al., 2003).

Another aspect of the plant community that directly affects the priming rate is rooting density. This dictates whether the input of fresh substrate is pulsed or continuous. In grassland soils, roots are very dense and have a multitude of tips, which leads to almost continuous biological activity throughout the rhizosphere (Kuzyakov, 2010). This is also true of fast-growing communities, where the high turnover of root tips, rapid exploration of roots and affiliations with mycorrhizae and other microbes potentially results in a large amount of SOM being exposed to priming. However, plant roots and mycorrhizae are also important in aggregate stability, and the balance between forming and breaking down aggregates is likely to be highly dynamic, so plant community characteristics could diminish the priming effect (Gould et al., 2016). Nonetheless, priming effects are considered so important that is has been asserted that if fresh C inputs are reduced to zero, the stability of SOC is maintained (Fontaine et al., 2007).

2.2.3 DECOMPOSITION

Plant tissue itself is a C-rich material that is added to the soil in the form of both aboveground and belowground litter, and is the material most likely to become a stable part of SOM. SOM is mostly composed of plant material that is metabolized by microbes and is thus in various stages of decomposition, making its molecular composition enormously complex. This decomposition mainly takes place on available substrate, regardless of chemical complexity (i.e., not chemically bound to clay mineralogy or otherwise unavailable; Lehmann and Kleber, 2015). This new understanding runs counter to traditional paradigms of "recalcitrant" and "labile" litter, which hypothesized that the larger and more complex the chemical structure, the more likely it would be to be retained in long-term C storage. Data from a range of plant community types appears to support the replacement of the recalcitrant-labile continuum (grasslands at Rothamsted, Jenkinson and Rayner, 1977, and a selection of deciduous and evergreen forest litters, Klotzbücher et al., 2011), but mineralogy and climate are often stronger drivers of decomposition rate. This is because clay tends to retain nutrients through ionic charges, which can alter grassland community composition toward a more exploitative sward (Fry et al., 2017). However, litter quality in a general sense has implications for decomposition rates. The more N rich the tissue, the more nutritious and the quicker it is likely to be targeted for decomposition (Fig. 2.1). It is therefore possible to infer both turnover and C stock under different vegetation types (Paul, 2016). A cross-latitude study of decomposition rates, using litter transplant techniques to disentangle climatic effects, revealed that litter species identity accounted for 34% of the variation in decomposition rates, across all biomes from tropical to subarctic (Makkonen et al., 2012). Further, the main determinants of decomposition could be further broken down into a number of indicator traits—magnesium concentration, water saturation capacity and tannin concentration (Makkonen et al., 2012), and the overall resource uptake strategy on a continuum of acquisitiveness is indicative of decomposability across biomes (Cornwell et al., 2008). The litter from Mediterranean forests was consistently the most decomposable, regardless of the biome it was transplanted to, while tropical rainforest litter was the least decomposable. This indicates that decomposer communities are relatively generalizable across biomes and are highly flexible. However, the work by Hobbie et al. (2010) showed that this is far from a simple correlation, and calcium, lignin, and other components are crucial determinants of decomposition.

Plant communities themselves can have a considerable effect on decomposition rate by generating a so-called "homefield" advantage (HFA). An HFA occurs when litter decomposes more rapidly under the same plant species from which the litter originated, indicating that the decomposer community is adapted for particular plant communities (Veen et al., 2015). This can occur either directly through altering litter biomass, root turnover, and rhizodeposition, or indirectly through favoring certain soil microbial and faunal guilds (Vivanco and Austin, 2008). This phenomenon has been observed in subtropical forests (Yu et al., 2015), old-growth Nothofagus forests in South America (Vivanco and Austin, 2008), and spruce (boreal) forests (Chomel et al., 2015). However, the HFA is by no means universal, as tropical rainforests do not show HFA, and Giesselmann et al. (2011), and Allison et al. (2013) found inconclusive results in grasslands, while St John et al. (2011) found no HFA with reciprocal transfer of grassland and forest litter. A preliminary acceleration of decomposition rates associated with HFA, and subsequent deceleration, means that the net effect on C storage is likely to be not only secondary to climate and litter quality, but also temporally controlled. This shift in decomposition rates could result in C storage rates that are no different from litter where no HFA is evident (Ayres et al., 2009a,b). More work is needed, however, to discover whether these effects are conserved across other vegetation types.

Furthermore, root turnover, and leaf and root senescence are seasonal additions to the soil, with fine root production estimated to account for as much as 33% of global primary productivity (Majdi and Andersson, 2005). Therefore, turnover of these structures is of huge significance to the global C cycle and research efforts to characterize these inputs have been considerable (Strand et al., 2008; Freschet et al., 2013; Solly et al., 2014). This is partly because turnover rates are difficult to generalize, varying according to soil nutrient heterogeneity, as shown in sugar maple forests (Burton et al., 2000). Further, fine root turnover is mediated by plant functional type, as demonstrated by the disparity in turnover rates in heather and grasses in heathlands (Aerts et al., 1992). Shifts within forest community composition are also expected to impact on fine root turnover over time, potentially leading to altered C sequestration capacity within ecosystems and at larger regional scales (McCormack et al., 2013). By contrast, leaf litter has a far lower C:N ratio and comparisons of leaf and root decomposition rates differ widely across ecosystems, with the majority of studies showing that roots are slower to decompose than leaf material (Chihuahuan desert; Kemp et al., 2003; temperate trees; Hobbie et al., 2010), although in some cases no differences have been detected (tropical forest; Cusack et al., 2009) and some leaves have been found to decompose slower (Hawaiian forests; Ostertag and Hobbie, 1999). This could be because of the contrast between fine and coarse roots—fine roots often seem to be highly decomposable, at a rate that is similar to leaves (Freschet et al., 2013). Interestingly, a comparison of fine root and pine needle decomposition showed that root decomposition resulted in 28% more C being retained in the soil, litter from roots were stabilized as N-rich molecules, while needles were retained over a short period in the C-rich SOM fraction of the soil (Bird et al., 2008). This contrast in decomposition pathways indicates that the longevity of C from litter in soils may be subject to different chemical processes, and consequences for C storage are complex.

At the molecular level, plant tissues make up a number of forms, comprising starch, hemicelluloses, and lignin molecules, which have varying levels of complexity and break down at varying rates (Aerts, 1997). Plant tissues also vary widely between above- and belowground organs in terms of C:N, which alters litter quality. For example, roots tend to be more C rich than shoots, and there has been little relationship found between root and shoot allometry in temperate tree species and grassland species. This is likely due to different morphologies, and also much higher variation in root chemistry relative to shoot chemistry, which alters decomposability (Hobbie et al., 2010). A meta-analysis comparing mean shoot and root litter residence times of a range of grass, forb, and crop species indicated that roots have a longer residence time in soil by up to 2.4 times (Rasse et al., 2005). Notably, this study showed that maize roots had a much higher residence time (i.e., 48−180 months) than the roots of other grasses, including *Festuca* spp. and *Stipa* spp. (i.e., 2 and 21 months, respectively) and legumes like *Vicia* spp. (5 months). This could partially explain why C stocks in croplands are higher than might be expected. In forests, the poor tissue quality and large amount of woody structures means that C cycling is slower than herbaceous systems. However, understory vegetation is likely to have more balanced tissue stoichiometry than trees, particularly in conifer dominated stands (Zhao et al., 2013), which could offer habitat for fauna and microbes. Increased N and C cycling in the evergreen understory layer under laurel forests in US mountain ranges strongly increased ecosystem C sequestration in a simulation of 50 years of nutrient cycling (Chastain et al., 2006). There is some evidence that understory vegetation enhances soil C storage because of improved litter quality compared with stands of only overstory trees. The overall C balance of inputs and C sink strength is still likely to be very high relative to nonwoody systems, so nutrient cycling remains relatively slow (De Deyn et al., 2008). Further, dead wood can take a number of years to decompose because of the complex chemical bonds that make up lignin, and the specialized nature of wood decomposers (Fukami et al., 2010). True organic matter accumulation in forests can therefore be anywhere in the order of years to millennia, which is why there is such concern over widespread deforestation and soil C loss (Baldrian and Banin, 2017).

2.3 PLANT COMMUNITIES AND MECHANISMS OF SOIL CARBON INCORPORATION

Plants synthesize and release a wide range of C-based molecules into the soil. In theory, the larger and more complex the molecule, the less labile it is. This continuum from recalcitrant to labile substrates can also be considered as a continuum from structural to metabolic-type molecules. However, the huge variability in chemical bond strength, coupled with external variables such as aeration, temperature, and soil microbial community composition means that predicting residence time of C in soil across contrasting plant communities is rarely straightforward (Amelung et al., 2008).

The most basic plant-derived C molecule is glucose, which is the first product of photosynthesis across all plant species. Glucose, which is a monosaccharide, consists of a ring of six C molecules with associated hydroxyl (OH) groups. In the plant, it has many roles including cell-cycle progression, signaling for genetic and protein expression, and growth and metabolism (Rolland et al., 2006). It is also a basic building block (i.e., C-skeleton) for amino acids such as glutamine (Zheng, 2009). The signaling is particularly important because it dictates which (and how much) sugar will be directed to different plant organs and directly into the soil. Glucose is a basic component of many other molecules, from two-glucose disaccharides such as sucrose to polyglucose molecules such as lignin. Experiments with various C-substrates show that glucose is the most abundant monomer released during decomposition of POM across all vegetation types (Gunina and Kuzyakov, 2015). It is also one of the preferred substrates for microbial uptake. Therefore, of all potential C forms glucose is the most vulnerable to metabolism. If glucose becomes part of the long-term C stock, rather than lost to mineralization as CO_2, it is likely to be as part of the microbial biomass (living and dead); (Dilly, 2001; Dungait et al., 2012). Plant functional group diversity (i.e., grass, legume, forb) significantly increases catabolic activity in temperate grassland systems, especially if legumes are present, which in some cases can reduce the amount of C incorporated into long-term storage (Stephan et al., 2000), although De Deyn et al. (2011) found that addition of a subordinate legume increased C sequestration. More research is needed into the plant community effects on glucose cycling because data are lacking to enable comprehensive comparisons, possibly because it is difficult to confidently differentiate between liquid exudates and mucilage in the soil.

Other simple plant-derived sugars including sucrose, fucose, xylose, cellobiose and galactose are formed from polymerized glucose molecules for various purposes such as signaling from the surface of plant cells and the formation of matrix polysaccharides for cell walls (Nguyen, 2009). Sucrose is the major transport sugar in plants. A disaccharide made up of two linked glucose molecules, it plays a key regulatory role in C allocation and sugar signaling. In soil, sucrose is hydrolyzed into glucose and/or fructose, and easily metabolized by microbes. It is then found as the main component of DOC in soil water. From here the C is either immobilized into microbial biomass or onto organomineral complexes, or incorporated into glycoproteins by AMF, which are also key agents in aggregation of grassland soils. Aggregation through glycoproteins enables the monomeric C forms to be stabilized over longer terms than if they were free in the soil (Singh et al., 2013). A cross-community database showed that sugars increase linearly with SOM in a ratio of approximately 1:4, and are also dependent on plant community composition (Gunina and Kuzyakov, 2015)—the sugar to SOM ratio is similar in grasslands and cropland, but in forests it is much lower (Guggenberger et al., 1994). Most of this is extremely transient, and only about 10% of sugars are incorporated into SOM.

Phenolic acids such as coumaric acid, vanillic acid, and sinapic acid are C-based compounds that are commonly released into soils through exudation from roots or decomposition of leaf tissues. The production and function of these acids can vary considerably between contrasting plant communities. Certain compounds may be predominantly synthesized in order to deter insect and pathogen attack in communities with high growth rates and high nutrient availability, such as oil seed rape (Siemens et al., 2002), while communities at high elevations might invest in phenolics that reduce UV damage (Khoddami et al., 2013). Polyphenols might also have allelopathic effects in soil (Inderjit et al., 2011). Allelopathic effects can be positive or negative on neighboring plants. These have knock-on effects on photosynthesis and respiration, and further, on plant community dynamics. For example, communities such as heathlands and boreal forests are very rich in phenolic compounds, which result in poor quality litter and decreased rates of nutrient cycling (i.e., the polyphenolic compounds inhibit decomposition; García-Palacios et al., 2016). This, in tandem with low annual temperatures, is responsible for the high C stocks stored in heathland and boreal soils (Bonan and Shugart, 1989; De Deyn et al., 2008). Free phenolic acids in the soil may be sorbed onto other soil particles or otherwise chemically altered through ionization or oxidation. They may also be included in SOM through polymerization.

Cellulose, hemicellulose, and cellobiose are umbrella terms for a wide range of polysaccharides such as xyloglucans, xylan, and glucomannan. Collectively, these compounds make up cell walls in nonwoody plants common in plant communities such as savanna, steppe, temperate grasslands, and understory herbs. Cellulose is the most abundant polymer in the soil and is mainly comprised of glucose monomers (Gunina and Kuzyakov, 2015). These polysaccharides are relatively simple to degrade—a wide range of bacterial and fungal groups can easily decompose them by selectively targeting the various chemical bonds (Zak et al., 2006). Hemicellulose is generally slower to break down, as shown in boreal forests (Sjöberg et al., 2004). However, the decomposition rate of hemicellulose increases as temperatures rise, so it probably breaks down much more quickly in tropical systems (McTiernan et al., 2003). In deciduous woodlands, cellulose and hemicellulose collectively make up about 69% of the wood portion and 29% of the leaves, with lignin making up much of the remainder (Swift et al., 1979). In the woody fraction, this cellulose and hemicellulose is largely protected by lignin (Talbot and Treseder, 2012), so it will often remain undecomposed until the lignin has been degraded. This ensures that C sequestration is a very slow process in woodlands, shrublands, and forests across the world (Zak et al., 2006). However, in grasses and forbs, the lignin fraction is minimal or absent in most species and so the cellulose cell walls are vulnerable to enzymatic attack by the fungal groups Ascomycetes, Basidiomycetes, and various bacterial groups. Therefore, as long as the plant community is not subject to abiotic stress that reduces microbial and faunal activity, these structures will quickly enter the POM fraction of soil and then be decomposed and incorporated into microbial or animal tissues, or SOM.

Finally, lignin comes in a wide range of forms. It is the group of C-based molecules produced by plants with the largest molecular weight and it is mostly found in woody stands. Accordingly, forests (particularly those dominated by evergreens) have DOC with the highest molecular weight of any plant community (Kiikkilä et al., 2013). There are a few guilds of fungi including white-rot that specifically depolymerize lignin using lignolytic enzymes such as lignin peroxidase (Zhang et al., 1991). However, these fungi are very specific, in limited abundance (Zak et al., 2006), and are inhibited by high soil N (Waldrop and Zak, 2006). Therefore, decomposers in forest systems with high C:nutrient ratios are much slower at breaking down plant inputs and C storage is consequently high. However, increasing evidence shows that the mere size and complexity of the molecule does not necessarily predict decomposability, particularly because molecules that fall under the lignin umbrella may have many soluble components (Lehmann and Kleber, 2015). Finally, climatic envelopes mainly dictate vegetation distribution patterns at global and regional scales, and, as a result, lignin synthesis and decomposition rates are likely to be impeded or accelerated based on water availability and temperature.

2.3.1 INCORPORATION OF MOLECULES INTO CARBON STOCKS

2.3.1.1 Physical mechanisms

Carbon can reside in soil from minutes to millennia. In soil, it can be "protected" in a number of ways from microbial metabolism and enzymatic attack, which substantially slows the rate of decomposition. This is mediated by plant community type (forest, grassland and so on), which plays a large role in determining the type of litter and resulting POM that is incorporated into C stocks.

Nonetheless, interactions between C-based molecules, clay minerals, and other physicochemical features of the soil should not be overlooked. Carbon protection could occur via inclusion into soil aggregates or sorption onto organomineral surfaces through a number of chemical processes (Jastrow et al., 2007). Soils with a high proportion of clay and silt (particle sizes $< 20\,\mu m$) have negatively charged particles and these are instrumental in sorbing C to minerals, thus protecting them from further decomposition (Baldock and Skjemstad, 2000). It should be noted though, that this "protection" is not permanent, but substantially slows the rate of decomposition (Baldock and Skjemstad, 2000; Lützow et al., 2006). Jastrow et al. (2007) suggest that optimizing this protection could involve preferentially planting perennial species over annuals, and reducing physical disturbance, so tillage of crops is not ideal (Jastrow et al., 2007), although subsequent management could reduce or reverse the benefit (Powlson et al., 2014). Chemical processes change plant residues into SOM, and are crucial to C sequestration (O'Brien and Jastrow, 2013). Highly complex molecular forms such as lignin can also offer a degree of protection, although this is context-dependent due to the varying forms of lignin and local soil N availability and climatic conditions (Talbot and Treseder, 2012).

Recent literature has begun to debate the classic paradigm of labile and recalcitrant molecules, and the perception of a gradient of decomposability and vulnerability to breakdown. This assumes that lignin, as the most biochemically large and complex, should be the most stable C form, while monomers such as fructose and glucose should persist in the soil for minutes to hours (Lehmann and Kleber, 2015). A new paradigm emerged in the literature that asserts that accessibility is the key feature in decomposition of organic matter (Kleber et al., 2011; Dungait et al., 2012; Kallenbach et al., 2016). Evidence is mounting that any exposed organic material is vulnerable to microbial and animal attack, as well as weathering, with multiple processes acting on the residues (Cotrufo et al., 2013). Thus, as the organic material becomes smaller, it undergoes multiple oscillations between formation and destruction of aggregates, and adsorption and desorption onto mineral surfaces until ultimately it is lost as CO_2 (Lehmann and Kleber, 2015). The idea of lability and recalcitrance may not be wholly obsolete, however. While molecules have been categorized in this manner based on their size and perceived stability in the face of decomposers, there may be a possibility to integrate molecular weight into the accessibility paradigm based on the strength of molecular bonds. For example, lignin has been traditionally perceived to be recalcitrant based on its size and the complexity of its components—some of its molecular linkages are highly resistant to degradation (in particular lignin forms with high densities of C−C linkages, known as condensed lignins), so a complex array of enzymes is required to fully depolymerize it (Munk et al., 2015). Lignin also exhibits hydrophobicity, which means that degradation in the soil is hampered by wet conditions (Laurichesse and Avérous, 2014). Therefore, improved understanding of the biochemistry of soil C compounds could offer scope for an integration between the labile-recalcitrant, and accessibility/C protection paradigms. At the time of writing, there has been little integration of this new accessibility/C protection paradigm into studies comparing different plant community types, but the basic principles are likely to be consistent across most vegetation types and climates.

2.3.1.2 Microbial mechanisms

Carbon may be incorporated into microbial biomass, where it is immobilized until the organism dies, emits it as dissolved waste, or via respiration. There is some debate concerning the role of microorganisms in the formation of C stocks in terms of nutrient stoichiometry. If soil nutrients such as N or P are limited, soil C may be in excess and therefore of little use to microbes (see Chapter 5: Climate,

Geography and Soil Abiotic Properties as Modulators of Soil C Storage). This is especially apparent in plant communities that are predisposed toward high C:N and C:P ratios, particularly in temperate and boreal forests, peat bogs, and calcareous grasslands (Stevenson et al., 2016). Many of these systems have shown decreased microbial C use efficiency (CUE) with higher C:nutrient ratios, which means that tropical forests and peat bogs are likely to have lower CUE because C is in such excess (Manzoni et al., 2012; Sinsabaugh et al., 2013). However, in terms of forest ecosystems, there is a large disparity of CUE across forest types, where tropical forests retain approximately 30% of photosynthetically acquired C, while in temperate forests it is closer to 50% (Chambers et al., 2004). This is likely due to sensitivity to temperature changes and water availability, which vary across biomes. More nutrient-rich areas such as temperate and alpine grasslands have much more available N, and this corresponds with better CUE and more C retained in the ecosystem.

Interestingly, high CUE also occurs when artificial inorganic nutrients are added, such as on croplands or grasslands cultivated for fodder, which results in an increase in soil C stock (Ramirez et al., 2010). Microbes have two possible courses of action under nutrient limited conditions, and which strategy is preferable is the focus of some debate (Van Der Heijden et al., 2008). One argument runs that from a stoichiometric point of view, the microbe would be better served to release excess C as respired CO_2. However, the other side of the debate suggests that energetically, the best outcome is to release the C as DOC or to immobilize it as SOC (Creamer et al., 2015). Much work has been published on this topic, especially using elevated CO_2 to increase C:N ratios artificially in a range of free-air CO_2 enrichment (FACE) experiments, and there is more empirical support for losses as CO_2, leading to its use as a marker for CUE (Spohn and Chodak, 2015). Further, increasing N deposition on peat bogs has been shown to vastly increase CO_2 emissions, but also increases enzymatic activity that results in a flush of leachable DOC (Bragazza et al., 2006). In contrast, Maaroufi et al. (2015) found that in boreal forests, long-term high levels of N deposition contributed to enhanced C sequestration via decreases in microbial biomass and respiration. Taken together, these findings illustrate the importance of considering how nutrient availability can affect microbial processes differently across contrasting plant communities, leading to changes in C sequestration.

2.4 PLANT COMMUNITY ATTRIBUTES, SOIL CARBON CYCLING AND STORAGE

2.4.1 GENERALIZED BIOMASS AND COMMUNITY EFFECTS

Certain physical characteristics of plant communities have implications for the amount of C entering the soil. If a community is very productive (i.e., grasslands), the soil is nearly completely covered by vegetation and is therefore less vulnerable to abiotic stresses such as drying, flash floods, and drought—all of which can erode the soil. This type of community generates a dense network of roots and considerable inputs of organic matter, which enables the build-up of SOM that tends to be more stable. In contrast, communities that are sparsely populated (i.e., desert) or subjected to frequent disturbance (i.e., agricultural) will often retain less C. This is particularly apparent in cereal crops (i.e., monoculture plant communities), where the soil is regularly exposed, turned over, and suffers large C losses from resulting mineralization and erosion (Fig. 2.4; Lal, 2006). Much of the C in ecosystems dominated by woody species (i.e., forests, shrublands) is stored in standing

FIGURE 2.5

Carbon storage can be affected by the plant community at different spatial scales. At the cellular or rhizosphere level, plant communities with a dominance of mycorrhizal species tend to promote higher C sequestration due to increased fungal bio- and necromass in the soil. Within a plant community, contrasting plant species might promote more or less C sequestration, depending on the functional groups present and which traits the individual members of these groups possess (Wright et al., 2004). For example, grassland species tend to be more acquisitive and boreal forest species tend to be more conservative, leading to lower versus higher C storage, respectively. Finally, C storage is highly dependent upon the biome considered, with subarctic tundra (Abisko, Sweden) promoting overall higher levels of C storage due to the conservative plant communities that grow there. Furthermore, C storage can occur in different compartments across contrasting biomes based upon the plant community. In nemoral forests (Nikko National Park, Japan), C is primarily stored aboveground (i.e., tree biomass), while in lowland fynbos (De Hoop, Western Cape, South Africa) it is sequestered belowground (i.e., grass biomass and organic matter inputs via rhizodeposition).

Photo credits: Jonathan R. De Long, Ellen Fry, Paul Kardol, Angela Straathof.

biomass. However, for more herbaceous or grassy swards, the C will be cycled much more rapidly, so there is a higher proportion of C in the soil and the likelihood of stable C storage is lower than in mature forests (De Deyn et al., 2008).

Many years of research have focused less on primary productivity and cover of various plant groups as indicators of soil C stocks, and more on the characteristics of individual species and plant types across landscapes (Tilman et al., 1996; Hooper and Vitousek, 1998). Impacts of plants on soil C storage can take place from the scale of a root hair all the way up to biome level (Fig. 2.5). Soil C storage in small vegetation patches is primarily subject to idiosyncratic effects of species, phenotypic plasticity of the plants in the patch, competition between individuals for space, light, and resources, and also rhizosphere community composition and dynamics. Accordingly, shifts in plant community diversity and composition can have implications for soil C storage. In grasslands, e.g., there is evidence that C storage increases when diversity increases (see Box 2.2; Fornara and Tilman, 2008; Steinbeiss et al., 2008; De Deyn et al., 2009; Lange et al., 2015), and that addition of legumes (De Deyn et al., 2011) or C_4 grasses (Fornara and Tilman, 2008) to high-diversity grasslands can enhance the effect. There is also a long tradition of using grasses, forbs, and legumes in C storage studies as they have greatly contrasting effects (Knops and Tilman, 2000; De Deyn et al., 2009). Mixtures of different functional groups have been shown to increase standing biomass C, although these simple taxonomic classifications have been largely replaced with targeted classifications (Fry et al., 2014; Shipley et al., 2016). Plant community composition is, however, also very important for decomposition and leachability of DOC, partly because a wide range of species can result in many different nutrient uptake requirements and therefore losses of C through leaching are

---

BOX 2.2 LIMITATIONS OF LINKING PLANT FUNCTIONAL DIVERSITY AND COMMUNITY TRAITS TO C SEQUESTRATION

Over the past three decades, there have been an increasing number of studies investigating the role of plant community diversity in soil C sequestration. Large-scale experiments in grasslands such as Cedar Creek (Fornara and Tilman, 2008; Reich et al., 2001) and Jena (Chen et al., 2017; Lange et al., 2015; Roscher et al., 2004; Steinbeiss et al., 2008) have demonstrated that increasing plant diversity increases soil C storage, both directly and indirectly. For example, Fornara and Tilman (2008) found that high-diversity species communities stored 500% more soil C than monoculture communities due to the presence of certain plant functional groups, namely legumes (i.e., N-fixing plants) and C4 grasses. Further, Chen et al. (2017) showed that root decomposition decreased with increasing plant community diversity as the result of higher root C to N ratios (i.e., contrasting belowground plant traits), which lead to more C storage in the soil.

  Despite this mounting body of evidence, there are a number of important caveats to the conclusions reached by these experiments. First, these experiments were relatively short-term (i.e., measurements were taken 1−12 years after establishment) and the relationships between soil C storage and plant diversity might not hold over the long-term. Second, the soils in these systems were relatively C-poor at the onset of the experiment, 0.5% and 2.4% for Cedar Creek and Jena, respectively (Fornara and Tilman, 2008; Hacker et al., 2015), which means more dramatic effects of increasing plant diversity were likely to be realized. Finally, these experiments have used highly artificial, manipulative approaches that might not reflect the impacts of plant diversity on soil C storage in more natural ecosystems.

  These limitations demonstrate that further long-term studies in contrasting ecosystems with more realistic, natural plant diversity treatments are necessary and some work has begun to emerge along this vein. Conti and Diaz (2013), e.g., found that soil C storage in semiarid forests was primarily driven by the relative abundance of plants with tall and densely woody stems, with leaf traits showing no link to soil C storage. Changes in the relative abundances of plant trait values were generated by long-term contrasting land-use patterns (i.e., centennial scale) and the over-riding effect of C stored in woody tissue. In contrast, Wardle et al. (2012) showed that belowground and total ecosystem C sequestration along a 5000-year boreal forest chronosequence increased with time since last disturbance (i.e., fire), but they attributed this finding to other biotic and abiotic factors and not plant community diversity. Taken collectively, these findings indicate that ecosystem C storage is strongly linked to the diversity of both plant functional groups and traits in some systems, but these diversity effects do not equate to universally applicable patterns. These knowledge gaps highlight the need to consider the impact of plant community diversity within each individual biome when making recommendations of potential CO_2 mitigation plans.

reduced. This has been shown in the large body of literature on semiarid systems (Maestre et al., 2012) and grasslands (De Deyn et al., 2009).

Nurse plants are found in many different ecosystems, in response to a wide range of stresses (Armas et al., 2011). Their net effect comes in aiding the growth, performance, and stability of a localized plant community, which in turn improves soil C cycling and accumulation of SOM. These nurse plants can offer shade, water, nutrients, or indirect protection from herbivores—all which might impact on ecosystem C storage potential (Molina-Montenegro et al., 2015). Tussock grasses are found in deserts, prairies, alpine, and wet grasslands across the world and could be considered as one type of nurse plant. They have been shown to collect nutrients, soil, and microbes underneath their roots, thus enhancing the tussock effect (Northup et al., 1999). Adjoining plant communities benefit from these "resource islands," and so the tussocks are usually surrounded by species that are highly acquisitive such as annual grasses. Tussock grasses are unusual in this, and are therefore worthy of highlighting here. The implications for C storage under these communities are that C is likely to be quickly metabolized or included in microbial biomass. While microbial biomass is a relatively stable source of C, even after the cells have died, the long-term prospects of C storage are poor under tussocks. Studies of invasive tussock grasses indicate that despite the high biomass and activity rates, C storage is low compared with other grasses (Christian and Wilson, 1999). In particular, a study that compared soil C under different grasses in combination with legumes, showed that while an increase in soil C did occur in the combinations, the increase was far less under the tussock than under the non-tussock grass (Fisher et al., 1994). More research is needed to ascertain why this should be, but their benefits to the ecosystem in terms of nursing neighboring species are offset by low soil C storage.

Other patch effects are observed in arid and semiarid communities, where some species exhibit hydraulic lift (e.g., Mediterranean biome shrublands; Prieto et al., 2010). Hydraulic traits are key in very water-limited locations, and by "lifting" water from deeper soil layers and redistributing it near the surface, clusters of plants may arise in isolation (Caldwell et al., 1998). Carbon storage may be relatively high under these small clusters, but the area would be very heterogeneous and long-term C storage is difficult to predict. Once the water dries out, or the key species dies, there may be local extinctions of the patches, with ensuing erosion and loss of SOM. These could be considered "nurse plants," although hydraulics is only one example of this phenomenon.

2.4.2 PLANT FUNCTIONAL TRAITS

Plant functional traits can be morphological, physiological, stoichiometric, or reproductive characteristics of a plant. These traits can be divided into two types: response traits and effect traits. Together, response traits can be used to mechanistically describe the manner in which a plant community responds to a change in its environment, while effect traits describe how a plant community directly affects the environment. The response-and-effect framework shows that in a plant community, there is a continuum of overlap between plants that respond to a given environmental change, and those that are influential in driving community-level responses (Suding et al., 2008). This has strong implications for ecosystem functioning, including soil C cycling, because when scaled up to the community level, plant functional traits can give insight into how resistant or resilient a plant community might be to a given change, with further implications for stability of soil C stocks and fluxes.

Plant functional traits can be broken up into "hard" and "soft" traits, to distinguish traits that are too labor-intensive or costly to use in large-scale screening, and characteristics that are cheap and simple to measure, respectively (Díaz et al., 2004). As such, soft traits are commonly used and well characterized, and are well linked to a specific function (Cornelissen et al., 2003). Six plant traits are most useful for predicting rates of resource use and covary tightly, and place all land plants on a "fast-slow" continuum (also known as the Leaf Economics Spectrum, Wright et al., 2004). These six traits are SLA, photosynthetic rate (A), leaf N content (LNC), leaf P content (LPC), dark respiration rate (R), and leaf lifespan. The spectrum offers a broad characterization of high C input, high investment tissues at the "slow" end, with cycling of leaf and root litter and associated nutrients and food webs also being "slow." Such "slow" systems are typically resource poor, while nutrients and C are retained in the system for longer periods of time. Slow-cycling vegetation types include blanket bog, boreal forests, calcareous grassland and tundra. At the "fast" end are very N-rich tissues with simpler structural carbohydrates, lower investment and longevity, and rapid tissue decomposition and cycling. Fast-cycling habitats include annual grasslands, tropical rainforests, and marshy grasslands. Intermediate systems include alpine meadows (dependent on temperature), nemoral forests, and Mediterranean biome shrublands.

Grime's mass-ratio hypothesis proposed that the influence of the traits of an individual plant or species on ecosystem processes is directly proportional to its mass and/or abundance, respectively (Grime, 1998). As such, trait values for each species in a given community can be weighted by abundance and/or biomass, summed and scaled up over an experimental plot, a landscape, or even a continent (Lavorel and Garnier, 2002; Fig. 2.6). This allows for mechanistic studies on C cycling based on the relative abundance of plant species possessing certain traits, as demonstrated by Enquist et al. (2015) and Manning et al. (2015). Studies have also demonstrated that community-weighted values of leaf traits, such as SLA, leaf dry matter content, and LNC can explain a reasonable portion of variation in rates of litter decomposition across sites (Garnier et al., 2004; Quested et al., 2007; Fortunel et al., 2009), pointing to the tractability of scaling from leaf traits to process rates at the landscape scale. Therefore, the linkages between function and traits have huge support, particularly along environmental gradients where a wide range of traits might be expected (De Long et al., 2016b).

While efforts are currently underway to characterize root traits, at the time of writing, root traits are still woefully underrepresented in the traits literature (Bardgett et al., 2014). This is a problem because: (1) Roots and shoots are not allometric; (2) They respond differently to stress such as drought; (3) They are difficult to characterize in situ without destructive harvesting; (4) They have different purposes—i.e., shoots harvest sunlight and CO_2, while roots are the primary sites of plant stability, anchoring, and water and nutrient uptake; and (5) They have different pressures, such as spatial requirements, herbivory, microbial pathogens and saprotrophs, and resource heterogeneity—see Hodge (2004) for a review of plant plasticity under differing nutrient abundances. These contrasting characteristics all have direct consequences for our ability to predict C sequestration and CUE using vegetation type as an informative parameter. Given that roots are key sites of C input into the system, researchers need to be able to account for such inputs in a meaningful way. For example, Clemmenson et al. (2013) estimate that as much as 70% of stored C in soil of a boreal forest comes from roots and their associated organisms.

In spite of these hindrances, there are some categorical root traits that could prove useful in predicting soil C dynamics within plant communities. From a system-wide perspective, prevailing root

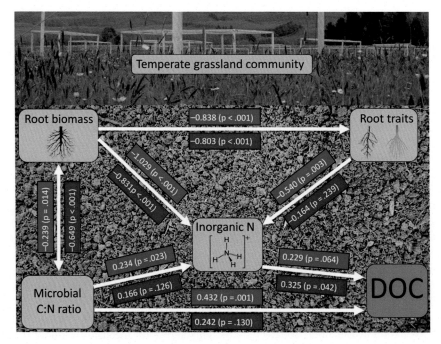

FIGURE 2.6

A structural equation model showing the impact of community-level root functional traits and biomass on dissolved organic soil C (DOC) in temperate grasslands, with soil microbial stoichiometry (i.e., microbial C:N ratio) and inorganic N as drivers. Under drought conditions, the relationship between root traits and soil C and N becomes decoupled (*red boxes*), whereas a close link can be observed in control conditions (*blue boxes*). This example illustrates how global change factors such as drought might alter relationships between different ecosystem processes within temperature grassland communities, resulting in changes to C lost from the system in the form of, e.g., DOC. The photo at the top shows drought shelters placed in a grassland field experiment in Selside, England.

The structural equation model was reproduced and adapted with permission from F.T. De Vries (De Vries, F.T., Brown, C., Stevens, C. J., 2016. Grassland species root response to drought: consequences for soil carbon and nitrogen availability. Plant Soil 409, 297–312 (De Vries, et al., 2016); http://creativecommons.org/licenses/by/4.0/).

structures could be sufficient to shed light on likely C relationships. For example, rooting depth will vary across plant systems (Canadell et al., 1996). As a simple comparison, forests have strong, thick, woody roots that tend to be distributed both along the surface and as much as 30 m deep (Pregitzer et al., 2002). Since woody structures have a high C:N ratio (C comprising about 50% of dry mass, Pregitzer et al., 2002), roots allocate large amounts of C to the formation of lignin, which ends up in the soil through exudates, microbial associations, and root sloughing. If the root is located in a very deep soil layer, microbial activity is minimized due to inhospitable abiotic conditions and mycorrhizal affiliation is limited (Clemmenson et al., 2013). Therefore, there is a good chance that the C will be preserved in these layers. Alternatively, grasslands have extensive, fibrous root structures that do not penetrate as deeply into the soil profile and are made of far more decomposable C compounds (Steen and Larsson, 1986). As a result, the C produced in the root systems of grassland communities may not reside in the soil as long as that which is produced in forests.

Plant communities dominated by species with greater investment in fine roots could experience more rapid C turnover and subsequent C loss from the soil than communities dominated by species that invest in longer-lived or storage roots. Understanding this can be a very simple method of applying root traits to C dynamics. Another potential root characteristic that could influence the rate of C acquisition into the system is architectural class or complexity. There is a great deal of literature that describes how the "order" of roots has a distinct set of traits (Pregitzer, 2002; Pregitzer et al., 2002; Guo et al., 2008). Specifically, zero order roots are the first roots that form—often the tap root in woody or herbaceous systems, but grassy plants with complex fibrous roots may have a different system (note that some literature, including Pregitzer et al., 2002 and McCormack et al., 2015 numbers the finest roots first order, with the tap or equivalent being the fifth order). First order roots are the roots that branch from the tap, and second order branch from them, and so on. There is a clear increase in nutrient uptake as the roots become finer (Roose and Fowler, 2004). However, Pregitzer et al. (2002) observed a marked decrease in C:N of root tissue as tree roots become finer and less woody, which correlates positively with respiration rates, so metabolism is high. These fine roots are also thought to be deciduous, but it is unknown whether the investment in these relatively costly structures is returned to the plant during the rapid turnover of fine roots (Gill and Jackson, 2000). This lack of N in tap roots also makes the roots less attractive from a decomposition perspective, especially compared with other plant systems (Aulen et al., 2012). Therefore, even a simple measurement of tissue C:N ratios could inform on the potential a plant community might have to sequester C.

2.5 PLANT–MYCORRHIZAL INTERACTIONS AND SOIL CARBON CYCLING

During the course of a plant's life, plant linkages to the microbial community can occur either inside or outside of the root. Carbon is traded with microbes in return for other nutrients such as nitrogen (N) and phosphorus (P) (Treseder, 2013). Plants exude low molecular weight C forms including sugars and amino acids in return for the service microbes provide in chemically changing N and P to plant available forms. This confers a number of benefits including protection from pathogen attack (Wehner et al., 2010), beneficial alterations in root tissue chemistry (Ouimette et al., 2013), and enhanced resource foraging capability either through root proliferation, or extended networks via mycorrhizal fungi (Comas and Eissenstat, 2009). By forming mutualistic associations with mycorrhizal fungi, plants can gain nutrients through mycelial networks and optimize foraging potential without having to use resources in costly root growth.

Transfer of C within roots can occur when mycorrhizal fungi form an association with the root. Mycorrhizae are specially adapted fungi that form mutualistic associates with plants by either coating the root with hyphae (ectomycorrhizal fungi (ECM)) or penetrating the root directly (AMF, orchid mycorrhiza), or both (ericoid mycorrhiza (ERM)). Carbon is tied up in the fungal mycelia and often greatly enhances the plant's root system. Therefore, mycorrhizae are an extremely strong sink for photosynthetically derived C. A portion of the C is fixed in the fungal mycelium, and this will not be turned over until it is either browsed on by fungal-feeders such as nematodes (Yeates, 1999) and collembolans (Johnson et al., 2005), or dies and is decomposed (Read and Perez-Moreno, 2003). The remainder is transferred to the soil via the mycelia in the form of glycoproteins, which bind the soil together, or other C-based molecules, the fate of which is still unclear (Högberg and Read, 2006; Finlay, 2008).

While an estimated 80% of land plants form mycorrhizal relationships (Smith and Read, 1997), different plant communities are often dominated by different types of mycorrhizae, and this has implications for soil C cycling and storage. For example, many woody plant communities are dominated by ECM because of their associations with important tree families across temperate, boreal and subtropical forests (Smith and Read, 1997; Landeweert et al., 2001). However, some tree species associate with AMF (De Deyn et al., 2008; Brzostek et al., 2013). These systems have vastly different C sink strengths depending on the types of plants in the community. In a boreal forest, ECM fungi have been estimated to receive about 15% of net primary production (Cairney, 2012). In contrast, AMF in grasslands receive 4%−20% of photosynthetic C, but directly stimulate higher photosynthetic rates from the hosts, so the actual value may be far higher relative to species uncolonized by mycorrhizae (Gaschuk et al., 2009). Further, ECM have been shown to reduce soil C loss, likely due to increased competition between ECM and free-living microbes for soil N (Averill et al., 2014; Averill and Hawkes, 2016). Tree root exudation is much higher in ECM-dominated systems, which tend to dominate in poor soils with mostly organic C, relative to AMF, which have access to more inorganic N in the soil. Brzostek and colleagues hypothesized that the higher exudation found in the more C-rich ECM communities is an attempt to liberate N from SOM using a range of catabolic enzymes (Brzostek et al., 2013). Heathland plant communities dominated by ERM produce high concentrations of phenolic and tannin compounds, which are difficult to break down and result in low decomposition rates, low soil N, and high amounts of SOM (Wurzburger and Hendrick, 2009). In contrast to AMF, ERM colonization actually decreases photosynthetic rates relative to uncolonized conspecifics (Woodward, 1999). This is because inorganic N is limiting and thus N uptake is low, and therefore heathland species tend to have very low relative growth rates, and tissue N and P, which means that C cycling rates are very low and ultimately C sink strength is high (Cornelissen et al., 2001). In short, the longest C residence times are associated with ERM, followed by ECM, and finally AMF which have the fastest cycling rates, most decomposable litter, and lowest organic matter fraction (Cornelissen et al., 2001). Consequently, there is a gradient of C sink strengths according to plant community type and mycorrhizal affiliation.

While much literature has focused on the beneficial aspects of the mycorrhizal-plant relationship, there is increasing evidence that the relationship can tip into parasitic or pathogenic effects, depending on some external shift in conditions (Johnson et al., 1997). The shift from mutualism to antagonism is particularly apparent when there is a drop in the supply of photosynthate, which causes the mycorrhizae to forage elsewhere for C. The supply of photosynthate could decrease as a consequence of high mineral nutrient availability from other sources, whereby the plant has less need of the nutrients offered by the fungus and thus withholds the photosynthate, leading to reduced development of mycelia (Johnson et al., 2003). This has been demonstrated in pine seedlings using both N (Newton and Pigott, 1991) and P (Agerer, 1993) fertilization. It could also be due to many other factors such as an increase in the plant's own requirements for the C, a drop in photosynthetic rate, the intrusion of a second plant that receives the nutrients, excessive demand for C from the fungus, or changes in fungal biomass (Wagg et al., 2015). This can result in loss of C from the system because the mycorrhizae may decompose SOM and thus cause a decrease in net soil C. This so-called "Plan B" hypothesis proposed by Talbot et al. (2008) has been observed experimentally in a wide range of systems with different types of mycorrhizae, including heathlands (Read et al., 2004) and temperate grasslands (Tu et al., 2006). However, the mechanisms and drivers of mycorrhizae switching from mutualists to antagonists have been understudied, and thus the effects on soil C storage are not well understood (Wagg et al., 2015).

2.6 PLANT-SOIL FEEDBACKS AND SOIL CARBON

Recently, a great deal of attention has been focused on PSFs (van der Putten et al., 2013; van der Putten, 2017), which have the potential to alter plant community composition and thereby impact on soil C cycling and storage. PSFs occur when plants change the soil abiotic and/or biotic environment, thereby altering the soil in a way that impacts on subsequent plant performance (Fig. 2.7;

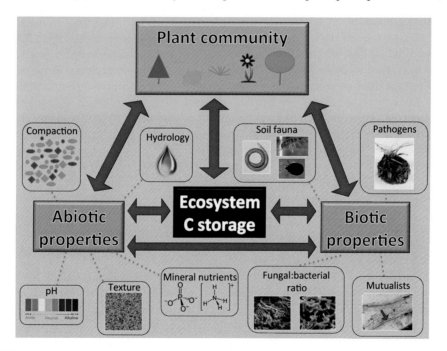

FIGURE 2.7

Abiotic and biotic soil drivers of C storage exist in a complex feedback relationship with the plant community. For example, plant communities with more acquisitive traits can promote higher availability of mineral nutrients and higher dominance of soil bacteria versus fungi, which in turn might further favors plants with acquisitive traits. Early successional acquisitive plants are known to harbor more pathogens (Kardol et al., 2007), which can promote successional development. Consequently, as succession proceeds, C sequestration typically increases (Post and Kwon, 2000; Pregitzer and Euskirchen, 2004). Broadly, the soil biotic community and abiotic properties may promote or inhibit certain plant species, thereby generating contrasting resistance or resilience within the plant community that might alter an ecosystem's capacity to sequester C.

Photo credits: Jonathan R. De Long, Dagmar Egelkraut, Ellen Fry, Paul Kardol.

Bever, 1994). If PSFs are negative, they will limit the growth and reproductive efforts of plants, while if such feedbacks are positive, they will enhance the fitness of the plants. Typically, negative feedback effects augment plant community diversity (Bennett et al., 2017), while positive PSFs promote the proliferation of limited diversity stands—i.e., monocultures (Dickie et al., 2014).

One important mechanism by which PSFs occur is via changes to the soil abiotic environment. For example, Bezemer et al. (2006) found that different plant species changed soil potassium availability over time, with grasses lowering potassium levels more dramatically than forbs. However, these findings depended strongly on the soil type considered—effects were pronounced in sandy soils, while negligible in chalk soils. These results demonstrate that contrasting plant functional groups might induce PSFs that alter the abiotic environment differently depending on soil type. Soil-type specific shifts in plant community composition could lead to different C storage potentials, especially if such shifts lead to dominance of plant species with more or less decomposition-resistant litter (Aerts, 1997). Furthermore, Bergmann et al. (2016) showed that soil structure could play an important role in tandem with plant functional traits (see Section 2.4.2) in driving PSFs. They found that plants with high specific root length (SRL)—i.e., more acquisitive traits—were better able to exploit resources in "disintegrated" soils—i.e., aggregates broken apart. On the other hand, plants with low SRL (i.e., more conservative traits) invested more in belowground structures and also were more susceptible to fungal pathogens. From a community perspective, this demonstrates that communities on more recently disturbed soils (i.e., early primary or secondary succession) might favor more acquisitive species due to a lack of soil aggregation. Consequently, lower soil C storage could result relative to plant communities composed of species with primarily conservative traits. Moreover, feedback generated by the entire plant community can alter soil hydrological processes. Crockett et al. (2016) discovered that *Eleocharis*-dominated fens generated lower hydraulic conductivity than *Carex*- and shrub-dominated fens. Reduced or accelerated water movement through these contrasting ecosystems could have implications for DOC release or sequestration.

Biotic changes to the soil incurred by PSFs that favor mutualists versus pathogens can also have consequences for plant community structure and soil C dynamics. Dickie et al. (2014) showed that in New Zealand the invasive tree *Pinus contorta* created a positive PSF that enhanced its own proliferation and that of other invasive tree species via associations with ECM mutualists. Given that invasive pines have been shown to increase C loss from the soil (Chapela et al., 2001), shifts in plant community structure that favor exotic pines will likely impact detrimentally on C storage within invaded ecosystems. On the other hand, negative pathogenic effects can be responsible for PSFs. Kardol et al. (2007) found that microbial-induced feedbacks accelerated succession in an old field, leading to the replacement of early successional with mid successional species. Typically, progressive successional changes will effectively shift the plant community to more conservative species that generate greater ecosystem C storage (Post and Kwon, 2000; Pregitzer and Euskirchen, 2004). However, despite these advances, it remains largely unknown if pathogens or mutualists are stronger drivers of PSFs in natural systems (van der Putten, 2017). Therefore, disentangling the relative roles the soil biota play is integral in understanding both the mechanisms behind PSFs and the resultant effects on soil C cycling and storage capacity.

Finally, research identifies the importance of PSFs in enhancing and maintaining plant community diversity across different ecosystems (Wubs and Bezemer, 2016; Bennett et al., 2017; Teste et al., 2017). Specifically, Bennett et al. (2017) showed that, when grown in conspecific soils, AMF tree species tended to experience negative PSFs, while ECM trees realized a positive feedback. Since mycorrhizae are known to control plant community productivity (van der Heijden et al., 2008), shifts in plant community composition that favor ECM trees due to positive PSFs could impact on C storage. In further support of this, Teste et al. (2017) found that ECM plants experienced greater positive, while N-fixing and nonmycorrhizal species experienced negative PSFs. They proposed that these feedbacks are responsible for maintaining the high level of diversity observed in Mediterranean plant communities. Given that enhanced plant diversity is generally associated with increasing C storage (De Deyn et al., 2011; Castro-Izaguirre et al., 2016), these findings have important implications for understanding the mechanisms by which plant communities may help or hinder C sequestration. Lastly, Wubs and Bezemer (2016) demonstrated that PSFs depend strongly on spatial scale, with more spatial heterogeneity generating greater negative feedbacks. This indicates that under field conditions, which are highly spatially variable due in part to diverse plant communities, PSFs may be more negative than predicted by glasshouse experiments. Consequently, more diverse communities with greater PSFs might further promote diversity and thereby bolster soil C storage.

2.7 **CLIMATE, PLANT COMMUNITIES, AND CARBON CYCLING**

The global climate is expected to drastically change in line with anthropogenic activity over the coming centuries (IPCC, 2014). While this is explored in other chapters (see Chapter 4: Leveraging a New Understanding of How Belowground Food Webs Stabilize Soil Organic Matter to Promote Ecological Intensification of Agriculture and Chapter 7: Agricultural Management Practices and Soil Organic Carbon Storage), here we will examine the specific effects of elevated CO_2, temperature, and precipitation change on a range of plant communities, and the resulting changes in C storage. There is compelling evidence that plant functional type modulates the effect of climate change variables in a systematic manner, and that plant diversity can reduce the severity of climatic events (Kreyling et al., 2008b).

Effects of elevated CO_2 (eCO_2) on plant physiology, performance, and soil C have been researched intensively in a global network of FACE experiments (Leakey et al., 2009). In most systems, while photosynthetic rate acclimates rapidly, C uptake increases overall (Ainsworth and Rogers, 2007). However, eCO_2 does not result in substantial gains in soil and wood biomass C sequestration when N or P are limited, which is usually the case in forests (Oren et al., 2001; Schlesinger and Lichter, 2001), and the extra C is returned to the atmosphere as respired CO_2 (Reich and Hobbie, 2013). Upon release of the constraint of low nutrient availability, e.g., if N fertilizer is added, plant productivity and soil C sequestration are vastly increased (Van Kessel et al., 2000; Oren et al., 2001). Much of the extra CO_2 taken up is allocated to short-lived tissues that are readily decomposable, such as leaves, and as such are more likely to become sequestered as part of the microbial biomass portion of SOM (Schlesinger and Lichter, 2001). Accordingly, decomposition rates are substantially higher with eCO_2 because of the relative lability of the inputs from the plant community, and priming rates are also increased, which results in a net decrease in stored C (van Groenigen et al., 2014). Some studies in temperate forests have shown that much of the labile C added to the soil from plants is sequestered, but this is insufficient to cover the deficit caused by the priming (Qiao et al., 2014; Sulman et al., 2014). There is also an indirect effect of eCO_2 that becomes apparent in warmer climates—eCO_2 improves water use efficiency, because of increased stomatal control. This results in wetter soils, which also enables microbial activity to increase (Battipaglia et al., 2013). Thus, the potential for widespread soil C loss to occur under a range of vegetation types as CO_2 increases is high.

Warming is projected to occur patchily across the globe in future decades, and the response of plant communities is likely to be rapid and nonlinear. Responses of soil C sequestration and storage to warming are likely to be mediated through responses of the plant and microbial communities. Plants may grow more rapidly under warming conditions and woody biomass will increase (Sistla et al, 2013). Germination rates are almost certain to change with higher temperatures, affecting recruitment and altering the plant community composition over time (Royal Botanic Gardens Kew, 2017). In a study conducted in the boreal forest, plant functional group was a stronger driver of decomposition than minimal warming (De Long et al., 2016a). Specifically, understory shrub cover was associated with slower decomposition rates, while moss cover was associated with faster decomposition rates. This indicates that C release from leaf litter is contingent on the presence of certain understory plant functional groups in boreal systems. In another example, vegetation composition in peatlands consists of one or two shrub, graminoid and bryophyte species, and studies have shown that the response of net ecosystem exchange to warming, and thus C dynamics are influenced by relative abundance of these groups (Ward et al., 2009, 2015; Dieleman et al., 2016). Also, in a long-term warming study, Walker et al. (2016) showed that when a complete plant community was present, ancient C (centuries to millennia) losses through respiration was at its lowest, compared to plots where specific plant functional groups had been removed and bare peat. Walker and colleagues suggested that this could be because intact vegetation increased root and microbial preference for recently fixed C, while a lack of certain vegetation groups resulted in preferential mineralization of older C.

Warming could also result in changes to growing season length, which will be particularly crucial for deciduous species. Temperature sensitivity of soil respiration is shown to be higher under deciduous species, and this is likely to increase vulnerability of stored C (Yuste et al., 2004). Warming could also lead to other growth-related limitations, particularly in arctic or tundra landscapes, where losses of soil C are likely to be far larger than potential gains through increased plant productivity in other regions (Crowther et al., 2016, although see Sistla et al., 2013). Other indirect warming effects through the plant community could arise through community effects on decomposition rates, which are often stronger than direct effects of warming on microbial activity (Ward et al., 2015). There is some evidence that high plant diversity can buffer the effects of warming on soil C storage, but warming could itself result in decreased species richness and thereby limit C storage. Plant community composition has consistently stronger effects on C cycling than warming alone across many communities, including grasslands (Steinauer et al., 2015), peatlands (Ward et al., 2015), and tundra (Sistla et al., 2013). Thus, atmospheric warming in the coming century will impact on soil C storage indirectly through the plant community (Dieleman et al., 2012).

Change to precipitation regimes is likely to be a key agent of future climate change, and this will have direct and indirect effects on soil C storage. Droughts and flash floods are expected to occur more frequently in magnitudes formerly described as "once-a-century" across temperate latitudes (IPCC, 2014). Direct effects will occur through altered diffusion of nutrients, more erosion and leaching of nutrients during heavy rainfall, and soil hydrophobicity during droughts and shifts

FIGURE 2.8

Massive erosion gullies in Western Kenya created by a combination of poor land management practices and extreme precipitation events driven in part by global climate change. As large quantities of topsoil are washed away, C sequestration is negatively impacted on both directly (via physical removal of the soil and organic matter) and indirectly (via changes to the plant community that lead to decreased primary productivity and C input into the system). Further, plants that colonize the exposed soil tend to have more acquisitive traits, fostering faster nutrient cycling and thereby limiting the C sequestration potential of the entire ecosystem.

Photo credit: Aida Bargués Tobella.

in osmotically-vulnerable microbial communities (Fig. 2.8; Kaisermann et al., 2015). All of these will impede the C storage capacity of soils, and physical and chemical stability. Sparse and disturbed vegetation types will exacerbate these effects. However, indirect responses through the plant community are likely to be just as important and less predictable. For example, long-term snow exclusion in the boreal forest was shown to indirectly negatively impact on the nematode community via shifts in understory plant community composition (De Long et al., 2016c). Further, evidence suggests that the deeper the soil, the less vulnerable the soil C is to extreme rainfall shifts, and this closely links with vegetation type as discussed above (Jobbagy and Jackson, 2000). Therefore, shrublands, forests, and some grassland types may have more stable C stocks (Knapp et al., 2008). Rainfall pattern (i.e., frequency and interval as opposed to quantity) changes are likely to directly reduce plant primary productivity and alter plant community composition, as demonstrated by Knapp et al. (2002) in prairie grasslands, which would lead to reductions in stored C. There are also likely to be changes in photosynthetic rate as a result of increased stomatal conductance, which would be most visible in shallower rooted systems that are dependent on smaller water pulses at the surface. In drylands, this has been shown to result in a net source of CO_2, with concomitant reductions in stabilization of C in the soil (Scott et al., 2009). In temperate acid grasslands, Fry et al. (2013) described a strong interaction effect between decreased summer rainfall and plant functional group—annual plants were less susceptible to drought overall compared with perennial species. Therefore, longer-lived species had lower C cycling rates under drought (i.e., both net ecosystem CO_2 exchange and decomposition were reduced) than annual species. They further showed that this was not directly attributable to microbial shifts, indicating that plant functional group was the key driver of these observed patterns (Fry et al., 2016). By contrast, the EVENT experiment in Bayreuth found very few interactive drought and diversity effects on acid grassland and heath as a result of extreme drought events, which indicates that in these systems, diversity is less important in mitigating drought effects than in other systems (Kreyling et al., 2008a; Mirzaei et al., 2008).

Drought is well characterized, but flooding is increasingly focused on, in light of recent problems in European countries. In sparser communities, runoff and erosion are likely to be key influences in loss of soil C (Knapp et al., 2008). Drought/flash flood cycles are likely to increase in temperate areas such as continental Europe and the United Kingdom, with severe consequences for lowland grasslands and heaths. There is still a dearth of information on the potential for plant communities to exacerbate or mitigate flooding effects on stored and dynamic C. Much of the research has taken place thus far on plant communities that are already adapted to periodic waterlogging or flooding, which have specialisms such as aerenchyma, which enable C assimilation and fixation to continue under flooding (Colmer and Voesenek, 2009). However, while the first and most pressing problem is likely to be CO_2 starvation restricting photosynthesis, there is some evidence that plant communities have a level of innate resilience to submergence (Mommer and Visser, 2005).

Increased temperatures and drought can trigger fire events. Fire is a natural feature of many systems, notably heathlands, savanna, and all five locations of Mediterranean biome species (the Mediterranean basin, Californian chaparral, South African Western Cape fynbos, Western Australian sclerophyll vegetation, and Chilean Matorral). Mediterranean biome species are characterized by shrubby, sparse communities. Fires in these areas burn fast and hot, and many of the plant species in these communities are adapted to regenerate after or withstand fire. The resulting ash layer is nutrient-rich, and high in partly burned biomass and "pyromorphic organic matter" (González-Pérez et al., 2011; Knicker, 2011), which is resistant to biological degradation, and aggregate stability of the underlying soil is largely unaffected (Mataix-Solera et al., 2011). Therefore, small natural burns in areas with fire-adapted vegetation can increase C storage (Fig. 2.8; Knicker, 2007). However, climate warming coupled with changes in management that increase fuel loads can often result in more severe, longer lasting fires that destroy the top layer of the soil, volatilizing nutrients and removing SOM, leading to widespread erosion (Gimeno-García et al., 2000). These increasingly occur in, or spread to, areas that do not have fire-adapted vegetation, such as grasslands (Knicker, 2011). The Mediterranean Basin is particularly vulnerable to this (Schröter et al., 2005). Fire results in strong competition between regrowing species, which will then have ensuing effects on soil C. For example, in a savanna system, more frequent fires maintain the grassy sward, but also lower C stocks. However, a decrease in fire frequency can lead to encroaching "woody weeds," which are a problem for the animals relying on the vegetation, but conversely C stocks are higher. The resulting increased fuel loads associated with woody species invariably means more intense fires, so C storage is actually more ephemeral under these conditions (Bird et al., 2000).

2.8 **CONCLUSIONS AND WAY FORWARD**

Plant community impacts on soil C cycling and storage have been the focus of much research, resulting in quantification of inputs, outputs, and stocks of C across a wide range of ecosystems (Paul, 2016). This comprehensive understanding has informed global C models and enhanced their ability to predict impacts of global change on soil C stocks. Nevertheless, there is still much to be done (see Box 2.3). Technological methods to identify the composition of SOM are improving the quantification of C in different soil pools, and new methods enable tracing of C through the system (Ellerbrock and Gerke, 2016; Shahbaz et al., 2017). However, there are still many avenues that require more research. The linkages between microbial community composition and C storage, while well characterized at an individual or community scale, remain difficult to use in C models (Wieder et al., 2013). Notably, the study of PSFs, while rapidly advancing, is still lacking wide-spread applications across plant communities in the real world, and thus it is still difficult to use them to predict shifting soil C dynamics over time. Also, there are many ecosystems that are vastly underrepresented in the literature—C storage potential for wetlands, e.g., and aquatic plant communities do not appear often in soil C cycling literature. Similarly marginalized are deserts, where a surprisingly large amount of C is stored (Li et al., 2015). Inclusion of these ecosystems, and understanding of the dynamics of transient systems, such as germination of desert annuals after rainfall pulses, could be important advances in the study of soil C storage and offer more holistic, global perspectives, especially in light of global change. Finally, climate and land-use change are accelerating in many parts of the world, and, there is a pressing need to develop management strategies

---

BOX 2.3 KNOWLEDGE GAPS IN THE ROLE PLANT COMMUNITIES PLAY IN THE MODELING OF C STORAGE

In recent years, models that predict global C cycling and sequestration in soil have made leaps forward due to advances in the resolution of plant community composition, soil parameters, and predicted climate change. For example, using Dynamic Global Vegetation Models (DGVMs), Sitch et al. (2008) showed that although such models accurately predicted current terrestrial C budgets, they differed markedly in their predictions under extreme environmental change. Despite these improvements, there are still numerous gaps in the predictive capacity of models. For example, most soil C stocks have only been measured in the top 1–2 m, or even shallower depths, of the soil profile (Canadell et al., 1996; Ward et al., 2016). Given that plant rooting depth and, as a result, the amount of C stored in the soil, varies dramatically across vegetation types and with plant community composition (Canadell et al., 1996; Jobbagy and Jackson, 2000), improving models depends on obtaining more accurate soil C stock figures for depths up to and in some cases beyond 1 m. In conjunction with this, feedbacks between plants and soil that could impact on C cycling must be considered in tandem with current DGVMs (Ostle et al., 2009). This is particularly important since nutrient limitation in soils is likely to limit the capacity of an ecosystem to continue to act as a C sink as CO_2 levels rise (Wieder et al., 2015). Additionally, there has been a call for DGVMs to further integrate plant functional traits (Atkin, 2016) and their heritability/response to environmental factors, as well as community assembly (Scheiter et al., 2013), all of which could be useful in improving global C models. Furthermore, it is well known that the plant community drives the microbial community (Bardgett and Wardle, 2010), and the microbial community (along with climate) has the potential to strongly control the decomposition of organic matter (Bardgett et al., 2008; Creamer et al., 2015). Therefore, it is critical to account for how interactions between the plant community, the microbial community and climate change can lead to substantial alterations in soil C sequestration or loss from terrestrial systems (Sofi et al., 2016; Wieder et al., 2013).

BOX 2.4 TAKE HOME MESSAGE AND FUTURE DIRECTIONS

- Plant communities contrast in their litter quality, climatic range and community composition. These factors directly influence the amount of C stored and its turnover rate in the soil.
- Higher plant diversity promotes soil C storage through improved resource use and better quality of litter.
- More diverse plant communities also increase soil C storage because they will have a range of plant functional traits, that optimize space and resources effectively compared with low diversity of traits.
- Croplands are at risk of losing C stocks because of removal of plant cover, poor plant diversity and tillage and fertilization, which contribute to erosion and aggregate breakdown.
- Factors such as precipitation, topography, mean annual temperature, nutrient availability, and the soil biotic community all interact with one another and the plant community to control soil C dynamics.

Ways forward:

- Forge coordinated, collaborative research efforts across ecosystems to determine global patterns in soil C storage (i.e., FACE network, ITEX, NUTNET, LTER).
- Investigate how climate change drivers such as increasing temperatures, altered precipitation regimes, increasing atmospheric CO_2 concentrations, land-use change, and N deposition interact with plant community composition to alter soil C dynamics.
- Explore how potential mismatches between plant communities and soil organisms impact on soil C dynamics as their respective ranges expand or contract due to climate change (Bardgett et al., 2013; Classen et al., 2015; van der Putten, 2012) will impact on soil C sequestration.
- Considering these interactions in the next generation of experiments will pave the way for more robust predictions of how C cycling will behave as the Earth's climate continues to change.

---

for increasing and maintaining soil C, both for climate mitigation and sustainable agriculture. Managing plant communities for soil C offers a potential contribution to achieving these goals (Box 2.4).

REFERENCES

Adame, M.F., Fry, B., 2016. Source and stability of soil carbon in mangrove and freshwater wetlands of the Mexican Pacific coast. Wetlands Ecol. Manage. 24, 129–137.

Aerts, R., 1997. Climate, leaf litter chemistry and leaf litter decomposition in terrestrial ecosystems: a triangular relationship. Oikos 79, 439–449.

Aerts, R., Bakker, C., De Caluwe, H., 1992. Root turnover as determinant of the cycling of C, N, and P in a dry heathland ecosystem. Biogeochemistry 15, 175–190.

Agerer, R., 1993. Mycorrhizae: ectomycorrhizae and ectendomycorrhizae. Prog. Bot. 54, 505–529.

Ainsworth, E.A., Rogers, A., 2007. The response of photosynthesis and stomatal conductance to rising [CO_2]: mechanisms and environmental interactions. Plant Cell Environ. 30, 258–270.

Allison, S.D., Lu, Y., Weihe, C., Goulden, M.L., Martiny, A.C., Treseder, K.K., et al., 2013. Microbial abundance and composition influence litter decomposition response to environmental change. Ecology 94, 714–725.

Amelung, W., Brodowski, S., Sandhage-Hofmann, A., Bol, R., 2008. Combining biomarker with stable isotope analyses for assessing the transformation and turnover of soil organic matter. Adv. Agron. 100, 155–250.

Armas, C., Rodríguez-Echeverría, S., Pugnaire, F.I., 2011. A field test of the stress-gradient hypothesis along an aridity gradient. J. Veg. Sci. 22, 818−827.

Atkin, O., 2016. New Phytologist: bridging the 'plant function − climate modelling divide'. New Phytol. 209, 1329−1332.

Aulen, M., Shipley, B., Bradley, R., 2012. Prediction of *in situ* root decomposition rates in an interspecific context from chemical and morphological traits. Ann. Bot. 109, 287−297.

Averill, C., Hawkes, C.V., 2016. Ectomycorrhizal fungi slow soil carbon cycling. Ecol. Lett. 19, 937−947.

Averill, C., Turner, B.L., Finzi, A.C., 2014. Mycorrhiza-mediated competition between plants and decomposers drives soil carbon storage. Nature 505, 543−545.

Ayres, E., Steltzer, H., Berg, S., Wall, D.H., 2009a. Soil biota accelerate decomposition in high-elevation forests by specializing in the breakdown of litter produced by the plant species above them. J. Ecol. 97, 901−912.

Ayres, E., Steltzer, H., Simmons, B.L., Simpson, R.T., Steinweg, J.M., Wallenstein, M.D., et al., 2009b. Home-field advantage accelerates leaf litter decomposition in forests. Soil Biol. Biochem. 41, 606−610.

Baldock, J.A., Skjemstad, J.O., 2000. Role of the soil matrix and minerals in protecting natural organic materials against biological attack. Org. Geochem. 31, 697−710.

Baldrian, P., Banin, E., 2017. The forest microbiome: diversity, complexity and dynamics. FEMS Microbiol. Rev. 41, 109−130.

Bansal, S., Nilsson, M.C., Wardle, D.A., 2012. Response of photosynthetic carbon gain to ecosystem retrogression of vascular plants and mosses in the boreal forest. Oecologia 169, 661−672.

Bardgett, R.D., Freeman, C., Ostle, N.J., 2008. Microbial contributions to climate change through carbon cycle feedbacks. ISME J. 2, 805−814.

Bardgett, R.D., Wardle, D.A., 2010. Aboveground-Belowground Linkages: Biotic Interactions, Ecosystem Processes, and Global Change Oxford Series in Ecology and Evolution. Oxford University Press, Oxford.

Bardgett, R.D., Manning, P., Morriën, E., De Vries, F.T., 2013. Hierarchical responses of plant−soil interactions to climate change: consequences for the global carbon cycle. J. Ecol. 101, 334−343.

Bardgett, R.D., Mommer, L., De Vries, F.T., 2014. Going underground: root traits as drivers of ecosystem processes. Trends Ecol. Evol. 29, 692−699.

Batjes, N.H., 1996. Total carbon and nitrogen in the soils of the world. Eur. J. Soil Sci. 47, 151−163.

Battipaglia, G., Saurer, M., Cherubini, P., Calfapietra, C., McCarthy, H.R., Norby, R.J., et al., 2013. Elevated CO_2 increases tree-level intrinsic water use efficiency: insights from carbon and oxygen isotope analyses in tree rings across three forest FACE sites. New Phytol. 197, 544−554.

Bengough, A.G., McKenzie, B.M., 1997. Sloughing of root cap cells decreases the frictional resistance to maize *Zea mays* L. root growth. J. Exp. Bot. 48, 885−893.

Bennett, J.A., Maherali, H., Reinhart, K.O., Lekberg, Y., Hart, M.M., Klironomos, J., 2017. Plant-soil feedbacks and mycorrhizal type influence temperate forest population dynamics. Science 355, 181−184.

Bergmann, J., Verbruggen, E., Heinze, J., Xiang, D., Chen, B.D., Joshi, J., et al., 2016. The interplay between soil structure, roots, and microbiota as a determinant of plant-soil feedback. Ecol. Evol. 6, 7633−7644.

Bever, J.D., 1994. Feedback between plants and their soil communities in an old field community. Ecology 75, 1965−1977.

Bezemer, T.M., Lawson, C.S., Hedlund, K., Edwards, A.R., Brook, A.J., Igual, J.M., et al., 2006. Plant species and functional group effects on abiotic and microbial soil properties and plant-soil feedback responses in two grasslands. J. Ecol. 94, 893−904.

Bird, J.A., Kleber, M., Torn, M.S., 2008. ^{13}C and ^{15}N stabilization dynamics in soil organic matter fractions during needle and fine root decomposition. Org. Geochem. 39, 476−477.

Bird, M.I., Veenendaal, E.M., Moyo, C., Lloyd, J., Frost, P., 2000. Effect of fire and soil texture on soil carbon in a sub-humid savanna Matopos, Zimbabwe. Geoderma 94, 71−90.

Bokhorst, S., Kardol, P., Bellingham, P.J., Kooyman, R.M., Richardson, S.J., Schmidt, S., et al., 2017. Responses of communities of soil organisms and plants to soil aging at two contrasting long-term chronosequences. Soil Biol. Biochem. 106, 69–79.

Bonan, G.B., Shugart, H.H., 1989. Environmental factors and ecological processes in boreal forests. Annu. Rev. Ecol. Syst. 20, 1–28.

Bond, W.J., 1989. The tortoise and the hare: ecology of angiosperm dominance and gymnosperm persistence. Biol. J. Linn. Soc. 36, 227–249.

Bragazza, L., Freeman, C., Jones, T., Rydin, H., Limpens, J., Fenner, N., et al., 2006. Atmospheric nitrogen deposition promotes carbon loss from peat bogs. Proc. Natl. Acad. Sci. U.S.A. 103, 19386–19389.

Brunner, A.M., Busov, V.B., Strauss, S.H., 2004. Poplar genome sequence: functional genomics in an ecologically dominant plant species. Trends Plant Sci. 9, 49–56.

Brzostek, E.R., Greco, A., Drake, J.E., Finzi, A.C., 2013. Root carbon inputs to the rhizosphere stimulate extracellular enzyme activity and increase nitrogen availability in temperate forest soils. Biogeochemistry 115, 65–76.

Burton, A., Pregitzer, K., Hendrick, R., 2000. Relationships between fine root dynamics and nitrogen availability in Michigan northern hardwood forests. Oecologia 125, 389–399.

Cairney, J.W.G., 2012. Extramatrical mycelia of ectomycorrhizal fungi as moderators of carbon dynamics in forest soil. Soil Biol. Biochem. 47, 198–208.

Caldwell, M., Dawson, T., Richards, J., 1998. Hydraulic lift: consequences of water efflux from the roots of plants. Oecologia 113, 151–161.

Canadell, J., Jackson, R.B., Ehleringer, J.R., Mooney, H.A., Sala, O.E., Schulze, E.D., 1996. Maximum rooting depth of vegetation types at the global scale. Oecologia 108, 583–595.

Castro-Izaguirre, N., Chi, X., Baruffol, M., Tang, Z., Ma, K., Schmid, B., et al., 2016. Tree diversity enhances stand carbon storage but not leaf area in a subtropical forest. PLoS One 11, e0167771.

Chambers, J.Q., Tribuzy, E.S., Toledo, L.C., Crispin, B.F., Higuchi, N., Santos, J.D., et al., 2004. Respiration from a tropical forest ecosystem: partitioning of sources and low carbon use efficiency. Ecol. Appl. 14, 72–88.

Chapela, I.H., Osher, L.J., Horton, T.R., Henn, M.R., 2001. Ectomycorrhizal fungi introduced with exotic pine plantations induce soil carbon depletion. Soil Biol. Biochem. 33, 1733–1740.

Chastain Jr, R.A., Currie, W.S., Townsend, P.A., 2006. Carbon sequestration and nutrient cycling implications of the evergreen understory layer in Appalachian forests. Forest Ecol. Manage. 231, 63–77.

Chen, H.M., Mommer, L., van Ruijven, J., de Kroon, H., Fischer, C., Gessler, A., et al., 2017. Plant species richness negatively affects root decomposition in grasslands. J. Ecol. 105, 209–218.

Cheng, W., Parton, W.J., Gonzalez-Meler, M.A., Phillips, R., Asao, S., McNickle, G.G., et al., 2014. Synthesis and modeling perspectives of rhizosphere priming. New Phytol. 201, 31–44.

Cheng, W.X., Johnson, D.W., Fu, S.L., 2003. Rhizosphere effects on decomposition: controls of plant species, phenology, and fertilization. Soil Sci. Soc. Am. J. 67, 1418–1427.

Chomel, M., Guittonny-Larchevêque, M., DesRochers, A., Baldy, V., 2015. Home field advantage of litter decomposition in pure and mixed plantations under boreal climate. Ecosystems 18, 1014–1028.

Christian, J.M., Wilson, S.D., 1999. Long-term ecosystem impacts of an introduced grass in the northern great plains. Ecology 80, 2397–2407.

Classen, A.T., Sundqvist, M.K., Henning, J.A., Newman, G.S., Moore, J.A.M., Cregger, M.A., et al., 2015. Direct and indirect effects of climate change on soil microbial and soil microbial-plant interactions: what lies ahead? Ecosphere 6, 21.

Clemmenson, K.E., Bahr, A., Ovaskainen, O., Dahlberg, A., Ekblad, A., Wallander, H., et al., 2013. Roots and associated fungi drive long-term carbon sequestration in boreal forest. Nature 339, 1615–1618.

Colmer, T.D., Voesenek, L.A.C.J., 2009. Flooding tolerance: suites of plant traits in variable environments. Funct. Plant Biol. 36, 665–681.

Comas, L.H., Eissenstat, D.M., 2009. Patterns in root trait variation among 25 co-existing North American forest species. New Phytol. 182, 919–928.

Cong, W.F., Hoffland, E., Li, L., Six, J., Sun, J.H., Bao, X.G., et al., 2015. Intercropping enhances soil carbon and nitrogen. Global Change Biol. 21, 1715–1726.

Conti, G., Diaz, S., 2013. Plant functional diversity and carbon storage – an empirical test in semi-arid forest ecosystems. J. Ecol. 101, 18–28.

Cornelissen, J.H.C., Aerts, R., Cerabolini, B., Werger, M., van der Heijden, M., 2001. Carbon cycling traits of plant species are linked with mycorrhizal strategy. Oecologia 129, 611–619.

Cornelissen, J.H.C., Lavorel, S., Garnier, E., Díaz, S., Buchmann, N., Gurvich, D.E., et al., 2003. A handbook of protocols for standardised and easy measurement of plant functional traits worldwide. Aust. J. Bot. 51, 335–380.

Cornwell, W.K., Cornelissen, J.H.C., Amatangelo, K., Dorrepaal, E., Eviner, V.T., Godoy, O., et al., 2008. Plant species traits are the predominant control on litter decomposition rates within biomes worldwide. Ecol. Lett. 11, 1065–1071.

Cotrufo, M.F., Wallenstein, M.D., Boot, C.M., Denef, K., Paul, E., 2013. The Microbial Efficiency-Matrix Stabilization MEMS framework integrates plant litter decomposition with soil organic matter stabilization: do labile plant inputs form stable soil organic matter? Global Change Biol. 19, 988–995.

Cotrufo, M.F., Soong, J.L., Horton, A.J., Campbell, E.E., Haddix, M.L., Wall, D.G., et al., 2015. Formation of soil organic matter via biochemical and physical pathways of litter mass loss. Nat. Geosci. 8, 776–779.

Creamer, C.A., Jones, D.L., Baldock, J.A., Farrell, M., 2015. Stoichiometric controls upon low molecular weight carbon decomposition. Soil Biol. Biochem. 79, 50–56.

Crockett, A.C., Ronayne, M.J., Cooper, D.J., 2016. Relationships between vegetation type, peat hydraulic conductivity, and water table dynamics in mountain fens. Ecohydrology 9, 1028–1038.

Crowther, T.W., Todd-Brown, K.E.O., Rowe, C.W., Wieder, W.R., Carey, J.C., Machmuller, M.B., et al., 2016. Quantifying global soil carbon losses in response to warming. Nature 540, 104–108.

Cusack, D.F., Chou, W.W., Yang, W.H., Harmon, M.E., Silver, W.L., The Lidet Team, 2009. Controls on long-term root and leaf litter decomposition in neotropical forests. Global Change Biol. 15, 1339–1355.

De Deyn, G.B., Cornelissen, J.H.C., Bardgett, R.D., 2008. Plant functional traits and soil carbon sequestration in contrasting biomes. Ecol. Lett. 11, 516–531.

De Deyn, G.B., Quirk, H., Yi, Z., Oakley, S., Ostle, N.J., Bardgett, R.D., 2009. Vegetation composition promotes carbon and nitrogen storage in model grassland communities of contrasting soil fertility. J. Ecol. 97, 864–875.

De Deyn, G.B., Shiel, R.S., Ostle, N.J., McNamara, N.P., Oakley, S., Young, I., et al., 2011. Additional carbon sequestration benefits of grassland diversity restoration. J. Appl. Ecol. 48, 600–608.

De Long, J.R., Kardol, P., Sundqvist, M.K., Veen, G.F., Wardle, D.A., 2015. Plant growth response to direct and indirect temperature effects varies by vegetation type and elevation in a subarctic tundra. Oikos 124, 772–783.

De Long, J.R., Dorrepaal, E., Kardol, P., Nilsson, M.-C.-, Teuber, L.M., Wardle, D.A., 2016a. Understory plant functional groups and litter species identity are stronger drivers of litter decomposition than warming along a boreal forest post-fire successional gradient. Soil Biol. Biochem. 98, 159–170.

De Long, J.R., Sundqvist, M.K., Gundale, M.J., Giesler, R., Wardle, D.A., 2016b. Effects of elevation and nitrogen and phosphorus fertilization on plant defence compounds in subarctic tundra heath vegetation. Funct. Ecol. 30, 314–325.

De Long, J.R., Laudon, H., Blume-Werry, G., Kardol, P., 2016c. Nematode community resistant to deep soil frost in boreal forest soils. Pedobiologia 59, 243–251.

De Vries, F.T., Brown, C., Stevens, C.J., 2016. Grassland species root response to drought: consequences for soil carbon and nitrogen availability. Plant Soil 409, 297–312.

Dennis, P.G., Miller, A.J., Hirsch, P.R., 2010. Are root exudates more important than other sources of rhizode-posits in structuring rhizosphere bacterial communities? FEMS Microbiol. Ecol. 72, 313−327.

Díaz, S., Hodgson, J.G., Thompson, K., Cabido, M., Cornelissen, J.H.C., Jalili, A., et al., 2004. The plant traits that drive ecosystems: evidence from three continents. J. Veg. Sci. 15, 295−304.

Dickie, I.A., St John, M.G., Yeates, G.W., Morse, C.W., Bonner, K.I., Orwin, K., et al., 2014. Belowground legacies of *Pinus contorta* invasion and removal result in multiple mechanisms of invasional meltdown. AOB Plants 6, 15.

Dieleman, W.I.J., Vicca, S., Dijkstra, F.A., Hagedorn, F., Hovenden, M.J., Larsen, K.S., et al., 2012. Simple additive effects are rare: a quantitative review of plant biomass and soil process responses to combined manipulations of CO_2 and temperature. Global Change Biol. 18, 2681−2693.

Dieleman, C.M., Branfireun, B.A., Lindo, Z., 2016. Northern peatland carbon dynamics driven by plant growth form—the role of graminoids. Plant Soil . Available from: http://dx.doi.org/10.1007/s11104-016-3099-3.

Dilly, O., 2001. Microbial respiratory quotient during basal metabolism and after glucose amendment in soils and litter. Soil Biol. Biochem. 33, 117−127.

Dungait, J.A.J., Hopkins, D.W., Gregory, A.S., Whitmore, A.P., 2012. Soil organic matter turnover is governed by accessibility not recalcitrance. Global Change Biol. 18, 1781−1796.

Edwards, E.J., Osbourne, C.P., Strömberg, C.A.E., Smith, S.A., C4 Grasses Consortium, 2010. The origins of C4 grasslands: integrating evolutionary and ecosystem science. Science 328, 587−591.

Ellerbrock, R.H., Gerke, H.H., 2016. Analysing management-induced dynamics of soluble organic matter composition in soils from long-term field experiments. Vadose Zone J. 15. Available from: http://dx.doi.org/10.2136/vzj2015.05.0074.

Enquist, B.J., Norberg, J., Bonser, S.P., Violle, C., Webb, C.T., Henderson, A., et al., 2015. Chapter Nine − Scaling from traits to ecosystems: developing a general trait driver theory via integrating trait-based and metabolic scaling theories. Adv. Ecol. Res. 52, 249−318.

Finlay, R.D., 2008. Ecological aspects of mycorrhizal symbiosis: with special emphasis on the functional diversity of interactions involving the extraradical mycelium. J. Exp. Bot. 59, 1115−1126.

Fisher, M.J., Rae, I.M., Ayarza, M.A., Lascano, C.E., Sanz, J.I., Thomas, R.J., et al., 1994. Carbon storage by introduced deep-rooted grasses in the South American savannas. Nature 371, 236−238.

Fontaine, S., Barot, S., Barré, P., Bdioui, N., Mary, B., Rumpel, C., 2007. Stability of organic carbon in deep soil layers controlled by fresh carbon supply. Nature 450, 277−280.

Fontaine, S., Henault, C., Aamor, A., Bdioui, N., Bloor, J.M.G., Maire, V., et al., 2011. Fungi mediate long term sequestration of carbon and nitrogen in soil through their priming effect. Soil Biol. Biochem. 43, 86−96.

Fornara, D.A., Tilman, D., 2008. Plant functional composition influences rates of soil carbon and nitrogen accumulation. J. Ecol. 96, 314−322.

Fortunel, C., Garnier, E., Joffre, R., Kazakou, E., Quested, H., Grigulis, K., et al., 2009. Leaf traits capture the effects of land use changes and climate on litter decomposability of grasslands across Europe. Ecology 90, 598−611.

Freschet, G.T., Cornwell, W.K., Wardle, D.A., Elumeeva, T.G., Liu, W., Jackson, B.G., et al., 2013. Linking litter decomposition of above- and below-ground organs to plant−soil feedbacks worldwide. J. Ecol. 101, 943−952.

Fry, E.L., Manning, P., Allen, D.G.P., Hurst, A., Everwand, G., Rimmler, M., et al., 2013. Plant functional group identity modifies precipitation change impacts on grassland ecosystem function. PLoS One. Available from: https://doi.org/10.1371/journal.pone.0057027.

Fry, E.L., Power, S.A., Manning, P., 2014. Trait based classification and manipulation of functional groups in diversity experiments. J. Veg. Sci. 25, 248−261.

Fry, E.L., Manning, P., Macdonald, C., Hasegawa, S., De Palma, A., Power, S.A., et al., 2016. Shifts in microbial communities do not explain the response of grassland ecosystem function to plant functional composition and rainfall change. Soil Biol. Biochem. 92, 199−210.

Fry, E.L., Pilgrim, E.S., Tallowin, J.R.B., Smith, R.S., Mortimer, S.R., et al., 2017. Plant, soil and microbial controls on grassland diversity restoration: a long-term, multi-site mesocosm experiment. J Appl. Ecol. Available from: https://doi.org/10.1111/1365-2664.12869.

Fukami, T., Dickie, I.A., Paula Wilkie, J., Paulus, B.C., Park, D., Roberts, A., et al., 2010. Assembly history dictates ecosystem functioning: evidence from wood decomposer communities. Ecol. Lett. 13, 675−684.

García-Palacios, P., Shaw, E.A., Wall, D.H., Hättenschwiler, S., 2016. Temporal dynamics of biotic and abiotic drivers of litter decomposition. Ecol. Lett. 19, 554−563.

Garnier, E., Cortez, J., Billès, G., Navas, M.-L., Roumet, C., Debussche, M., et al., 2004. Plant functional markers capture ecosystem properties during secondary succession. Ecology 85, 2630−2637.

Gaschuk, G., Kuyper, T.W., Leffelaar, P.A., Hungria, M., Giller, K.E., 2009. Are the rates of photosynthesis stimulated by the carbon sink strength of rhizobial and arbuscular mycorrhizal species? Soil Biol. Boichem. 41, 1233−1244.

Geiger, D.R., Servaites, J.C., 1994. Diurnal regulation of photosynthetic carbon metabolism in C3 plants. Annu. Rev. Plant Physiol. 45, 235−256.

Giesselmann, U.C., Martins, K.G., Brandle, M., Schadler, M., Marques, R., Brandi, R., 2011. Lack of home-field advantage in the decomposition of leaf litter in the Atlantic rainforest of Brazil. Appl. Soil Ecol. 49, 5−10.

Gill, R.A., Jackson, R.B., 2000. Global patterns of root turnover for terrestrial ecosystems. New Phytol. 147, 13−31.

Gimeno-García, E., Andreu, V., Rubio, J.L., 2000. Changes in organic matter, nitrogen, phosphorus and cations in soil as a result of fire and water erosion in a Mediterranean landscape. Eur. J. Soil Sci. 51, 201−210.

Gitelson, A.A., Peng, Y., Arkebauer, T.J., Suyker, A.E., 2015. Productivity, absorbed photosynthetically active radiation, and light use efficiency in crops: implications for remote sensing of crop primary production. J. Plant Physiol. 177, 100−109.

González-Pérez, J.A., González-Vila, F.J., Almendros, G., Knicker, H., 2011. The effect of fire on soil organic matter − a review. Environ. Int. 30, 855−870.

Gould, I.J., Quinton, J.N., Weigelt, A., De Deyn, G.B., Bardgett, R.D., 2016. Plant diversity and root traits benefit physical properties key to soil function in grasslands. Ecol. Lett. 19, 1140−1149.

Grassein, F., Till-Bottraud, I., Lavorel, S., 2010. Plant resource-use strategies: the importance of phenotypic plasticity in response to a productivity gradient for two subalpine species. Ann. Bot. 106, 637−645.

Grime, J.P., 1998. Benefits of plant diversity to ecosystems: immediate, filter and founder effects. J. Ecol. 86, 902−910.

van Groenigen, K.J., Qi, X., Osenberg, C.W., Luo, Y., Hungate, B.A., 2014. Faster decomposition under increased atmospheric CO_2 limits soil carbon storage. Science 344, 508−509.

Guggenberger, G., Christensen, B.T., Zech, W., 1994. Land-use effects on the composition of organic matter in particle-size separates of soil: I. Lignin and carbohydrate signature. Eur. J. Soil Sci. 45, 449−458.

Gunina, A., Kuzyakov, Y., 2015. Sugars in soil and sweets for microorganisms: review of origin, content, composition and fate. Soil Biol. Biochem. 90, 87−100.

Guo, L.W., Xu, D.Q., Shen, Y.K., 1994. The causes of midday decline of photosynthetic efficiency in cotton leaves under field conditions. Acta Phytophys. Sin. 20, 360−366.

Guo, D.L., Xia, M.X., Wei, X., Chang, W.J., Liu, Y., Wang, Z.Q., 2008. Anatomical traits associated with absorption and mycorrhizal colonization are linked to root branch order in twenty-three Chinese temperate tree species. New Phytol. 180, 673−683.

Hacker, N., Ebeling, A., Gessler, A., Gleixner, G., Mace, O.G., de Kroon, H., et al., 2015. Plant diversity shapes microbe-rhizosphere effects on P mobilisation from organic matter in soil. Ecol. Lett. 18, 1356−1365.

Hanley, S.J., Karp, A., 2014. Genetic strategies for dissecting complex traits in biomass willows *Salix* spp. Tree Physiol. 34, 1167−1180.

Hardiman, B.S., Gough, C.M., Halperin, A., Hofmeister, K.L., Nave, L.E., Bohrer, G., et al., 2013. Maintaining high rates of carbon storage in old forests: a mechanism linking canopy structure to forest function. For. Ecol. Manage. 298, 111−119.

Hatch, M.D., Slack, C.R., 1966. Photosynthesis by sugar-cane leaves—a new carboxylation reaction and pathway of sugar formation. Biochem. J 101, 103−111.

Hicks Pries, C.E., Schuur, E.A.G., Crummer, K.G., 2013. Thawing permafrost increases old soil and autotrophic respiration in tundra: partitioning ecosystem respiration using $\delta 13C$ and $\Delta 14C$. Global Change Biol. 19, 649−661.

Hobbie, S.E., Oleksyn, J., Eissenstat, D.M., Reich, P.B., 2010. Fine root decomposition rates do not mirror those of leaf litter among temperate tree species. Oecologia 162, 505−513.

Hodge, A., 2004. The plastic plant: root responses to heterogeneous supplies of nutrients. New Phytol. 162, 9−24.

Högberg, P., Read, D.J., 2006. Towards a more plant physiological perspective on soil ecology. Trends Ecol. Evol. 21, 548−554.

Hooper, D.U., Vitousek, P.M., 1998. Effects of plant composition and diversity on nutrient cycling. Ecol. Monogr. 68, 121−149.

Iason, G., Moore, B., Lennon, J., Stockan, J., Osler, G., Campbell, C., et al., 2012. Plant secondary metabolite polymorphisms and the extended chemical phenotype. In: Iason, G.R., Dicke, M., Hartley, S.E. (Eds.), The Ecology of Plant Secondary Metabolites: From Genes to Global Processes. Cambridge University Press, New York, NY, pp. 247−268.

Inderjit, Wardle, D.A., Karban, R., Callaway, R.M., 2011. The ecosystem and evolutionary contexts of allelopathy. Trends Ecol. Evol. 26, 655−662.

IPCC, 2014. In: Core Writing Team, Pachauri, R.K., Meyer, L.A. (Eds.), Climate Change 2014: Synthesis Report. Contribution of Working Groups I, II and III to the Fifth Assessment Report of the Intergovernmental Panel on Climate Change. IPCC, Geneva, 151 pp.

Janzen, H.H., 2004. Carbon cycling in earth systems − a soil science perspective. Agric. Ecosyst. Environ. 104, 399−417.

Jastrow, J.D., Amonette, J.E., Bailey, V.L., 2007. Mechanisms controlling soil carbon turnover and their potential application for enhancing carbon sequestration. Clim. Change 80, 5.

Jenkinson, D.S., Rayner, J.H., 1977. The turnover of soil organic matter in some of the Rothamsted classical experiments. Soil Sci. 123, 298−305.

Jobbagy, E.G., Jackson, R.B., 2000. The vertical distribution of soil organic carbon and its relation to climate and vegetation. Ecol. Appl. 10, 423−436.

Johnson, D., Krsek, M., Wellington, E.M.H., Stott, A.W., Cole, L., Bardgett, R.D., et al., 2005. Soil invertebrates disrupt carbon flow through fungal networks. Science 309, 1047.

Johnson, N.C., Graham, J.H., Smith, F.A., 1997. Functioning of mycorrhizal associations along the mutualism-parasitism continuum. New Phytol. 135, 575−586.

Johnson, N.C., Rowland, D.L., Corkidi, L., Egerton-Warburton, L.M., Allen, E.B., 2003. Nitrogen enrichment alters mycorrhizal allocation at five mesic to semiarid grasslands. Ecology 84, 1895−1908.

Jones, D.L., 1998. Organic acids in the rhizosphere − a critical review. Plant Soil 205, 25−44.

Jones, D.L., Nguyen, C., Finlay, R.D., 2009. Carbon flow in the rhizosphere: carbon trading at the soil−root interface. Plant Soil 321, 5−33.

Kaisermann, A., Maron, P.A., Beaumelle, L., Lata, J.C., 2015. Fungal communities are more sensitive indicators to non-extreme soil moisture variations than bacterial communities. Appl. Soil Ecol. 86, 158−164.

Kallenbach, C.M., Frey, S.D., Grandy, A.S., 2016. Direct evidence for microbial-derived soil organic matter formation and its ecophysiological controls. Nat. Commun. 7, 13630−13640.

Kardol, P., Cornips, N.J., van Kempen, M.M.L., Bakx-Schotman, J.M.T., van der Putten, W.H., 2007. Microbe-mediated plant-soil feedback causes historical contingency effects in plant community assembly. Ecol. Monogr. 77, 147−162.

Keesstra, S.D., Bouma, J., Wallinga, J., Tittonell, P., Smith, P., Cerdà, A., et al., 2016. The significance of soils and soil science towards realization of the United Nations Sustainable Development Goals. Soil 2, 111–128.

Kemp, P.R., Reynolds, J.F., Virginia, R.A., Whitford, W.G., 2003. Decomposition of leaf and root litter of Chihuahuan desert shrubs: effects of three years of summer drought. J. Arid Environ. 53, 21–39.

Kho, L.K., Jepsen, M.R., 2015. Carbon stock of oil palm plantations and tropical forests in Malaysia: a review. Singap. J. Trop. Geogr. 36, 249–266.

Khoddami, A., Wilkes, M.A., Roberts, T.H., 2013. Techniques for analysis of plant phenolic compounds. Molecules 18, 2328–2375.

Kiikkilä, O., Smolander, A., Kitunen, V., 2013. Degradability, molecular weight and adsorption properties of dissolved organic carbon and nitrogen leached from different types of decomposing litter. Plant Soil 373, 787–798.

Kleber, M., Nico, P., Plante, A.F., Filley, T., Kramer, M., Swanston, C., et al., 2011. Old and stable soil organic matter is not necessarily chemically recalcitrant: implications for modeling concepts and temperature sensitivity. Global Change Biol. 17, 1097–1107.

Klotzbücher, T., Kaiser, K., Guggenberger, G., Gatzek, C., Kalbitz, K., 2011. A new conceptual model for the fate of lignin in decomposing plant litter. Ecology 92, 1052–1062.

Knapp, A.K., Fay, P.A., Blair, J.M., Collins, S.L., Smith, M.D., Carlisle, J.D., et al., 2002. Rainfall variability, carbon cycling, and plant species diversity in a mesic grassland. Science 298, 2202–2205.

Knapp, A.K., Beier, C., Briske, D.D., Classen, A.T., Luo, Y., Reichstein, M., et al., 2008. Consequences of more extreme precipitation regimes for terrestrial ecosystems. Bioscience 58, 811–821.

Knicker, H., 2007. How does fire affect the nature and stability of soil organic nitrogen and carbon? A review. Biogeochemistry 85, 91–118.

Knicker, H., 2011. Pyrogenic organic matter in soil: its origin and occurrence, its chemistry and survival in soil environments. Quat. Int. 243, 251–263.

Knops, J.M.H., Tilman, D., 2000. Dynamics of soil nitrogen and carbon accumulation for 61 years after agricultural abandonment. Ecology 81, 88–98.

Kreyling, J., Beierkuhnlein, C., Elmer, M., Pritsch, K., Radovski, M., Schloter, M., et al., 2008a. Soil biotic processes remain remarkably stable after 100-year extreme weather events in experimental grassland and heath. Plant Soil 308, 175.

Kreyling, J., Wenigmann, M., Beierkuhnlein, C., et al., 2008b. Effects of extreme weather events on plant productivity and tissue die-back are modified by community composition. Ecosystems 11, 752–763.

Kuzyakov, Y., 2010. Priming effects: interactions between living and dead organic matter. Soil Biol. Biochem. 42, 1363–1371.

Lal, R., 2006. Enhancing crop yields in the developing countries through restoration of the soil organic carbon pool in agricultural lands. Land Degrad. Dev. 17, 197–209.

Landeweert, R., Hoffland, E., Finlay, R.D., Kuyper, T.W., van Breeman, N., 2001. Linking plants to rocks: ectomycorrhizal fungi mobilise nutrients from minerals. Trends Ecol. Evol. 16, 248–254.

Lange, M., Eisenhauer, N., Sierra, C.A., Bessler, H., Engels, C., Griffiths, R.I., et al., 2015. Plant diversity increases soil microbial activity and soil carbon storage. Nat. Commun. 6, 6707.

Laurichesse, S., Avérous, L., 2014. Chemical modification of lignins: towards biobased polymers. Prog. Polym. Sci. 39, 1266–1290.

Lavorel, S., Garnier, E., 2002. Predicting changes in community composition and ecosystem functioning from plant traits: revisiting the Holy Grail. Funct. Ecol. 16, 545–556.

Leakey, A.D.B., Ainsworth, E.A., Bernacchi, C.J., Rogers, A., Long, S.P., Ort, D.R., 2009. Elevated CO_2 effects on plant carbon, nitrogen, and water relations: six important lessons from FACE. J. Exp. Bot. 60, 2859–2876.

Lehmann, J., Kleber, M., 2015. The contentious nature of soil organic matter. Nature 528, 60–68.

Li, Y., Wang, Y.-G., Houghton, R.A., Tang, L.-S., 2015. Hidden carbon sink beneath desert. Geophys. Res. Lett. 42, 5880−5887.

Lützow, M.V., Kögel-Knabner, I., Ekschmitt, K., Matzner, E., Guggenberger, G., Marschner, B., et al., 2006. Stabilization of organic matter in temperate soils: mechanisms and their relevance under different soil conditions − a review. Eur. J. Soil Sci. 57, 426−445.

Maaroufi, N.I., Nordin, A., Hasselquist, N.J., Bach, L.H., Palmqvist, K., Gundale, M.J., 2015. Anthropogenic nitrogen deposition enhances carbon sequestration in boreal soils. Global Change Biol. 21, 3169−3180.

Maestre, F.T., Quero, J.L., Gotelli, N.J., Escudero, A., Ochoa, V., Delgado-Baquerizo, M., et al., 2012. Plant species richness and ecosystem multifunctionality in global drylands. Science 335, 214−218.

Maier, M., Schack-Kirchner, H., Hildebrand, E.E., Holst, J., 2010. Pore-space CO_2 dynamics in a deep, well-aerated soil. European J. Soil Sci. 61, 877−887.

Majdi, H., Andersson, P., 2005. Fine root production and turnover in a Norway spruce stand in northern Sweden: effects of nitrogen and water manipulation. Ecosystems 8, 191−199.

Makkonen, M., Berg, M.P., Handa, I.T., Hättenschwiler, S., van Ruijven, J., van Bodegom, P.M., et al., 2012. Highly consistent effects of plant litter identity and functional traits on decomposition across a latitudinal gradient. Ecol. Lett. 15, 1033−1041.

Manning, P., de Vries, F.T., Tallowin, J.R.B., Smith, R., Mortimer, S.R., Pilgrim, E.S., et al., 2015. Simple measures of climate, soil properties and plant traits predict national-scale grassland soil carbon stocks. J. Appl. Ecol. 52, 1188−1196.

Manzoni, S., Taylor, P., Richter, A., Porporato, A., Ågren, G.I., 2012. Environmental and stoichiometric controls on microbial carbon-use efficiency in soils. New Phytol. 196, 79−91.

Marschner, H., 1995. Mineral Nutrition of Higher Plants, second ed. Academic Press, London.

Mataix-Solera, J., Cerdà, A., Arcenegui, V., Jordán, A., Zavala, L.M., 2011. Fire effects on soil aggregation: a review. Earth-Sci. Rev. 109, 44−60.

Matthews, E., 1997. Global litter production, pools, and turnover times: estimates from measurement data and regression models. J. Geophys. Res. 102, 18771−18800.

McCormack, M.L., Eissenstat, D.M., Prasad, A.M., Smithwick, E.A.H., 2013. Regional scale patterns of fine root lifespan and turnover under current and future climate. Global Change Biol. 19, 1697−1708.

McCormack, M.L., Dickie, I.A., Eissenstat, D.M., Fahey, T.J., Fernandez, C.W., Guo, D., et al., 2015. Redefining fine roots improves understanding of below-ground contributions to terrestrial biosphere processes. New Phytol. 207, 505−518.

McNaughton, S.J., 1983. Compensatory plant-growth as a response to herbivory. Oikos 40, 329−336.

McTiernan, K.B., Coûteaux, M.-M., Berg, B., Berg, M.P., Calvo de Anta, R., Gallardo, A., et al., 2003. Changes in chemical composition of *Pinus sylvestris* needle litter during decomposition along a European coniferous forest climatic transect. Soil Biol. Biochem. 35, 801−812.

Mirzaei, H., Kreyling, J., Zaman Hussain, M., Li, Y., Tenhunen, J., Beierkuhnlein, C., et al., 2008. A single drought event of 100-year recurrence enhances subsequent carbon uptake and changes carbon allocation in experimental grassland communities. J. Plant Nutr. Soil Sci. 171, 681−689.

Molina-Montenegro, M.A., Oses, R., Torres-Díaz, C., Atala, C., Núñez, M.A., Armas, C., 2015. Fungal endophytes associated with roots of nurse cushion species have positive effects on native and invasive beneficiary plants in an alpine ecosystem. Perspect. Plant Ecol. Evol. Syst. 17, 218−226.

Mommer, L., Visser, E.J.W., 2005. Underwater photosynthesis in flooded terrestrial plants: a matter of leaf plasticity. Ann. Bot. 96, 581−589.

Munk, L., Sitarz, A.K., Kalyani, D.C., Mikkelsen, J.D., Meyer, A.S., 2015. Can laccases catalyse bond cleavage in lignin? Biotechnol. Adv. 33, 13−24.

Murphy, C.A., Foster, B.L., Gao, C., 2016. Temporal dynamics in rhizosphere bacterial communities of three perennial grassland species. Agronomy 6, 17.

Murty, D., Kirschbaum, M.U.F., McMurtrie, R.E., McGilvray, H., 2002. Does conversion of forest to agricultural land change soil carbon and nitrogen? A review of the literature. Global Change Biol. 8, 105−123.

Newton, A.C., Pigott, C.D., 1991. Mineral nutrition and mycorrhizal infection of seedling oak and birch. New Phytol. 117, 37−44.

Nguyen, C., 2009. Rhizodeposition of organic C by plant: mechanisms and controls. In: Lichtfouse, E., Navarrete, M., Debaeke, P., Véronique, S., Alberola, C. (Eds.), Sustainable Agriculture. Springer, Netherlands, pp. 97−123.

Northup, B.K., Brown, J.R., Holt, J.A., 1999. Grazing impacts on the spatial distribution of soil microbial biomass around tussock grasses in a tropical grassland. Appl. Soil Ecol. 13, 259−270.

O'Brien, S.L., Jastrow, J., 2013. Physical and chemical protection in hierarchical soil aggregates regulates soil carbon and nitrogen recovery in restored perennial grasslands. Soil Biol. Biochem. 61, 1−13.

Oren, R., Ellsworth, D.S., Johnsen, K.H., Phillips, N., Ewers, B.E., Maier, C., et al., 2001. Soil fertility limits carbon sequestration by forest ecosystems in a CO_2-enriched atmosphere. Nature 411, 469−472.

Ostertag, R., Hobbie, S., 1999. Early stages of root and leaf decomposition in Hawaiian forests: effects of nutrient availability. Oecologia 121, 564−573.

Ostle, N.J., Smith, P., Fisher, R., Woodward, F.I., Fisher, J.B., Smith, J.U., et al., 2009. Integrating plant-soil interactions into global carbon cycle models. J. Ecol. 97, 851−863.

Ouimette, A., Guo, D., Hobbie, E., Gu, J., 2013. Insights into root growth, function, and mycorrhizal abundance from chemical and isotopic data across root orders. Plant Soil 367, 313−326.

Panagos, P., Borelli, P., Poesen, J., Ballabio, C., Lugato, E., Meusburger, K., et al., 2015. The new assessment of soil loss by water erosion in Europe. Environ. Sci. Policy 54, 438−447.

Paterson, E., Sim, A., 2013. Soil-specific response functions of organic matter mineralization to the availability of labile carbon. Global Change Biol. 19, 1562−1571.

Paul, E.A., 2016. The nature and dynamics of organic matter: plant inputs, microbial transformations, and organic matter stabilisation. Soil Biol. Biochem. 98, 109−126.

Pearcy, R.W., Calkin, H., 1983. Carbon dioxide exchange of C_3 and C_4 tree species in the understory of a Hawaiian forest. Oecologia 58, 26−32.

Pérez-Harguindeguy, N., Díaz, S., Cornelissen, J.H.C., Vendramini, F., Cabido, M., Castellanos, A., 2000. Chemistry and toughness predict leaf litter decomposition rates over a wide spectrum of functional types and taxa in central Argentina. Plant Soil 218, 21−30.

Phillips, D.A., 1992. Flavonoids: plant signals to soil microbes. In: Stafford, H.A., Ibrahim, R.K. (Eds.), Phenolic Metabolism in Plants. Recent Advances in Phytochemistry, vol. 26. Springer, Boston, MA.

Phillips, R.P., Meier, I.C., Bernhardt, E.S., Grandy, A.S., Wickings, K., Finzi, A.C., 2012. Roots and fungi accelerate carbon and nitrogen cycling in forests exposed to elevated CO_2. Ecol. Lett. 15, 1042−1049.

Poll, C., Marhan, S., Ingwersen, J., Kandeler, E., 2008. Dynamics of litter carbon turnover and microbial abundance in a rye detritusphere. Soil Biol. Biochem. 40, 1306−1321.

Post, W.M., Kwon, K.C., 2000. Soil carbon sequestration and land-use change: processes and potential. Global Change Biol. 6, 317−327.

Powers, J., Veldkamp, E., 2005. Regional variation in soil carbon and δ13C in forests and pastures of northeastern Costa Rica. Biogeochemistry 72, 315−336.

Powlson, D.S., Stirling, C.M., Jat, M.L., Gerard, B.G., Palm, C.A., Snachez, P.A., et al., 2014. Limited potential of no-till agriculture for climate change mitigation. Nat. Clim. Change 4, 678−683.

Pregitzer, K.S., 2002. Fine roots of trees − a new perspective. New Phytol. 154, 267−270.

Pregitzer, K.S., Euskirchen, E.S., 2004. Carbon cycling and storage in world forests: biome patterns related to forest age. Global Change Biol. 10, 2052−2077.

Pregitzer, K.S., DeForest, J.L., Burton, A.J., Allen, M.F., Ruess, R.W., Hendrick, R.L., 2002. Fine root architecture of nine North American trees. Ecol. Monogr. 72, 293−309.

Prieto, I., Martínez-Tillería, K., Martínez-Manchego, L., Montecinos, S., Pugnaire, F.I., Squeo, F., 2010. Hydraulic lift through transpiration suppression in shrubs from two arid ecosystems: patterns and control mechanisms. Oecologia 163, 855–865.

Puijalon, S., Piola, F., Bornette, G., 2008. Abiotic stresses increase plant regeneration ability. Evol. Ecol. 22, 493–506.

van der Putten, W.H., 2012. Climate change, aboveground-belowground interactions, and species' range shifts. Annu. Rev. Ecol. Evol. Syst. 43, 365–383.

van der Putten, W.H., 2017. Belowground drivers of plant diversity. Science 355, 134–135.

van der Putten, W.H., Bardgett, R.D., Bever, J.D., Bezemer, T.M., Casper, B.B., Fukami, T., et al., 2013. Plant-soil feedbacks: the past, the present and future challenges. J. Ecol. 101, 265–276.

Qiao, N., Schaefer, D., Blagodatskaya, E., Zou, X., Xu, X., Kuzyakov, Y., 2014. Labile carbon retention compensates for CO_2 released by priming in forest soils. Global Change Biol. 20, 1943–1954.

Quested, H., Eriksson, O., Fortunel, C., Garnier, E., 2007. Plant traits relate to whole-community litter quality and decomposition following land use change. Funct. Ecol. 21, 1016–1026.

Ramirez, K.S., Craine, J.M., Fierer, N., 2010. Nitrogen fertilization inhibits soil microbial respiration regardless of the form of nitrogen applied. Soil Biol. Biochem. 42, 2336–2338.

Rasse, D.P., Rumpel, C., Dignac, M.F., 2005. Is soil carbon mostly root carbon? Mechanisms for a specific stabilisation. Plant Soil 269, 341–356.

Read, D.J., Perez-Moreno, J., 2003. Mycorrhizas and nutrient cycling in ecosystems — a journey towards relevance? New Phytol. 157, 475–492.

Read, D.J., Leake, J.R., Perez-Moreno, J., 2004. Mycorrhizal fungi as drivers of ecosystem processes in heathland and boreal forest biomes. Can. J. Bot. 82, 1243–1263.

Reich, P.B., Hobbie, S.E., 2013. Decade-long soil nitrogen constraint on the CO_2 fertilisation of plant biomass. Nat. Clim. Change 3, 278–282.

Reich, P.B., Knops, J., Tilman, D., Craine, J., Ellsworth, D., Tjoelker, M., et al., 2001. Plant diversity enhances ecosystem responses to elevated CO_2 and nitrogen deposition. Nature 410, 809–812.

Rolland, F., Baena-Gonzalez, E., Sheen, J., 2006. Sugar sensing and signalling in plants: conserved and novel mechanisms. Plant Biol. 57, 675–709.

Roose, T., Fowler, A.C., 2004. A mathematical model for water and nutrient uptake by plant root systems. J. Theor. Biol. 228, 173–184.

Roscher, C., Schumacher, J., Baade, J., Wilcke, W., Gleixner, G., Weisser, W.W., et al., 2004. The role of biodiversity for element cycling and trophic interactions: an experimental approach in a grassland community. Basic Appl. Ecol. 5, 107–121.

Royal Botanic Gardens Kew, 2017. Seed Information Database (SID). Version 7.1. Available from: <http://data.kew.org/sid/> (May 2017).

Ryan, M.G., Lavigne, M.B., Gower, S.T., 1997. Annual carbon cost of autotrophic respiration in boreal forest ecosystems in relation to species and climate. J. Geophys. Res. 102, 28871–28883.

Rytter, R.M., 2012. The potential of willow and poplar plantations as carbon sinks in Sweden. Biomass Bioenergy 36, 86–95.

Sage, R.F., 2016. A portrait of the C_4 photosynthetic family on the 50th anniversary of its discovery: species number, evolutionary lineages, and Hall of Fame. J. Exp. Bot. 67, 4039–4056.

Scharlemann, J.P.W., Tanner, E.V.J., Hiederer, R., Kapos, V., 2014. Global soil carbon: understanding and managing the largest terrestrial carbon pool. Carbon Manage. 5, 81–91.

Scheiter, S., Langan, L., Higgins, S.I., 2013. Next-generation dynamic global vegetation models: learning from community ecology. New Phytol. 198, 957–969.

Schlesinger, W.H., 1990. Evidence from chronosequence studies for a low carbon-storage potential of soils. Nature 348, 232–234.

Schlesinger, W.H., Lichter, J., 2001. Limited carbon storage in soil and litter of experimental forest plots under increased atmospheric CO_2. Nature 411, 466–469.

Schröter, D., Cramer, W., Leemans, R., Prentice, I.P., Araújo, M.B., Arnell, N.W., et al., 2005. Ecosystem service supply and vulnerability to global change in Europe. Science 310, 1333−1337.

Scott, R.L., Jenerette, G.D., Potts, D.L., Huxman, T.E., 2009. Effects of seasonal drought on net carbon dioxide exchange from a woody-plant-encroached semiarid grassland. J. Geophys. Res. 114, G04004.

Shahbaz, M., Kuzyakov, Y., Heitkamp, F., 2017. Decrease of soil organic matter stabilisation with increasing inputs: mechanisms and controls. Geoderma 304, 76−82.

Shipley, B., De Bello, F., Cornelissen, J.H.C., Laliberté, E., Laughlin, D.C., Reich, P.B., 2016. Reinforcing loose foundation stones in trait-based plant ecology. Oecologia 180, 923−931.

Siemens, D.H., Garner, S.H., Mitchell-Olds, T., Callaway, R.M., 2002. Cost of defense in the context of plant competition: *Brassica rapa* may grow and defend. Ecology 83, 505−517.

Singh, P.K., Singh, M., Tripathi, B.N., 2013. Glomalin: an arbuscular mycorrhizal fungal soil protein. Protoplasma 250, 663−669.

Sinsabaugh, R.L., Manzoni, S., Moorhead, D.L., Richter, A., 2013. Carbon use efficiency of microbial communities: stoichiometry, methodology and modelling. Ecol. Lett. 16, 930−939.

Sistla, S.A., Moore, J.C., Simpson, R.T., Gough, L., Shaver, G.R., Schimel, J.P., 2013. Long-term warming restructures Arctic tundra without changing net carbon storage. Nature 497, 615−618.

Sitch, S., Huntingford, C., Gedney, N., Levy, P.E., Lomas, M., Piao, S.L., et al., 2008. Evaluation of the terrestrial carbon cycle, future plant geography and climate-carbon cycle feedbacks using five Dynamic Global Vegetation Models DGVMs. Global Change Biol. 14, 2015−2039.

Six, J., Frey, S.D., Thiet, R.K., Batten, K.M., 2006. Bacterial and fungal contributions to carbon sequestration in agroecosystems. Soil Sci. Soc. Am. J. 70, 555−569.

Sjöberg, G., Nilsson, S.I., Persson, P., Karlsson, P., 2004. Degradation of hemicellulose, cellulose and lignin in decomposing spruce needle litter in relation to N. Soil Biol. Biochem. 36, 1761−1768.

Smith, S.E., Read, D.J., 1997. The role of mycorrhizas in ecosystems, Mycorrhizal Symbiosis, second ed. Academic Press, London, pp. 409−452.

Smith, P., Martino, D., Cai, Z., Gwary, D., Janzen, H., Kumar, P., et al., 2008. Greenhouse gas mitigation in agriculture. Philos. Trans. R. Soc. Lond. Ser. B. 363, 789−813.

Sofi, J.A., Lone, A.H., Ganie, M.A., Dar, N.A., Bhat, S.A., Mukhtar, M., et al., 2016. Soil microbiological activity and carbon dynamics in the current climate change scenarios: a review. Pedosphere 26, 577−591.

Solly, E.F., Schöning, I., Boch, S., Kandeler, E., Marhan, S., Michalzik, B., et al., 2014. Factors controlling decomposition rates of fine root litter in temperate forests and grasslands. Plant Soil 382, 203−218.

Soong, J.L., Parton, W.J., Calderon, F., Campbell, E.E., Cotrufo, M.F., 2015. A new conceptual model on the fate and controls of fresh and pyrolized plant litter decomposition. Biogeochemistry 124, 27−44.

Spohn, M., Chodak, M., 2015. Microbial respiration per unit biomass increases with carbon-to-nutrient ratios in forest soils. Soil Biol. Biochem. 81, 128−133.

St John, M.G., Orwin, K.H., Dickie, I.A., 2011. No 'home' versus 'away' effects of decomposition found in a grassland-forest reciprocal litter transplant study. Soil Biol. Biochem. 43, 1482−1489.

Steen, E., Larsson, K., 1986. Carbohydrates in roots and rhizomes of perennial grasses. New Phytol. 104, 339−346.

Steinauer, K., Tilman, D., Wragg, P.D., Cesarz, S., Cowles, J.M., Pritsch, K., et al., 2015. Plant diversity effects on soil microbial functions and enzymes are stronger than warming in a grassland experiment. Ecology 96, 99−112.

Steinbeiss, S., Beßler, H., Engels, C., Temperton, V.M., Buchmann, N., Roscher, C., et al., 2008. Plant diversity positively affects short-term soil carbon storage in experimental grasslands. Global Change Biol. 14, 2937−2949.

Stephan, A., Meyer, A.H., Schmid, B., 2000. Plant diversity affects culturable soil bacteria in experimental grassland communities. J. Ecol. 88, 988−998.

Stevenson, B.A., Sarmah, A.K., Smernik, R., Hunter, D.W.F., Fraser, S., 2016. Soil carbon characterisation and nutrient ratios across land uses on two contrasting soils: their relationships to microbial biomass and function. Soil Biol. Biochem. 97, 50−62.

Stockmann, U., Adams, M.A., Crawford, J.W., Field, D.J., Henkaarchi, N., Jenkins, M., et al., 2013. The knowns, known unknowns and unknowns of sequestration of soil organic carbon. Agric. Ecosyst. Environ. 164, 80–99.

Strack, M., Zuback, Y., McCarter, C., Price, J., 2015. Changes in dissolved organic carbon quality in soils and discharge 10 years after peatland restoration. J. Hydrol. 527, 345–354.

Strand, A.E., Pritchard, S.G., McCormack, M.L., Davis, M.A., Oren, R., 2008. Irreconcilable differences: fine-root life spans and soil carbon persistence. Science 319, 456–458.

Suding, K.N., Lavorel, S., Chapin, F.S., Cornelissen, J.H.C., Díaz, S., Garnier, E., et al., 2008. Scaling environmental change through the community-level: a trait-based response-and-effect framework for plants. Global Change Biol. 14, 1125–1140.

Sulman, B.N., Phillips, R.P., Oishi, A.C., Shevliakova, E., Pacala, S.W., 2014. Microbe-driven turnover offsets mineral-mediated storage of soil carbon under elevated CO2. Nat. Clim. Change 4, 1099–1102.

Swift, M.J., Heal, O.W., Anderson, J.M., 1979. Decomposition in Terrestrial Ecosystems. University of California Press, Berkeley, CA.

Talbot, J.M., Treseder, K.K., 2012. Interactions among lignin, cellulose, and nitrogen drive litter chemistry–decay relationships. Ecology 93, 345–354.

Talbot, J.M., Allison, S.D., Treseder, K.K., 2008. Decomposers in disguise: mycorrhizal fungi as regulators of soil C dynamics in ecosystems under global change. Funct. Ecol. 22, 955–963.

Teste, F.P., Kardol, P., Turner, B.L., Wardle, D.A., Zemunik, G., Renton, M., et al., 2017. Plant-soil feedback and the maintenance of diversity in Mediterranean-climate shrublands. Science 355, 173–176.

Tilman, D., Wedin, D., Knops, J., 1996. Productivity and sustainability influenced by biodiversity in grassland ecosystems. Nature 379, 718–720.

Treseder, K.K., 2013. The extent of mycorrhizal colonization of roots and its influence on plant growth and phosphorus content. Plant Soil 371, 1–13.

Trumbore, S., 2000. Age of soil organic matter and soil respiration: radiocarbon constraints on belowground C dynamics. Ecol. Appl. 10, 399–411.

Tu, C., Booker, F.L., Watson, D.M., Chen, X., Rufty, T.W., Shi, W., et al., 2006. Mycorrhizal mediation of plant N acquisition and residue decomposition: impact of mineral N inputs. Global Change Biol. 12, 793–803.

Van Der Heijden, M.G.A., Bardgett, R.D., van Straalen, N.M., 2008. The unseen majority: soil microbes as drivers of plant diversity and productivity in terrestrial ecosystems. Ecol. Lett. 11, 296–310.

Van Kessel, C., Horwath, W.R., Hartwig, U., Harris, D., LÜscher, A., 2000. Net soil carbon input under ambient and elevated CO_2 concentrations: isotopic evidence after 4 years. Global Change Biol. 6, 435–444.

Veen, G.F., Freschet, G.T., Ordonez, A., Wardle, D.A., 2015. Litter quality and environmental controls of home-field advantage effects on litter decomposition. Oikos 124, 187–195.

Vilmundardóttir, O.K., Gísladóttir, G., Lal, R., 2015. Soil carbon accretion along an age chronosequence formed by the retreat of the Skaftafellsjokull glacier, SE-Iceland. Geomorphology 228, 124–133.

Vivanco, L., Austin, A.T., 2008. Tree species identity alters forest litter decomposition through long-term Plant Soil interactions in Patagonia, Argentina. J. Ecol. 96, 727–736.

Wagg, C., Veiga, R., van der Heijden, M.G.A., 2015. Facilitation and antagonism in mycorrhizal networks. In: Horton, T.R. (Ed.), Mycorrhizal Networks, Ecological Studies, vol. 224. Springer, Netherlands.

Waldrop, M.P., Zak, D.R., 2006. Response of oxidative enzyme activities to nitrogen deposition affects soil concentrations of dissolved organic carbon. Ecosystems 9, 921–933.

Walker, T.N., Garnett, M.H., Ward, S.E., Oakley, S., Bardgett, R.D., Ostle, N.J., 2016. Vascular plants promote ancient peatland carbon loss with climate warming. Global Change Biol. 22, 1880–1889.

Walker, T.S., Bais, H.P., Grotewald, E., Vivanco, J.M., 2003. Root exudation and rhizosphere biology. Plant Physiol. 132, 44–51.

Ward, S.E., Bardgett, R.D., McNamara, N.P., Ostle, N.J., 2009. Plant functional group identity influences short-term peatland ecosystem carbon flux: evidence from a plant removal experiment. Funct. Ecol. 23, 454–462.

Ward, S.E., Orwin, K.H., Ostle, N.J., Briones, M.J.I., Thomson, B.C., Griffiths, R.I., et al., 2015. Vegetation exerts a greater control on litter decomposition than climate warming in peatlands. Ecology 96, 113−123.

Ward, S.E., Smart, S.M., Quirk, H., Tallowin, J.R.B., Mortimer, S.R., Shiel, R.S., et al., 2016. Legacy effects of grassland management on soil carbon to depth. Global Change Biol. 22, 2929−2938.

Wardle, D.A., Jonsson, M., Bansal, S., Bardgett, R.D., Gundale, M.J., Metcalfe, D.B., 2012. Linking vegetation change, carbon sequestration and biodiversity: insights from island ecosystems in a long-term natural experiment. J. Ecol. 100, 16−30.

Watt, M., McCully, M.E., Jefree, C.E., 1993. Plant and bacterial mucilages of the maize rhizosphere: comparison of their soil binding properties and histochemistry in a model system. Plant Soil 151, 151−165.

Wehner, J., Antunes, P.M., Powell, J.R., Mazukatow, J., Rillig, M.C., 2010. Plant pathogen protection by arbuscular mycorrhizas: a role for fungal diversity? Pedobiologia 53, 197−201.

West, T.O., Post, W.M., 2002. Soil organic carbon sequestration rates by tillage and crop rotation: a global data analysis. Soil Sci. Soc. Am. J. 66, 1930−1946.

Wieder, W.R., Bonan, G.B., Allison, S.D., 2013. Global soil carbon projections are improved by modelling microbial processes. Nat. Clim. Change 3, 909−912.

Wieder, W.R., Cleveland, C.C., Smith, W.K., Todd-Brown, K., 2015. Future productivity and carbon storage limited by terrestrial nutrient availability. Nat. Geosci. 8, 441−445.

Wright, I.J., Reich, P.B., Westoby, M., Ackerly, D.D., Baruch, Z., Bongers, F., et al., 2004. The worldwide leaf economics spectrum. Nature 428, 821−827.

Woodward, F.I., 1999. Chapter 22: Issues when scaling from plants to globe. In: Press, M.C., Scholes, J.D., Barker, M.G. (Eds.), Physiological Plant Ecology: 39th Symposium of the British Ecological Society. Blackwell Science, Oxford.

Wubs, E.R.J., Bezemer, T.M., 2016. Effects of spatial plant-soil feedback heterogeneity on plant performance in monocultures. J. Ecol. 104, 364−376.

Wurzburger, N., Hendrick, R.L., 2009. Plant litter chemistry and mycorrhizal roots promote a nitrogen feedback in a temperate forest. J. Ecol. 97, 528−536.

Xu, X., Schimel, J.P., Thornton, P.E., Song, X., Yuan, F., Goswami, S., 2014. Substrate and environmental controls on microbial assimilation of soil organic carbon: a framework for Earth system models. Ecol. Lett. 17, 547−555.

Yeates, G.W., 1999. Effects of plants on nematode community structure. Annu. Rev. Phytopathol. 37, 127−149.

Yin, H., Wheeler, E., Phillips, R.P., 2014. Root-induced changes in nutrient cycling in forests depend on exudation rates. Soil Biol. Biochem. 78, 213−221.

Yu, Z., Huang, Z., Wang, M., Liu, R., Zheng, L., Wan, X., et al., 2015. Nitrogen addition enhances home-field advantage during litter decomposition in subtropical forest plantations. Soil Biol. Biochem. 90, 188−196.

Yuste, J., Janssens, I.A., Carrara, A., Ceulemans, R., 2004. Annual Q10 of soil respiration reflects plant phenological patterns as well as temperature sensitivity. Global Change Biol. 10, 161−169.

Zak, D.R., Blackwood, C.B., Waldrop, M.P., 2006. A molecular dawn for biogeochemistry. Trends Ecol. Evol. 21, 288−295.

Zhang, Y.Z., Reddy, C.A., Rasooly, A., 1991. Cloning of several lignin peroxidase lip encoding genes: sequence analysis of the lip6 gcne from the white-rot basidiomycete *Phanerochaete chrysosporium*. Gene 97, 191−198.

Zhao, L., Hu, Y.-L., Lin, G.-G., Gao, Y.-C., Fang, Y.-T., Zeng, D.-H., 2013. Mixing effects of understory plant litter on decomposition and nutrient release of tree litter in two plantations in Northeast China. PLoS One. Available from: https://doi.org/10.1371/journal.pone.0076334.

Zheng, Z.-L., 2009. Carbon and nitrogen nutrient balance signalling in plants. Plant Signal. Behav. 4, 584−591.

MICROBIAL MODULATORS AND MECHANISMS OF SOIL CARBON STORAGE

3

Pankaj Trivedi[1], Matthew D. Wallenstein[1], Manuel Delgado-Baquerizo[2] and Brajesh K. Singh[3]

[1]*Colorado State University, Fort Collins, CO, United States* [2]*University of Colorado, Boulder, CO, United States*
[3]*Western Sydney University, Penrith, NSW, Australia*

3.1 INTRODUCTION

Globally, soils contain more carbon (C) than the atmosphere and vegetation combined, and play a vital role in regulating climate, nutrient cycling and biodiversity. These fundamental ecosystem services are essential to human and environmental well-being (Schmidt et al., 2011; Trivedi et al., 2013). The C in terrestrial soils originates predominantly from decomposing plant litter. But, almost all of this plant-derived C is transformed by soil fauna and microbes, and the efficiency of these transformations is a critical determinant of net ecosystem C storage (Singh et al., 2010; Schmidt et al., 2011). In terrestrial ecosystems, the uptake of CO_2 from the atmosphere by net primary production is dominated by higher plants, but microorganisms contribute greatly to the ecosystem C budgets through their roles as detritivores, plant symbionts, or pathogens, thereby modifying nutrient availability and influencing C turnover and retention in soil (Bardgett et al., 2008; Singh et al., 2010; Schimel and Schaeffer, 2012; King, 2011; Trivedi et al., 2013; Luo et al., 2016). Given the key role of soil biota to ecosystem functioning, there is continued and growing interest in elucidating the explicit relationships between vegetation and belowground processes (Fig. 3.1; Wardle et al., 2004; Xiao et al., 2007; De Deyn et al., 2008; Jin et al., 2010). Soil microbes regulate processes related to soil C dynamics through their survival strategies at the organismal level, competitive and synergistic interactions at the community level, and feedback control mechanisms at the ecosystem level. A comprehensive, mechanistic understanding of ecosystem processes requires that responses at the organism level be directly related to responses of the genome, and these in turn be linked to higher levels of organization. Challenges in manipulating soil microbial community for C storage include identification of major players in C storage and mobilization; determination of genetic basis of the mechanisms involved in C sequestration, understanding complex interactions and feedbacks between physical environment, plants and soil microbes over large spatial and temporal scales, and incorporation of microbial community patterns and process rates to a rigorous ecological framework when most ecosystem models "black box" microbiology (King, 2011; Schimel and Schaeffer, 2012; Trivedi et al., 2013; Wieder et al., 2015).

While the role of soil microbial communities in soil C transformation and stabilization is now widely recognized (Liang and Balser, 2011; Miltner et al., 2012; Box 3.1), the efficiency of these

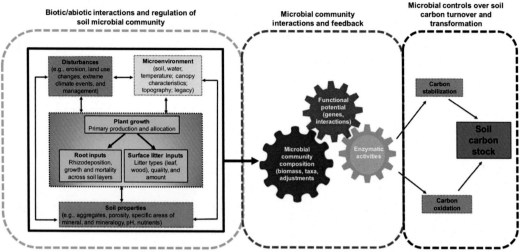

Linkage between microbial community structure and functions

FIGURE 3.1

A conceptual model illustrating the interactions between site biotic and abiotic factors and regulations of soil microbial community, linkage between soil microbial structure and function, and microbial controls over soil carbon stabilization and turnover via effects on enzymatic activities.

BOX 3.1 MYCORRHIZAL FUNGI AS MODULATORS OF SOIL ORGANIC CARBON (SOC) DYNAMICS

Mycorrhizal fungi are one of the major modulators of SOC dynamics as their interactions with plants represent a major link between atmospheric CO_2 and C stored in terrestrial ecosystems. The overall contribution of mycorrhizal fungi to soil C storage function of the kinds of hyphae produced, the residence time of accumulated hyphal residues, the production of glomalin, and the role played by mycorrhiza in the stabilization of soil aggregates. Mycorrhizal fungi channel significant amounts of recently fixed plant C through the soil, affecting a number of soil processes including the composition and activity of microbial communities and SOM mineralization. Some of the mycorrhizal fungi (particularly those establishing ecto-, ericoid and orchid-mycorrhizas) directly mineralize SOM with a purpose to utilize N and/or P without utilization of its C. Through effective scavenging of essential nutrients, these fungi create localized nutrient deficiency for other C decomposers consequently reducing their biomass and rates of decomposition. On the other hand, arbuscular mycorrhizal fungi lack efficient C degrading exoenzymes and thus possibly interact with the associated microbes for SOM mineralization. Several studies have provided empirical evidence regarding the involvement of mycorrhizal fungi in soil C storage processes under global change. The effect of different mycorrhizal types on soil C pools have far greater consequence than impacts associated with an ecosystem's productivity, climate, or the physical properties of its soil. The functional traits of mycorrhizal fungi have been linked to C storage at ecosystem-to-global scales, suggesting that plant—decomposer competition for nutrients exerts a fundamental control over the terrestrial C cycle. A thorough understanding of the ecology of mycorrhizal fungi available in public datasets along with its constrained diversity leads to the possibility of incorporating mycorrhizal fungi community composition and function in a trait-based framework that can inform Earth system models.

processes varies with environmental factors and climate, and can be altered by management practices (Fig. 3.1). These factors can induce structural shifts in soil microbial communities that lead to alterations in their function (Hackl et al., 2005; Brockett et al., 2012). Although significant advancements have been made to elucidate the taxonomic composition of soil microbial communities (Högberg et al., 2007; Thoms et al., 2010; Burns et al., 2015), our ability to directly link phylogeny to function is limited. This knowledge gap is a barrier to explicitly including microbes in global biogeochemical models such as Earth system models (ESMs) that inform citizens and policy makers of C dynamics and exchange between the biosphere and the atmosphere (Wieder et al., 2013, 2015). The most fundamental problems for microbiologists to resolve in this regard include identifying the processes, mechanisms, and groups of organisms that are most important for sequestration and determining whether they can be managed at large enough spatial, and long enough temporal scales, to achieve desired levels of C storage. This chapter aims to provide a mechanistic understanding of the role microbes play in C storage and propose pathways to include these data to modify existing ecological models.

3.2 KEY MICROBIAL MODULATORS OF SOIL C STORAGE IN TERRESTRIAL ECOSYSTEMS

3.2.1 COMMUNITY STRUCTURE

Environmental parameters such as temperature and moisture are typically the dominant controls on microbial process rates. Given the incredible diversity of the soil microbiome, scientists have long hypothesized that microbial community composition was also an additional control on processes such as decomposition. Comparative studies of C substrate decomposition by soil microbial communities, generally only conducted on a few soil types (Brant et al., 2006; Hanson et al., 2008; Steinweg et al., 2008; Rinnan and Bååth, 2009; Eilers et al., 2010), failed to observe a strong role for microbial community composition in driving decomposition rates (Strickland et al., 2009). However, other studies that focused on more narrowly defined processes or specific functional responses, found that the composition of soil microbiomes seems to exert an additional level of control on process rates and their sensitivity to environmental drivers (Schimel, 1995; Schimel and Gulledge, 1998; Cavigelli and Robertson, 2000). Studies that have used advanced "omics" analysis to provide an explicit understanding on the relationship between microbial community and function have demonstrated significant control of microbial communities on soil organic carbon (SOC) transformation and turnover (Weand et al., 2010; Trivedi et al., 2016). In particular, these studies have identified specific soil microbial attributes driving SOC dynamics, which is a significant step forward for gaining mechanistic understanding on the control of microbial communities on C storage potential in terrestrial ecosystems (Lucas et al., 2007; Acosta-Martinez et al., 2010).

Different microbes have different physiologies and, as a result, change in composition or relative abundance are likely to result in a shift in the rate of processes including soil C turnover (Waldrop and Firestone, 2006). For example, shifts in microbial community composition that produce hydrolytic enzymes facilitate C acquisition in support of soil microbial primary metabolism (Cusack et al., 2011), whereas shifts toward microbial communities producing oxidative enzymes (e.g., saprophytic fungi) help with degrading recalcitrant C substrates such as lingo-cellulose

(Sylvia et al., 2005; Berg and Laskowski, 2006). It is well known that fungi are major facilitators of the rate limiting step in litter decay in which cell wall lignin is completely mineralized to CO_2 (Eisenlord et al., 2013). Higher C accumulation in fungal-dominated soils can be attributed to higher C use efficiency (CUE), longer retention of C in living biomass, and recalcitrant necromass resulting in longer resident time of C (Strickland and Rousk, 2010; Malik et al., 2016). Accordingly, higher fungal:bacterial ratios seem to lead to greater C accumulation (Strickland and Rousk, 2010; Trivedi et al., 2013; Malik et al., 2016). This generally accepted belief, however, does not take into consideration the role of external factors such as disturbance that can alter the community structure and functions of decomposer community. Shifts in bacterial community composition from oligotrophs (slower growing) to copiotrophs (fast growing) have also been related to greater SOC turnover in terrestrial ecosystems (Trivedi et al., 2013, 2015; Fierer et al., 2007; Leff et al., 2015; see Section 3.2.3 for more details on the ecological characterization of bacteria based on lifestyle). Trivedi et al. (2016) found that a strong relationship between different functional genes and their corresponding enzyme activities was maintained in cropping soils from three major grain producing regions in Australia after considering microbial community structure, total C and soil pH using structural equation modeling. Results showed that the variations in the activity of enzymes involved in C degradation were predicted by the functional gene abundance of the soil microbial community.

3.2.2 DIVERSITY

Diversity is generally described by indices taking into account a number of entities such as genotypes or species and their relative distribution in a community—richness, evenness, Shannon index etc. Although microbial diversity is directly related to ecosystem multifunctionality and specialized functions (Levine et al., 2011; Philippot et al., 2013; Powell et al., 2015; Delgado-Baquerizo et al., 2016a, b), various reports have suggested that resource availability and microbial abundance rather than diversity matrices are related to aggregate functions such as decomposition (Trivedi et al., 2016; Louis et al., 2016). Soil C modeling approaches have integrated microbial diversity by representing different functional groups of microorganisms according to their differing affinities for organic substrates (Moorhead and Sinsabaugh, 2006; Perveen et al., 2014) or enzyme-production strategies (Allison et al., 2013). For example, a trait-based approach that links microbial community composition with physiological and enzymatic traits predicts that interactions between different functional groups influence microbial-enzyme production that controls the litter degradation rates (Allison, 2012). However, our understanding of the functional traits associated with the microbial groups used in these models is incomplete, which limits validation of these models with empirical data (Waring et al., 2013). Recent demonstrations of the microbial diversity-SOM (soil organic matter) dynamics relationship suggests that overall microbial diversity modulates SOM mineralization and diversity indexes could be important variables explaining SOC transformation (Baumann et al., 2009; Juarez et al., 2013; Mau et al., 2015; Louis et al., 2016). Tardy et al. (2015) showed that soil microbial diversity indices were the best predictors of C-cycling activities, wherein higher variations in labile and recalcitrant organic matter mineralization were explained by bacterial and fungal diversity, respectively. Use of combined the dual isotope probing-technique with amplicon sequencing demonstrated direct linkages of priming effect with microbial diversity and composition (Mau et al., 2015). Recent evidence suggests that microbial diversity can explain mineralization of native SOC, while the priming effect is independent of microbial diversity. Overall, soil microbial diversity appears to affect soil processes involved in C dynamics (Louis et al., 2016). But, we are still lacking a full understanding of the underlying changes in soil community functioning with shifts in soil diversity (Nielsen et al., 2011; De Graaff et al., 2015). It seems that a change in taxonomic diversity could lead to a change in the functional traits of the whole community, and thus to a change in soil functioning.

3.2.3 **FUNCTIONAL TRAITS**

Resource economic theory provides a framework to understand how tradeoffs in life-history strategies result in growth trait variation among life forms. This theory assumes that growth traits develop from the allocation of limited resources to competing metabolic purposes—namely, growth, reproduction, or maintenance functions (Litchman and Klausmeier, 2008). To achieve a better understanding of the linkages between soil C turnover and the structure and function of microbial communities, Fierer et al. (2007) categorized soil microbes based on their trophic lifestyles and trait-tradeoff particularly in relation to C utilization. This classification categorizes bacterial communities across a continuum that goes from fast-growing bacteria that utilize labile forms of C and dominate nutrient-rich environments as "copiotrophs" (members of α-*Proteobacteria*, β-*Proteobacteria*, γ-*Proteobacteria*, Bacteroidetes) to slow-growing bacteria using relatively recalcitrant forms of C, and that dominate nutrient poor areas as "oligotrophs" (members of Acidobacteria and δ-*Proteobacteria*). Some other major soil bacterium groups such as Firmicutes and Actinobacteria do not fall within this copiotrophic-to-oligotrophic gradient and are often described as facultative copiotrophs/oligotrophs (Fierer et al., 2007; Trivedi et al., 2013). Traits that determine copiotrophic or oligotrophic lifestyles in soil bacteria are contrastingly different (Table 3.1). Functional groups based on these life-history traits are instrumental in determining the relative abundance of certain organisms in a given environment, influencing the outcome of many ecosystem processes depending on which growth strategy dominates (Follows et al., 2007). Finn et al. (2017) presented evidence whereby management practices that increases the bioavailability of SOC alter community structure and function to favor microbial species (e.g., copiotrophs) likely to be associated with increased rates of SOC loss compared with natural ecosystems. The study suggested manipulation of soil microbial community that reduce SOC loss through management practices can improve sequestration of SOC stocks and improve ecosystem services in agricultural soils.

To date, the copiotroph/oligotroph classification remains one of the most successful functional groupings of soil microbes and has helped to interpret how global environmental changes such as climate change (DeAngelis et al., 2015), nutrient deposition (Leff et al., 2015), and changes in management practices and land use (Trivedi et al., 2015) indirectly alter C cycling via microbial community composition. Recent studies have derived the genetic traits that determine the signatures of copiotrophy/oligotrophy in soil bacterial communities (Trivedi et al., 2013; Spring et al., 2015). These analyses have shown that microbial diversity of enzymes involved in the degradation of C substrates may be associated with lifestyle strategies, thus influencing SOC retention in terrestrial ecosystems. It has been postulated that copiotrophs have an increased ability to produce extracellular enzymes involved in the degradation of labile forms of C, while oligotrophs will have a higher potential to produce extracellular enzymes involved in the degradation of recalcitrant forms of C (Singh et al., 2010; Trivedi et al., 2013). Also, copiotrophs would be expected to have a higher

Table 3.1 Physiological, Genomic, and Environmental Response Traits That Determine Copiotrophic or Oligotrophic Lifestyles in Soil Bacteria

Traits	Copiotrophs	Oligotrophs
Physiological		
Growth rates	High	Low
Growth yield	Low	High
Growth strategy	Fast and famine	Equilibrium
Cell maintenance requirements	High	Low
Ease of cultivation	High	Low
Resistance to stress	Low	High
Genomic		
rRNA copy numbers	High	Low
Prophages	Many	Few
Repeats with CRISPRs	Many	Few
COG category N (cell motility)	High	Low
COG category T (signal transduction)	High	Low
COG category V (defense mechanisms)	High	Low
COG category Q (secondary metabolites biosynthesis, transport, and catabolism)	High	Low
Ability to produce enzymes for labile C degradation	High	Low
Ability to produce enzymes for recalcitrant C degradation	Low	High
Flagella production	High	Low
Membrane transporters	High	Low
ATP-dependent transporters	High	Low
Substrate specific high affinity transporters	Low	High
Stress response genes	Low	High
Response to Management Practices		
Nutrient additions	Rapid increase	Equilibrium
Extreme environmental conditions	Rapid decline	Equilibrium

proportion of transporters than slower growing organisms such as oligotrophs to support fast growth rates and their nutrient requirements (Trivedi et al., 2013).

While the physiological foundation for copiotroph/oligotroph bacteria is established and there is evidence that we can use taxonomic information to place soil bacteria along this continuum in life-history strategies (Fierer et al., 2007), much less is known on whether such trait-based classification exist in soil fungi. Garrett (1951) proposed a conceptual framework wherein "allochthonous" fungi carry out initial stages of decomposition on freshly added organic materials, while the "autochthonous" fungi attack more recalcitrant materials—the latter comprise more stable communities than the former, and are little affected by additions of fresh organic matter. While conceptually tidy, the terms bear little relation to the actual behavior of decomposer fungi and several researchers have questioned the existence of a well-defined autochthonous mycoflora (Kjøller and Struwe, 1992; Brandl and Andersen, 2015). Some researchers have further suggested that basidiomycetes that produce a range of enzymes to degrade recalcitrant C can be categorized as oligotrophs, while ascomycetes that grow fast and have lower capacity to degrade recalcitrant C substrates—especially lignin—are copiotrophs (Morrison et al., 2016). However, fungi are known to have broader enzymatic capabilities (McGuire et al., 2010), slower biomass turnover rates (Rousk and Bååth, 2011), and potentially greater CUE (Six et al., 2006) than bacteria, and therefore it is difficult to categorize fungal communities into well-defined traits based groups that differentially influence SOC dynamics. At this time, limited availability of fungal genomes from terrestrial environments only partially supports the broad hypothesis of trait-based groupings in relation to C turnover. In a report, Treseder and Lennon (2015) synthesized information from 157 whole fungal genomes and showed that traits associated with C mineralization varied most at the phylum level. Traits that regulate the decomposition of complex organic matter-lignin peroxidases, cellobiohydrolases, and crystalline cellulases were positively related, but they were more strongly associated with free-living filamentous fungi. More sequencing efforts will be required to provide insights into the relationship between taxonomy and traits among soil fungi that will be critical for defining ecological coherent units and predicting/interpreting community functions related to key aspects of C cycling. It should, however, be noted that linking the phylogenetic and trait-based patterns of biogeography of soil microbial communities based on enzyme production is difficult considering the role of "cheaters," horizontal gene transfer, and different ecological function(s) of a single enzyme. Also considering the vast physiological and metabolic differences within different species of a single phylogenetic group (e.g., family and genus) it is difficult to assign similar roles in ecosystem process dynamics, particularly those related to SOC turnover.

3.2.4 BACTERIAL:FUNGAL RATIO

A widely applied proxy microbial indicator to relate microbial community to SOC dynamics is based on the sub-division of microbes into major decomposer groups, namely the fungi and bacteria, indexed as fungal:bacterial (F:B) ratio (Bardgett et al., 1996; Strickland and Rousk, 2010 and references within). The ratio has been extensively used in soil ecology, particularly in the context of land management and its effects on soil carbon sequestration (Bardgett et al., 1996; Strickland and Rousk, 2010; Bailey et al., 2002; de Vries et al., 2006; Fierer et al., 2009; Sinsabaugh et al., 2013). It has been postulated that soil microbial communities' F:B ratio is linked to its capacity to store C—with higher F:B dominance being linked to higher carbon sequestration. A shift in a fungal dominance is thought to enhance SOC accumulation due to: (1) The ability of soil fungi to produce a broader array of enzymes that are capable of transforming and stabilizing inputs (de Boer et al., 2005); (2) Enhanced soil aggregation (Six et al., 2006); (3) Slower biomass turnover rates (Rousk and Bååth, 2011); and (4) Potentially greater CUE due to higher C:N ratio of fungal biomass (Six et al., 2006; Wallenstein et al., 2007). Although correlations between F:B and environmental change factors such as land use change are well established, there is still a lack of experimental studies demonstrating the underlying physiological and genetic mechanisms underpinning these field observations (Strickland and Rousk, 2010).

To bridge these knowledge gaps, studies have employed metaanalysis, manipulative experimentation, modeling and next-generation "omics" approaches to advance current understanding of the role of fungal dominance in enhancing soil C storage. Using metaanalysis and an enzyme-driven biogeochemical model, Waring et al. (2013) showed that soil F:B was largely controlled by resource stoichiometry, wherein the relationship between F:B and C:N is consistent with the idea of a threshold elemental ratio at which microbial metabolism switches from dependence upon energy (C) availability to nutrient availability (Sterner and Elser, 2002; Allen and Gillooly, 2009). Differences in the threshold elemental ratio between the fungi and bacteria will result in variable relationships between F:B ratios and soil C:N along a nutrient gradient that will change when both fungi and bacteria are C-limited, when both are N-limited, and when the two groups differ in element limitation (Waring et al., 2013). Therefore, responses of F:B to fertilization with C or N may not be fully predictable unless the nutrient limitation status of each functional group is known. Recently, Malik et al. (2016) employed RNA sequencing, protein profiling and isotope tracer techniques in a litter degradation experiment with two soils with similar physicochemical properties, but different F:B ratios, to show that soil F:B ratios are linked to altered C cycling wherein high F:B soil demonstrates high C storage. It has been suggested that it would be advantageous when relating to SOC turnover with F:B ratio, that a distinction between traits of bacterial and fungal subgroups should be made (Strickland and Rousk, 2010). For example, RNA-seq analysis showed that the ratio between Actinobacteria to the rest of the bacteria increased only in low F:B soils suggesting the Actinobacteria may take up the role of fungi in litter degradation when fungal abundance is low (Malik et al., 2016). It may, therefore, be necessary to tailor the F:B ratio to specific functional guilds within the bacterial and fungal subgroups in order to accurately develop the indicators for SOC dynamics in terrestrial ecosystems.

3.2.5 CARBON USE EFFICIENCY

CUE (biomass-C synthesized per substrate consumed) of soil microbial communities influences what proportion of organic C utilized is released into the atmosphere as CO_2, potentially remains in the soil as organic matter in living cells, or as dead SOM (Billings and Ballantyne, 2013; Bradford and Crowther, 2013; Hagerty et al., 2014; Geyer et al., 2016) and therefore is a critical control on C storage in ecosystems (Allison et al., 2010; Manzoni et al., 2012). The CUE is a function of the cellular demand for energy and biosynthesis, and thus a function of the physiological state and the type of compounds that are being produced (Dijkstra et al., 2015). A higher value of CUE translates to greater growth per unit C acquired, meaning that, if all else is equal, more C is available to higher trophic levels, detrital pathways, and potentially for ecosystem storage (Bradford and Crowther, 2013). It is expected that fungi, on average, have higher CUE than do bacteria, leading to higher C stored in environments with greater F:B ratios (Strickland and Rousk, 2010; see also Section 3.2.4). In bacteria, oligotrophs have a higher CUE compared to copiotrophs because of their lower maintenance energy requirements (Trivedi et al., 2013). Although, CUE increases for both copiotroph and oligotroph as the energy content of the resource is increased, maximal growth efficiency of oligotrophs is higher than copiotrophs, and that it is reached at a lower concentration of limiting resources (Trivedi et al., 2013; Roller and Schmidt, 2015).

Substrate chemistry can alter CUE due to its direct effect on cellular metabolism, but may also drive CUE indirectly by selecting for distinct microbial communities with different prevailing life histories (Trivedi et al., 2013; Kallenbach et al., 2016; Geyer et al., 2016). For instance, highly reduced substrate-C (high free energy) typically promotes higher CUE within a community, but may also select for a copiotrophic-dominated community with an inherently lower CUE (Geyer et al., 2016). In general, soil microbial communities appear to have high CUE (approximate average of 0.55) in soils, seem in an apparent contradiction to the idea that soil is a C-limited environment where most microbes are not growing, or only grow slowly, and where maintenance energy demand dominates substrate use (Blagodatskaya et al., 2014; Reischke et al., 2015). High CUE in soil environments has important and potential far-reaching consequences as it affects how we model microbial activity in soils, and think about the relative importance of key process that stabilize microbial population size and community composition (Hagerty et al., 2014; Dijkstra et al., 2015).

While the concept of CUE is relatively straightforward, there are important methodological challenges to measuring it in soils. Typically, [13]C (and sometimes ^{14}C) labeled substrates are added to soils, and traced into respiration, biomass, and soil. But the apparent CUE will decrease with time as the labeled C is reprocessed. Thus, factors like temperature which appear to affect CUE may simply be altering microbial turnover, resulting in the appearance of lower CUE due to faster reprocessing (Hagerty et al., 2014). Future explorations of physiological, ecological, and evolutionary processes shaping microbial communities will provide insights into the variations in CUE among diverse microbes and will improve our knowledge of how microbial communities impact carbon flux, from the small scale of host microbiome interactions to large-scale annual CO_2 flux from an ecosystem. In this respect, Geyer et al. (2016) have provided a framework that enhances data interpretation and theoretical advances by unifying the existing concepts and terminology underlying CUE by accounting for population, community, and ecosystem-scale controls over the fate of metabolized organic matter.

3.3 FACTORS IMPACTING MICROBIAL MODULATION OF C STORAGE
3.3.1 SOIL STRUCTURE (AGGREGATES)

A particularly important variable controlling SOC dynamics and soil fertility is the process of soil aggregation (Gupta and Germida, 1988; Six et al., 2004, 2006; Trivedi et al., 2015, 2017). Experiments and theory propose hierarchical models for the aggregation process wherein the mechanisms that form and stabilize aggregates are the orientation of the mineral particles, the attractions between these particles, the interactions between these particles and the soil microorganisms, and the production of aggregating agents by these microorganisms (Oades and Waters, 1991). Microbes play a major role in aggregate stabilization by modulating the interactions among organic matter, metals, and mineral surfaces through the production of specific proteins. Glomalin, considered the most important protein for aggregate stabilization, is produced by a phylogenetically narrow range of arbuscular mycorrhizal fungi (AMF) within the order Glomales (King, 2011) and is reported to directly increase that stability of the aggregates by impacting hydration (water penetration and hydrophobicity) and therefore the accessibility of hydrolytic enzymes to bulk organic matter (Rillig et al., 2010). However, the evidence for the importance of glomalins in soil aggregation is weak and further research is required to elucidate the role of such an operationally defined suite of proteins in soil stabilization processes. In addition to glomalin, other proteins, such as the hydrophobins and chaplins, are also reported to play important roles in microaggregate formation (King, 2011). Despite the important role of microbial proteins in aggregate stabilization, we have a limited understanding on their interactions with SOM turnover within different sized aggregates. Also, in soils with heterogeneous habitat conditions and mixed microbial communities, it may be difficult to isolate the role of individual proteins in SOC dynamics in different soil aggregates. A better understanding of glomalin, hydrophobins, and chaplins could directly lead to new or modified management practices for SOC retention in agroecosystems and other managed systems.

Soil aggregates are generally divided into three or four classes based on their size—e.g., large mega- (>250 µm), macro- (250−50 µm), and microaggregates (<50 µm) (Tiemann et al., 2015; Trivedi et al., 2015). Mega- and macroaggregates protect plant-derived and microbial-derived SOM, contain high amounts of fungal biomass, and are enriched with labile C and N originating predominately from plant residues (Six et al., 2006; Tiemann et al., 2015; Trivedi et al., 2015). On the other hand, microaggregates are formed by microbial-induced bonding of clay particles, polyvalent metals, and organometal complexes as well as being characterized by lower concentrations of new and labile carbon (lower C:N ratio) and increased amounts of physically protected and biochemically more recalcitrant C compared to macroaggregates (Elliott, 1986). SOM turnover in different sized soil aggregates is influenced by several variables including management practices, organic matter loading, mechanical disturbance, temperature, and moisture, along with functional groups of soil microbes (Six et al., 2006; Tiemann et al., 2015; Trivedi et al., 2015, 2017). It has been postulated that spatial distribution of resources within different sized soil aggregates provides niches for the heterogeneous distribution of contrasting functional groups of microorganisms that regulates ecosystem-level decomposition and nutrient mineralization (Davinic et al., 2012; Tiemann and Grandy, 2015; Trivedi et al., 2015, 2017). Furthermore, the location of microorganisms within aggregates and pores influences their functionality due to accessibility of substrates and microclimatic properties such as oxygen concentration, gas diffusivity, moisture content, and competition (Gupta and Germida, 2015). For example, it has been shown that pore geometry and connectivity, oxygen diffusion rates, and bacterial community structure have more pronounced effects on the mechanics of soil C dynamics in microaggregates compared with larger sized aggregates (Vos et al., 2013; Rabbi et al., 2016). Conceptually, microaggregate conditions are understood to be responsible for the protection of soil C, however quantitative knowledge on the specific factors driving SOC dynamics in microaggregates is still lacking. Overall, we have a limited understanding on the mechanisms and relative contributions of different microbial groups to SOM turnover in different aggregates. Knowledge of microbial communities and their activities within different microenvironments (i.e., aggregate size) is currently poor but essential for understanding the regulation of soil C process rates and, therefore, C sequestration at the cropping system scale.

In general, the bacterial community composition within soil varies with aggregate size, and microaggregates have a higher microbial biomass, abundance, and diversity compared to macroaggregates (Monreal and Kodama, 1997; Neumann et al., 2013). The interaction between bacteria, organic matter and its forms, metal ions, and clay affects the survival of bacteria, as SOM, minerals, and aggregate sizes may offer different levels of nutrients and habitat to bacteria (Van Gestel et al., 1996; Sessitsch et al., 2001). Increased organic matter recalcitrance and stabilization via organomineral interactions with the decreasing size of aggregates may explain why microaggregates have higher bacterial abundances than larger aggregates. Studies have further suggested that distinct aggregate size classes support distinct microbial habitats which supports the colonization of different microbial communities (Davinic et al., 2012; Trivedi et al., 2015). Results from microbial community composition in aggregates are generally consistent with the hypothesized shifts in general life-history strategies with bacterial taxa that are faster-growing and more copiotrophic being favored in large aggregates with higher amounts of labile C, while slow-growing and more oligotrophic dominate microaggregates wherein recalcitrant C is present in higher levels (Trivedi et al., 2015). Microbial community functions such as the production of extracellular enzymes, decomposition, and production of aggregate binding agents have previously been liked to aggregate stabilization and SOC persistence/accrual (Six et al., 2006; Tiemann and Grandy, 2015). Studies have demonstrated increased activity of enzymes involved in degradation of labile C with a decrease in aggregate size (Qin et al., 2010; Nie et al., 2014; Bach and Hofmockel, 2015). On the other hand, the activity of NAG (i.e., chitin degradation) has been reported to increase with decreased aggregate size (Tiemann et al., 2015; Tiemann and Grandy, 2015). Increased enzyme activities related to C decomposition with decreasing aggregate size may be due to a higher OC concentration in microaggregates compared with macroaggregates (Nie et al., 2014; Trivedi et al., 2015, 2017). Recent research findings have highlighted a differential response of management practices and microbial control on the C turnover in macroaggregates (wherein higher amount of labile C was present) and microaggregate (wherein higher amount of recalcitrant C was present) should be explicitly considered when accounting for management impacts on soil functions including C storage (Trivedi et al., 2015, 2017, Fig. 3.2). Quantifying the relative importance of different microbial groups for C storage and functions in aggregate-size fractions will significantly advance this area of science by providing the empirical data required for models that predict global soil C sequestration (Wieder et al., 2015).

3.3.2 LAND USE AND MANAGEMENT PRACTICES

A range of manipulation approaches for land management to influence soil decomposition processes has been proposed to impact soil C turnover. Recent studies have provided strong evidence that microbial community composition responds significantly to land use and management practices and are correlated with soil processes, including SOC turnover (Geisseler and Scow, 2014; Trivedi et al., 2015; Leff et al., 2015). Conservation tillage (including no-tillage), high crop residue retention, and crop rotation practices are known to improve SOC storage, which results in greater productivity and sustainability of agroecosystems (Six et al., 2006; Hobbs et al., 2008). Interestingly, long-term experiments have revealed that conservational agricultural practices have predictable effects on soil microbial structures regardless of cross-site variations (Frey et al., 1999; Sipilä et al., 2012). In no-till agroecosystems, the location of crop residue on the surface favors fungi as the primary decomposers, whereas bacteria are more important in conventional tillage (Hendrix et al., 1986; Frey et al., 1999). Crop rotations change the microbial community structure and activity, with positive effects on aggregate formation and SOM accrual (Tiemann et al., 2015; Trivedi et al., 2017). In addition, compared to the conventional farming systems, low-inputs systems had a high functional gene diversity demonstrating that low input systems may help in sustaining higher soil microbial community diversity and potentially enhanced functions mediated by soil microbes in nutrient cycling and C storage (Xue et al., 2013).

Nitrogen (N) management in soils has been considered as a key to the sustainable use of terrestrial ecosystems (Ollivier et al., 2011). The impact of N availability on soil microbial communities and detailed microbial mechanisms that drive the impact on SOC turnover in response to N

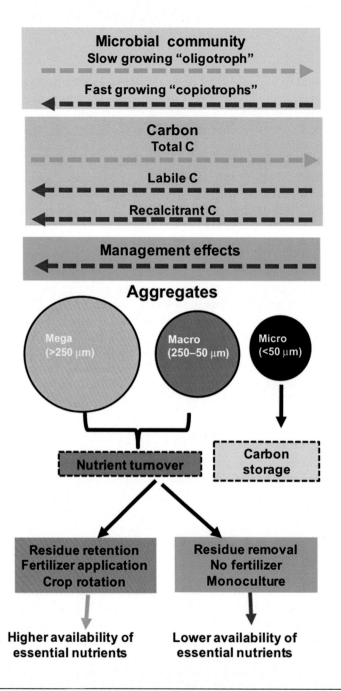

FIGURE 3.2

Trajectory of microbial community composition, soil C and management practices (*colored boxes*) and the agro-ecosystem processes taking place within different sized aggregates (*dashed box*). The conclusions are drawn from multiple studies that have described the microbial communities in different sized aggregates and their impact of SOC turnover and ecological processes (Trivedi et al., 2015, 2017; Wood et al., 2016; Tiemann et al., 2015; Tiemann and Grandy, 2015; Davinic et al., 2012).

addition is discussed by Macdonald et al. (Chapter 6: Soil Nutrients and Soil Carbon Storage: Modulators and Mechanisms). Briefly, across the wide array of soil types, soil respiration and microbial biomass were consistently suppressed under N amendments (Treseder, 2008) and these belowground responses are not solely related to the C inputs to the soil by the vegetation (Ramirez et al., 2012). N enrichment decreases extracellular enzymes that are associated with the degradation of recalcitrant C (Ramirez et al., 2012) which correlates with the decreases in the richness and diversity of genes involved in the decay process (Eisenlord et al., 2013). Similarly to conservational management practices, N amendment causes a consistent shift in the predominant microbial life-history strategies, favoring a more active, copiotrophic microbial community (Campbell et al., 2010; Ramirez et al., 2010, 2012; Fierer et al., 2012; Eisenlord et al., 2013). Shifts in the catabolic capabilities of communities across N gradients were significantly correlated with phylogenetic and metagenomic responses, indicating possible linkages between the structure and functioning of soil microbial communities (Fierer et al., 2012). In addition to N management, other land use change and management such as the conversion of forest, pastured, and cropped land have ecosystem scale impacts in terms of soil cycling of organic compounds, microbial community, and soil nutrient dynamics (Bonan, 2008; Hartmann et al., 2012).

Together, results from various metagenomics-based studies have demonstrated that although microbial community composition varied considerably across the diverse land sites, management practices that increase the availability of dissolved organic carbon (DOC; proxy for easily utilizable labile C source) elicits changes to the composition of microbial communities in consistent ways across sites by selecting for microbial groups that have certain functional traits (Ramirez et al., 2012; Fierer et al., 2012; Geisseler and Scow, 2014; Trivedi et al., 2015; Leff et al., 2015; see Fig. 3.3). These results suggest generalizable patterns in these responses across a wide range of climatic and edaphic environments and confirm their existence, despite large cross-site differences in microbial community structure. For example, our results on several long-term soil management trials with contradicting treatments have a clear distinction in the structure of their bacterial community with an increase of taxa belonging to Actinobacteria, alpha-beta-gamma Proteobacteria, Bacteroidetes, and a decrease in the relative abundance of taxa belonging to Acidobacteria, delta-Proteobacteria across land use strata that increase the availability of readily utilizable C substrates (Fig. 3.3A). Families belonging to these taxa are categorized as copio- and oligotrophs at class levels thus suggesting that management induced shifts in tropic traits could have important implications for soil C cycling. These observed patterns correspond to broader ecological theory and set the stage for more targeted hypothesis testing to explore the relative contribution of different functional microbial groups in SOC dynamics and turnover.

Several studies show that soil management practices have a differential effect on genes involved in carbon degradation (Xue et al., 2013; Kuramae et al., 2014; Trivedi et al., 2015; Leff et al., 2015; also see Fig. 3.3B). A significant increase in the abundance of genes involved in the degradation of relatively labile carbon (e.g., starch and cellulose), was observed in management practices that increase SOC content. On the other hand, genes involved in the degradation of recalcitrant forms of carbon were higher in control treatments. These results provide evidence that microbial community functional gene structure could significantly affect SOC turnover and storage. The increase in the gene abundance and corresponding enzyme activities involved in the degradation of labile forms of C in treatments with higher DOC, may be a consequence of the stimulation of both microbial growth and activity by improved nutrient

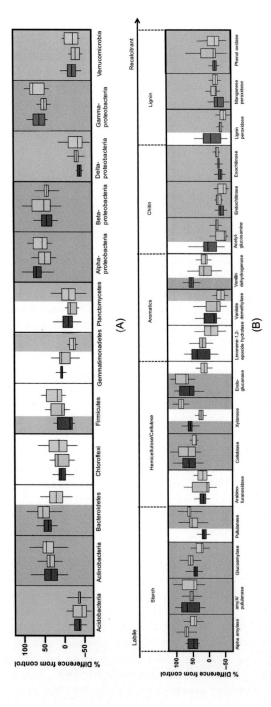

Differences in the relative abundance of higher-level taxa (A) and key gene families involved in carbon degradation (B) between control and crop rotation (*gray colored boxplots*); N-addition (*green colored boxplots*); and residue retained (*yellow colored boxplots*) plots at three long-term soil management trials at Narrabri; Karoonda; and Cunderdin respectively. Boxplots show quartile values for each taxon. *Blue and orange backgrounds* show significant decreases and increases in the relative abundances of specific taxa or gene families, respectively (false discovery rate-corrected $P < .05$). Differences are comparisons to cotton-wheat-vetch rotation versus continuous cotton (Control) at Narrabri; Optimal variable nutrient application (nitrogen (urea) and phosphorous (DAP) was applied at a rate of 80 and 10 kg ha^{-1}, respectively) versus no nutrient application (Control) at Karoonda; and cereal-legume-brassica "residue retained" versus cereal-legume-brassica "residue removed" (Control) at Cunderdin sites. Bacterial community composition and functional gene families were determined by 454 pyrosequencing and Geochip, respectively (details for techniques and analysis are provided in Trivedi et al., 2015). In the Geochip analysis (B) the complexity of carbon is presented in the order from labile to recalcitrant.

availability as well as changes in microbial community composition (Fig. 3.3A and B). These findings are supported by a large-scale study on the response of soil microbial communities to experimentally added nitrogen and phosphorous at 25 grassland sites across globe (Leff et al., 2015) which showed similar trends in the shifts in the ecological attributes of microbial community compositions and functions. Future research exploring direct linkage, in real time, for soil processes and metabolic properties of soil microbial community will help to identify the ecological consequences and in identifying management practices that could increase net productivity and C storage.

3.3.3 **SOIL MOISTURE**

Soil water content is a key control on soil microbial activities, limiting both soil respiration and enzymatic activities and thus influencing SOC decomposition and soil CO_2 efflux (Xu and Qi, 2001; Reichstein et al., 2003; Brockett et al., 2012). As soil moisture decreases, substrate diffusion becomes limited, resulting in the physiological adaptation of microbes to lower water potentials, thus slowing down biochemical process rates and hence soil respiration (Schimel et al., 2007; Bouskill et al., 2013). Changes in soil water availability can alter soil microbial communities by causing shifts in community composition through the local extinction of certain taxa, or by shifting the abundance from one taxon to another (Evans and Wallenstein, 2012, 2014). In general, bacteria tend to be more prominent in wet or disturbed systems, such as tilled agricultural land and marshlands, while the abundance of fungi increases in drier, less disturbed areas like forests, no-till agroecosystems (García-Orenes et al., 2013). These effects are also evident within bacterial community, e.g., in tropical humid forests, lower soil moisture seemingly selects for Actinomycetes and Planctomycetes (Bouskill et al., 2013), both of which are adapted to grow at lower water potentials and are widely distributed in arid soils.

Microbial responses to soil moisture can be highly variable. Low water availability limits the physiological performance of microbes and the diffusion of nutrients in the soil pore space (Robertson et al., 1997). However, broad consistency in microbial functional responses to moisture allows microbially-driven soil C processes to acclimatize leading to similar levels of microbial activities under low- and high-water availability as in normal conditions (Schimel et al., 2007; Curiel Yuste et al., 2007). Furthermore, unique physiological responses of individual microbial taxa can also result in process variability in responses to water availability (Lennon et al., 2012). Specific microbial moisture responses are reported to result in historic contingencies that modifies SOC dynamics from microsite to ecosystem scale. For example, Barnard et al. (2013) reported that adaptation of soil bacterial communities to extreme soil desiccation and rapid rewetting is expressed through different life-strategies, conferring a spectacular resilience to the potentially active bacterial communities across sites, despite site-specific community composition and structure. In contrast, the fungal community was not detectably different among sites, and was largely unaffected, showing marked resistance to desiccation.

Transient rain events may contribute disproportionately to soil C cycling, particularly when microbial responses are both rapid and large, or when short-duration rain pulses allow older soil C pools to be accessed (Collins et al., 2008; Placella et al., 2012). Microbial response patterns displayed phylogenetic clustering and were primarily conserved at the subphylum level, suggesting that resuscitation strategies after wet-up of dry soil may be a phylogenetically conserved ecological trait (Placella et al., 2012). Changes in precipitation altered the potential for bacterial carbohydrate degradation that directly impacts the SOC dynamics (Martiny et al., 2017). Climate models project that precipitation patterns are likely altered in the future, resulting in increased duration of droughts and increased frequency of large soil rewetting events, which could expose microorganisms to more stressful periods. This has implications for the microbial-mediated turnover of SOC, therefore, understanding how changes in soil moisture content shape microbial communities and their functions is important for predicting C feedbacks to global climate change (Cregger et al., 2012). For example, Fuchslueger et al. (2016) reported that the legacy of drought history strongly affects microbial responses related to turnover of recent plant-derived C during further drought events with repercussions on ecosystem carbon dynamics in a changing climate. Additional studies are essential to fully characterize the effects of soil moisture and substrate supply on microbial physiology and responses toward SOC turnover, so that biogeochemical models can accurately predict the effects of climate change on soil processes.

3.3.4 SOIL PH

Soil bacterial communities are strongly influenced by pH, which explains a large proportion of the variance in soil bacterial diversity and community composition at local (Rousk et al., 2010), regional (Kuramae et al., 2012; Griffiths et al., 2011), and continental scales (Lauber et al., 2009; Delgado-Baquerizo et al., 2016a,b). In contrast, the relative abundance of fungi remains unaffected by pH, and fungal diversity is reported to be only weekly related to pH (Rousk et al., 2009). The differential effect of pH on the abundance and diversity of bacterial and fungal community, might be related to the differences in optimal growth range. Bacterial taxa have a narrow pH range for optimal growth, while fungi generally exhibit wider pH ranges for optimal growth (Rousk et al., 2010). Using a continuous soil pH gradient, Rousk et al. (2009) have demonstrated functional redundancy in C mineralization due to the contrasting effects of soil pH on the growth of bacterial and fungal community. This observation is supported by a study that showed that soil pH has a direct strong impact on the composition of soil microbial community, but not on the production of enzymes involved in C degradation (Trivedi et al., 2016).

3.3.5 PLANT COMMUNITY

Vegetation type and structure can influence microbial activity through direct and indirect mechanisms such as: (1) Modification of microclimate; (2) Alterations in the quantity and quality of litter; (3) The supply of root exudates, and/or; (4) The above-below ground allocation patterns of organic matter (De Deyn et al., 2008; Wardle et al., 2012; Jassey et al., 2013). Many studies show that SOC is closely related to soil microbial community structure and function under different types of vegetation (Grayston and Prescott, 2005; Franklin and Mills, 2009; You et al., 2014). Increased C storage in ecosystems with high plant diversity is a direct function of soil microbial community, indicating that the increase in C storage is mainly influenced by the integration of new C into soil and less by the decomposition of existing soil C (Lange et al., 2015). Data from long-term field experiments revealed that plant species richness had direct positive effects on root C inputs as well as metabolic activity of soil microorganisms, measured as basal respiration (Lange et al., 2015). There was no direct relationship between plant species richness or the presence of legumes and soil C storage, showing that fine root carbon inputs and microbial activity are the main drivers explaining the positive effect of plant diversity on soil C storage. A shift in the metabolic activity of soil microorganisms toward anabolic activity with plant diversity resulted in more microbial C per unit root C in high-diversity plots over time (Lange et al., 2015; Liang et al., 2015). Greater plant diversity facilitates increased microbial activity that results in increments of turnover rates of root litter and exudates as indicated by increased microbial respiration (Lange et al., 2015; Liang et al., 2015). Thus, microbial products associated with increased microbial respiration—such as microbial necromass—end up in slow-cycling SOM pools in the form of reduced organic material (Schmidt et al., 2011).

Higher plant diversity alters decomposition dynamics of new residue inputs via positive impacts on soil microbes (particularly Gram- and AMF), which may be linked to the benefits on soil nutrient cycling and SOM dynamics. For example, plant diversity in a converted grassland with Eutric Fluvison soil, increase the CUE of soil microbes while having no impact of exudation efficiency on phosphorous acquiring enzymes (Phosphatase) suggesting the need for efficient C, rather than P, cycling underlying plant species richness and phosphatase activity (Hacker et al., 2015). In grasslands, microbial biomass is determined by plant diversity that can influence soil C dynamics irrespective of global environmental change factors (Thakur et al., 2015). Similarly, tree mortality and subsequent secondary succession alter soil microbial taxonomic and metabolic activities resulting in higher soil respiration rates together with higher bacterial diversity and anomalously high representation of copiotrophic bacteria (Yuste et al., 2012). Overall results on the relationships between plant diversity, soil microbes, and C storage in several independent experimental settings have provided strong evidence that microbial communities can serve as proxy for C transfer into stable forms, and that plant diversity and associated microbial communities can significantly contribute to C sequestration. This leads us to reconsider the role of soil microorganisms as sources rather than sinks for slow-cycling organic matter (Lange et al., 2015).

3.3.6 MICROBE-FAUNA INTERACTIONS

Decomposer food web of soils is complex wherein the interactions between microbes and soil fauna and their associated functions can directly or indirectly influence soil C dynamics (Grandy et al., 2016; more detailed by Wood and Bradford, Chapter 4: Leveraging a New Understanding of How Belowground Food Webs Stabilize Soil Organic Matter to Promote Ecological Intensification of Agriculture). Soil fauna exerts multiple effects on soil C turnover through their role in shredding and redistribution of litter, altering soil aggregate structure, predation, and accelerating nutrient cycling in soil and litter (Verhoef and Brussaard, 1990; Brussaard et al., 2007; Coleman, 2008). Shredding of plant litter by soil meso- and macroinvertebrates represents an important control on decomposition rates as it can increase the surface area of litter that facilitates decomposition by soil microbial community (Chamberlain et al., 2006; Soong et al., 2016). Litter chemistry is altered during gut passage in saprotrophic fauna that enhances microbial activity during the early stages of litter decomposition (Wickings and Grandy, 2011). Bioturbator fauna impact soil aggregation and pore space structure thus influencing the dynamics of decomposition by altering the resource availability to soil microbes (David, 2014; Wickings et al., 2012; Soong et al., 2016). Several studies have shown that microbial grazing by soil meso- and macrofauna across a wide range of taxonomic groups and size classes can modify microbial community structure and functions (Wickings and Grandy, 2011; A'Bear et al., 2014), which are the proximal controls over SOM dynamics. In spite of their postulated importance in SOC dynamics in terrestrial ecosystems, we have a limited understanding on microbe-fauna interactions and how these might be influenced by land management practices and/or climate change. In future, a mechanistic understanding on the relationships between microbes and soil fauna and how they may relate to SOC formation and persistence will be required for conceptual exploration of food web interactions related to SOC dynamics (Grandy et al., 2016).

3.3.7 TEMPERATURE

Temperature regulates the rate of biogeochemical cycling through the control on microbial metabolism (Bradford and Crowther, 2013). Experimental and observational studies have shown that temperature directly affects microbial physiology, biomass, and composition, and functions particularly to those related to the production of extracellular enzymes involved in C degradation (Zhou et al., 2012; Frey et al., 2013; Rousk et al., 2012; German et al., 2012). Increase in temperature could initially stimulate decomposition by enhancing the metabolism of soil microbes, provoking increases in microbial CO_2 production (Lloyd and Taylor, 1994) leading to soil C losses, higher soil respiration rates, and an overall positive feedback to global warming (Jenkinson et al., 1991). However, this response can be transient (Luo et al., 2001) and the warming effect can be declined over time and eventually became nonsignificant (Melillo et al., 2002).

Recent studies have provided a mechanistic understanding of the effects of warming on the roles played by microorganisms on ecosystem C storage, including: (1) Shifting microbial community composition, most likely led to the reduced temperature sensitivity of heterotrophic soil respiration (Barcenas-Moreno et al., 2009; Rousk et al., 2012; Wei et al., 2014), (2) Differentially stimulating genes for degrading labile, but not recalcitrant C so as to maintain long-term soil C stability (Frey et al., 2013), (3) Enhancing nutrient cycling processes to promote plant nutrient use efficiency and hence plant growth; and (4) Evolutionary adaptation (Romero-Olivares et al., 2015) and acclimatization of individual microbes or communities to adapt to changes in temperature thus maintaining status-quo in metabolism and physiological responses (Allison et al., 2010; Bradford and Crowther, 2013). These mechanisms are nonexclusive, and their influence may vary among seasons (Contosta et al., 2015), ecosystems, and across time scales. Overall, a broad array of indirect mechanisms under warming will likely affect microbial physiology and community composition, and these effects will likely cooccur with the direct effects of warming on microbial activities (detailed mechanisms are discussed in Chapter 8: Impact of Global Changes on Soil C Storage—Possible Mechanisms and Modeling Approaches). We postulate that the effect of warming on microbial communities and their related functions will be variable and will depend on the local climate history wherein the overall changes in temperature are likely to have strong effects on microbial communities and decomposition processes in alpine and arctic regions. For example, Karhu et al. (2014) observed the strongest responses of microbial communities to soil respiration in soils from cold climatic regions and suggested that the substantial carbon stores in Arctic and boreal soils could be more vulnerable to climate warming than currently predicted. Most studies till date have focused on either temperature or water effects on microbial controls SOM decomposition, but only a few have explored the combined effect of both (Curiel Yuste et al., 2007). Given the projected decreases in precipitation and increases in temperatures for most of the terrestrial ecosystems, it is particularly important to understand how the interaction of both factors may affect microbial regulation of SOM decomposition in a changing world particularly under extreme climate events.

3.3.8 PRIMING

The priming effect, i.e., the enhanced SOM decomposition due to amendment of fresh SOM or mineral nitrogen (Jenkinson et al., 1985; Kuzyakov, 2010), that provides a link between the quantity or quality of more thermodynamically labile inputs and the potential destabilization of soil C (Wild et al., 2014), has also been related to changes in microbial community structure and function (Fontaine et al., 2004; De Graaff et al., 2010; Blagodatskaya et al., 2014). The role of soil microbial communities in priming has received considerable attention because of its potential relevance for understanding SOM dynamics and feedback to climate warming. Soil priming in warming climates has been attributed to the activity of certain microbes, particularly gram-positive bacteria and fungi (De Graaff et al., 2010; Fontaine et al., 2011; Dijkstra et al., 2013; Creamer et al., 2015). However, the direction of priming may vary with the litter additions (negative with fresh litter, positive with preincubated litter) and is related to differences in the composition of microbial communities degrading soil-C, particularly gram-positive and gram-negative bacteria, resulting from litter addition (Pascault et al., 2013; Creamer et al., 2015). Links between SOM priming effects and microbial community dynamics have recently been reported for a variety of situations where labile C was present (Garcia-Pausas and Paterson, 2011; Pascault et al., 2013; Blagodatskaya et al., 2014). This probably underlies the widely-held rationale that successional dynamics among copiotrophs and oligotrophs within the microbial community explain the SOM priming induced by labile C additions. Chen et al. (2014) demonstrated that both copiotrophs and oligotrophs were beneficial for priming effects, with an increased contribution of oligotrophs under N-limitation. Thus, the priming phenomenon described in "microbial N mining" theory can be ascribed to oligotrophs, whereas "stoichiometric decomposition" theory, i.e., accelerated SOM mineralization due to balanced microbial growth, is explained by domination of copiotrophic species. However, the findings of few other studies directed at testing the connection between microbial successional dynamics (from copiotroph to oligotrophy switch) and priming did not translate into a change in the priming effect (Reinsch et al., 2013; Rousk et al., 2015). This calls for the identification for mechanisms other than general changes in microbial growth dynamics most often highlighted as a likely explanation for the priming of SOM (e.g., Kuzyakov, 2010; Garcia-Pausas and Paterson, 2011). Climate-dependent priming mechanisms where plant exudates counteract the strong protective effect of mineral-organic associations and facilitate the loss of C from soil systems by shifts in the activity and composition of soil microbial communities, challenge the long-held conceptual framework that suggest mineral-organic associations protect SOM from loss processes for millennia, or longer, by making it permanently inaccessible to microbes (Keiluweit et al., 2015). We postulate that although microbes play a major role in priming, the overall effects are not a purely biotic phenomenon and should be viewed as the sum of direct biotic cometabolic and indirect C mobilization mechanisms. Future investigations into the causes of "priming effects" should therefore consider both biotic and abiotic (e.g., fertilization, moisture content, temperature) drivers simultaneously, and focus on quantifying their relative contributions in different soil ecosystems (Keiluweit et al., 2015).

3.4 MICROBIAL REGULATION OF SOC DYNAMICS

It is now well-established that turnover of SOM in terrestrial ecosystems is governed by the cata-lytic activities of soil microbes. While the dependency of the SOM dynamics on individual envi-ronmental conditions such as pH, moisture, and temperature have been frequently studied, we lack a mechanistic and quantitative understanding on the controls of soil microbial communities on the SOM conversion and the feedback mechanisms that alter growth, activity, and interaction between primary producers and soil microbial community (Wieder et al., 2013; Luo et al., 2016). One of the major constrains in our ability to incorporate microbial data into ESMs, is the lack of direct evidence that soil microbes regulate the processes involved in soil C turnover (Schimel and Schaeffer, 2012; Trivedi et al., 2016). Process-level analysis such as respiration or enzymatic transformation of added substrate have been used as a proxy of microbial functions used in microbial models although it does provide insight into the overall microbial-mediated SOC turn-over yet does not provide concrete links on mechanisms, functional microbial community compo-sition and diversity that underpins process-level differences (Reeve et al., 2010; Talbot et al., 2014; You et al., 2014). While the relationship between changes in the microbial community and SOC turnover is well-established, empirical evidence of direct and individual regulation of SOC dynamics by soil microbial communities is lacking. This knowledge gap limits the coupling between microbial community functional traits, the environment context, and the ecosystem pro-cesses. Because of the lack of this evidence, it is generally assumed that the link between com-munity composition and the functional and metabolic responses is indirect (Comte et al., 2013). This is a significant constraint in our ability to develop a framework to directly incorporate microbial data into ESMs.

Here we provide three case studies that address bottom up (gene → genome → ecosystem based approaches) analysis to address questions regarding regulatory and metabolic networks of microbes involved in C turnover at community, aggregate, plot, and continental scales. First, we consider the phylogenetic level at which microbes form meaningful guilds, based on overall life-history strate-gies, and suggest phylogenetic conservation of traits related to SOM turnover. We then consider the controls on microbial communities on SOM dynamics at different sized aggregates and argue that the regulatory mechanisms differ with aggregate size and depend on the specialized microbial community present in different sized aggregates. We then provide evidence of the microbial regula-tion of enzymes involved in C degradation at continental scale. Lastly, we summarize results from recent studies that provide empirical evidence that soil microbes modulate soil processes directly linked to soil C dynamics.

3.4.1 ECOLOGICAL ROLE AND FITNESS: DOES GENOMIC BLUEPRINT PROVIDES INSIGHTS?

For decades, linking microbial community composition and function has been a difficult task. However, the availability of an increasing number of sequenced genomes and consistent annota-tion systems provides one way to link phylogeny and distribution of functional traits (Lauro et al., 2009; Zimmerman et al., 2013; Trivedi et al., 2013). A few studies have used the information from the whole genome sequences to study soil microbial community from different perspectives and has significantly advanced our understanding on the metabolic capabilities of prokaryotic taxa, their adaptation to various environmental niches, and their potential to influence regional- and global-level ecological phenomenon (Lauro et al., 2009; Zimmerman et al., 2013; Trivedi et al., 2013).

In soil microbial community, the study of trophic lifestyles (copiotroph vs oligotrophs, also see previous section) has been impaired by a lack of understanding of the molecular basis of this classification. Here we analyzed complete genome sequences of 1100 soil microbes to provide insights into bacterial traits that may contribute to survival and growth in soil, including their transporters and their ability to influence carbon storage and cycling. We didn't attempt to perform exhaustive comparison of all bacterial genomes, but provide trends which may be useful for linking the genomic structure to functional traits which could later be scaled to ecosystem-level processes. Our analysis showed that the number of genes per genome for the production of enzymes involved in the degradation of labile C were higher in copiotrophs. However, the number of genes per genome involved in the degradation of recalcitrant forms of C seems to be higher in oligotrophs and other typical soil inhabiting bacteria belonging to Actinomycetes and Firmicutes (Fig. 3.4). Our results thus provide evidence that the gene-diversity of C degrading enzymes may also be associated with lifestyle strategy and can also influence C storage in terrestrial ecosystems.

Our analysis further showed that in addition to resource acquisition, soil bacteria have evolved specific genomic structures and expression patterns to meet the requirements of its fitness and competitiveness under selective pressures imposed on the bacterial strains in an environmental niche. The number of genes that produce proteins involved in providing relief from oxidative stress arising from the rapid fluctuations in nutrient availability and environmental conditions (e.g., those involved in the production of catalase, oxidoreductase, and superoxidase dismutase) were higher in copiotrophs. On the other hand, oligotrophs and Actinomycetes in general have a greater number of genes involved in the production of trehalose (also higher in) beta-proteobacteria and glycogen synthetic branching enzyme that can provide protection under low nutrient and water limitations.

Transport proteins are a primary mechanism for nutrient uptake and assimilation in microbial cells, therefore, the identification and characterization of transport proteins may be important toward understanding the broad range of substrates available for assimilation and utilization by microbes (Jiao and Zheng, 2011). Copiotrophs and members of Firmicutes have a higher number of total transporters, ABC transporters, and Feo transporters that could import or export a broad range of compounds, such as carbohydrate, drugs, proteins, amino acids, inorganic anions, metal ions, lipids, and hydrocarbon. These features support fast growth rates, higher nutrient requirements of copiotrophs, and their preferences to the environments with pulses of nutrient availability such as the rhizosphere. Our analysis shows that oligotrophs may have evolved to minimize the number of "energetically expensive" transporters and instead rely on a relatively smaller number of broad-specificity transporters (Fig. 3.4) allowing them to thrive in low nutrient concentrations, but become saturated at high nutrient concentration leading to their selective exclusion by fast growers in nutrient-rich environments (Lauro et al., 2009; Roller and Schmidt, 2015). Overall, our

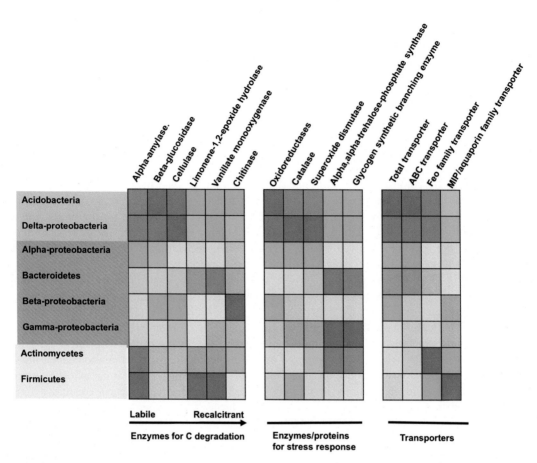

FIGURE 3.4

Fraction of genomes containing genes involved in the production of extracellular enzymes capable of degrading different C sources, enzymes/proteins for stress response, and transporters in different groups of bacteria associated with terrestrial ecosystems. For C degrading enzymes, the sugar substrates are shown in order of increasing recalcitrance. The gradations from *dark red-green-yellow* in for each enzyme represent lowest-intermediate-highest values. Oligotrophs, copiotrophs, and noncategorized bacterial phylum/groups are shown in *blue*, *gray* and *white boxes*, respectively. We selected complete genome sequences of 1100 bacteria associated with terrestrial ecosystems with the "Genome Browser" available at Microbial Genome and Metagenome Data Analysis pipeline of the DOE Joint Genome Institute (JGI) site (https://img.jgi.doe.gov/cgi-bin/m/main.cgi). The selected bacterial genomes were added to the "analysis cart" and analyzed for the presence of genes encoding enzymes and transporters using the "find function" option.

analysis demonstrates that trophic strategy is strongly reflected in genomic content and genomic signatures which can be used as a proxy for determining the ecological characteristics of uncultured microorganisms, thereby allowing the assessment of trophic life strategies from bacterial genome sequences.

3.4.2 MICROBIAL REGULATION OF SOIL C: DOES SOIL AGGREGATE SIZE MATTER?

The majority of soil microbial analyses are performed on bulk soil samples using methods that tend to standardize variability and result in a loss of information regarding the spatial distribution of microorganisms and, therefore, their controls on SOC turnover at aggregate scale (Six et al., 2006; Trivedi et al., 2015, 2017). Physical separation of the soil into distinct aggregate fractions (Elliott, 1986) and subsequent molecular analysis and statistical analysis allows investigation of, at appropriate scales, the relationships between microbial communities, and soil microhabitats. We have used a statistical modeling approach to determine the control of different factors in modulating the amount of soil C in different soil aggregates with the aim to develop a framework for how soil structure interacts with microbial composition and activity, and management practices—in particular how these interactions affect soil carbon storage. These analyses clearly showed the greater microbial control of C storage in microaggregates compared to megaaggregates for two long-term management sites with contrasting soil types and management (Fig. 3.5). In the macroaggregates, the amount of C is explained mainly by the interaction effects of residue management, soil microbial community, and extracellular enzymes. For example, the greater availability of relatively degradable particulate organic matter in macroaggregates would drive soil microbial community structure and function (Trivedi et al., 2015; Tiemann et al., 2015; Tiemann and Grandy, 2015). These results demonstrate that microbial responsiveness to crop management practices declined in smaller aggregates.

The variation in plant materials entering the soil after harvest may explain differences in total C content and biochemical processes between the residue management practices—the occlusion and retention of microbial metabolites in microaggregates is more likely to enhance the recalcitrance of C in the soil. Interestingly, the variations explained by the enzymatic activities decreased significantly in the microaggregates as compared to mega- and macroaggregates (Trivedi et al., 2015). These findings suggest the impact of different crop management regimes on soil C and microbial communities are mediated by aggregate size distribution, and these impacts are more pronounced in macroaggregates compared to microaggregates. Our study provides new evidence that soil aggregate size and their associated microbial communities (i.e., activity and composition of bacteria) modulate the effects of agricultural management on soil C storage and thus should be explicitly considered to determine the impact of management practices on soil health. Considering the relative importance of microaggregates in regulating C storage, greater attention is required to further understand both the structure and function of soil microbial community in microaggregates.

3.4.3 MICROBIAL REGULATION OF ENZYMATIC ACTIVITIES INVOLVED IN DEGRADATION OF SOIL C AT CONTINENTAL SCALE

Identifying the structural/functional relationships for microbial organisms is particularly critical to determine the importance of the soil microbial community in regulating ecosystem processes, and thus there is keen interest in developing theoretical and experimental approaches to disentangle the microbial regulation of soil functions from other biotic and abiotic drivers (e.g., Wallenstein and Hall, 2012; Talbot et al., 2014; You et al., 2014; Wood et al., 2016; Trivedi et al., 2016). Here, we evaluated the linkage between microbial functional genes and extracellular enzyme activity at continental scales from soil samples collected across three geographical regions of Australia (details provided in Trivedi et al., 2016). We characterized bacterial and fungal communities with next-generation sequencing and functional diversity using GeoChip 4.0, including abundance of known

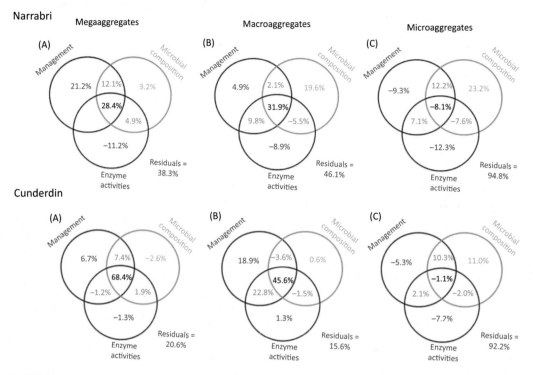

FIGURE 3.5

Variation partitioning (percent variance explained by land management, microbial composition, and enzymatic activities) of total C in megaaggregate (A), macroaggregate (B); and microaggregates (C) from two long-term soil management experiments. Shared effects of these variable groups are indicated by the overlap of circles. "Cropping System Experiment" at Narrabri, New South Wales (NSW), Australia (149°47′E, 30°13′S) in conducted on alkaline, self-mulching, grey Vertosol and has four crop rotation trials (Details provided in Trivedi et al., 2015). "Conservation agriculture cropping system experiments" at Cunderdin, Western Australia (117°14′E, 31°38′S) is conducted on red sandy clay loam soil and is based on four different cropping philosophies (details provided in Trivedi et al., 2017). Enzyme activities represent activities of α-Glucosidase (AG); β-D-Xylosidase (XYL); N-acetyl-β-Δ-glucosaminidase (NAG), and β-D-celluliosidase (CB). Microbial was determined by using MiSeq Illumina of soil bacterial community using 16S rDNA gene (details provided in Trivedi et al., 2017).

protein-coding genes related to the production of four key enzymes involved in C degradation. These enzymes were selected based on their general occurrence in different soil types, key role in soil C degradation (Baldrian, 2014; Trivedi et al., 2015) and the availability of well-established methods for determining their activity (Bell et al., 2013). To determine whether microbial structural composition and functional gene abundances enhance predictive models linking environmental parameters to the activity of enzymes involved in C degradation, we used a multimodel inference approach based on information theory (Burnham et al., 2011).

Our results clearly showed that microbial structural and functional community composition are the most important predictors of the activity of studied enzymes (Table 3.2). The models

Table 3.2 Best-Fitting Regression Models of the Activity of Enzymes Involved in C Degradation

Enzyme	Functional gene abundance	Microbial community composition – Bacteria	Microbial community composition – Fungi	Abiotic variables – pH	Abiotic variables – Total C	r^2	AIC_c	ΔAIC_c	W_i
N-acetyl-β-Glucosaminidase (NAG)		δ-Proteobacteria				0.886	-126.206	0	0.195
		Actinobacteria, δ-Proteobacteria				0.887	-124.362	1.844	0.078
		δ-Proteobacteria	Classiculomycetes			0.886	-124.036	2.17	0.066
		δ-Proteobacteria	Eurotiomycetes			0.886	-123.888	2.318	0.061
		δ-Proteobacteria				0.886	-123.777	2.429	0.058
		δ-Proteobacteria	Classiculomycetes, Eurotiomycetes			0.774	-86.38	39.826	<0.001
						0.635	-66.976	59.229	<0.001
α-Glucosidase (AG)			Fungi unclassified			0.903	-1.457	0	0.103
			Classiculomycetes			0.903	-1.238	0.219	0.093
			Classiculomycetes			0.907	-0.817	0.64	0.075
			Classiculomycetes, Fungi unclassified			0.907	-0.727	0.73	0.072
			Fungi unclassified			0.896	0.03	1.488	0.049
						0.205	101.194	102.651	<0.001
						0.157	106.555	108.013	<0.001
β-Xylosidase (XYL)			Classiculomycetes			0.941	232.382	0	0.061
			Classiculomycetes			0.937	232.552	0.17	0.056
			Classiculomycetes, Leotiomycetes			0.94	232.881	0.499	0.048
			Classiculomycetes, Leotiomycetes			0.942	234.141	1.758	0.025
		δ-Proteobacteria	Classiculomycetes			0.935	234.479	2.097	0.021
						0.888	264.768	32.386	<0.001
						0.815	285.228	52.846	<0.001
β-D-cellulosidase (CB)		Acidobacteria	Tremellomycetes,			0.973	241.159	0	0.081
		Acidobacteria,				0.971	241.684	0.525	0.062
		δ-Proteobacteria	Leotiomycetes			0.972	241.91	0.751	0.055
		Acidobacteria	Leotiomycetes			0.972	241.969	0.81	0.054
		Acidobacteria	Tremellomycetes, Leotiomycetes			0.973	242.215	1.057	0.048
		Acidobacteria, δ-Proteobacteria				0.827	332.684	91.525	<0.001
						0.435	390.49	149.331	<0.001

Each column represents different predictor variable(s). A total of 255, 63, 511, and 127 possible models which included all the possible combinations of independent predictor variables were obtained for NAG, AG, XYL, and CB, respectively. Of all the possible models, the first five are the best models which included the abundance of genes responsible for the enzyme production. The sixth model is the top model excluding the functional gene, while the seventh model is the best model with only abiotic data. The most parsimonious model is shown as the yellow color shading in the observed parameters. The best models are presented, ranked according to AIC_c value. ΔAIC_c = difference between the AIC_c of each model and that of the best model; $AICcW_i$ = Akaike weights.

incorporating the abundance of functional genes were better predictors of the enzymatic activities. Models incorporating the composition of microbial community explained the activity of enzymes better than those containing only the abiotic variables. We also noticed that inclusion of different microbial taxa provides better predictions for different enzymatic activities. Using the model-based approach designed to link functional composition to the activity of soil enzymes and to compare the importance of the microbial community composition to other abiotic drivers, we demonstrated that variations in microbial communities can impact the activities of enzymes involved in C degradation. Our results provide evidence for links between microbial community and ecosystem process, demonstrating that incorporation of microbial community structure and function in ESMs could potentially improve predictions. In accordance with our results, You et al. (2014) showed that the abundance of soil bacterial communities is strongly linked with the extracellular enzymes involved in carbon transformation, whereas the abundance of saprophytic fungi is associated with activities of extracellular enzymes driving carbon oxidation. Similarly, Trivedi et al. (2016) showed that variation in microbial community composition leads to differences in functional gene abundance, which in turn has consequences for the activity of enzymes directly linked to C degradation at field to regional scales. Altogether, our results along with previous studies, demonstrate the complex interactions and linkage among microenvironment, biotic variables and soil physiochemical properties in affecting SOC via microbial regulations.

3.4.4 EMPIRICAL EVIDENCE ON THE REGULATION OF SOM DYNAMICS BY SOIL MICROBIAL COMMUNITIES

A number of reports provide evidence for direct control of soil microbes on SOM dynamics. For example, Kallenbach et al. (2016) provided direct evidence that soil microbes produce chemically diverse, stable SOM and that SOM accumulation is driven by distinct microbial communities more so than clay mineralogy, where microbial-derived SOM accumulation is greatest in soils with higher fungal abundances and more efficient microbial biomass production. The results from the study points to new approaches to rebuild soil C contents that emphasize the influence of substrate—microbe interactions on the synthesis of novel SOM constituents that become mineral stabilized highlighting the importance of microbial residues in SOM dynamics. It is also evident that social dynamics among microbes producing extracellular enzymes ("decomposers") and microbes exploiting the catalytic activities of others ("cheaters") regulate organic matter turnover SOM dynamics. For example, it was demonstrated that the presence of cheaters increases nitrogen retention and organic matter build-up by downregulating the ratio of extracellular enzymes to total microbial biomass, allowing nitrogen-rich microbial necromass to accumulate (Kaiser et al., 2015). Moreover, increasing catalytic efficiencies of enzymes are outbalanced by a strong negative feedback on enzyme producers, leading to less enzymes being produced at the community level. Similarly, Lange et al. (2015) showed that the positive plant diversity effect on carbon storage is driven by root carbon inputs, but mediated by the soil microbial community.

Current understanding suggests that although microbial community composition is delineated strongly by the sampling regions—enzyme activities across regions are driven by the selection of common active members of microbial community (Trivedi et al., 2016). Using multivariate analysis techniques, researchers established the relative linkage between soil microbial community types and function-specific soil extracellular enzymes involved in C degradation (You et al., 2014; Trivedi et al., 2016). These findings further suggest that complex interactions between the biotic and abiotic site factors may further propagate in affecting the soil microbial community function involved in SOC transformation and/or turnover through modification of the bacterial and saprophytic fungal community structure, shifts in the allocation of plant-derived organic carbon to microbial biomass, and regulation of microbial activities. Moreover, factors affecting soil microbial community structure may also impose direct and/or indirect impacts on the microbial function, enhancing or weakening the overall controls of soil microbial community function by individual site factors. Overall these recent findings provide empirical evidence that soil microbial communities modulate soil processes directly linked to SOM dynamics and demonstrate the importance of explicit incorporation of microbial community attributes to improve quantitative predictions of the global C balance.

3.5 INCORPORATING MICROBIAL COMMUNITY AND FUNCTIONS IN PREDICTIVE AND MECHANISTIC SOIL C MODELS: CHALLENGES AND ADVANCEMENTS

Despite the significance of terrestrial carbon storage in the overall global C sequestration, global ecosystem models persistently diverge on fundamental predictions of the size and magnitude of feedbacks (Friend et al., 2014; Hoffman et al., 2014), contributing substantial uncertainty to the overall accuracy of Earth system prediction (Bodman et al., 2016). Some reviews have indicated that current ESMs do not represent the suite of mechanisms necessary for prediction of future SOM stocks (Conant et al., 2011; Dungait et al., 2012; Schmidt et al., 2011). These models vary widely in their predictions of SOC pools and numerical simulations, on average, tend to underestimate the global carbon turnover time by 36% (Carvalhais et al., 2014). Gaps in theory contribute to the failure of models (Wieder et al., 2013), but a lack of critical observation slows the pace of development of theory, and its implementation into models (Todd-Brown et al., 2012). Fundamentally, model projections of SOC dynamics rely on three components: Model structure, parameterization, and external variables (Luo et al., 2016). The uncertainties in ESMs predictions can result from biases in any one of three components across time, space, and soil depth.

Classical ESMs describe organic matter dynamics as first-order decay function formulated with multiple SOM pools, continuous driving variables (e.g., temperature, moisture), and controls on decomposition and stabilization factors (e.g., McGill et al., 1981). These models implicitly include microbial physiology wherein microbial responses are represented by rate modifying functions determined by the dynamics of environmental and management conditions, but the actual quantity or biomass of decomposer organisms are not included as a rate control. While useful, the strong feedback between microbial communities and organic matter pools and high sensitivity of the models to poorly-understood parameters controlling microbial growth and mortality, is postulated to result in massive uncertainties in the projections. Several studies have challenged the dominant paradigm of the "black box" approach toward microbial community structure and function in ESMs, and have explored ways to represent microbial controls on SOM transformations more explicitly. These efforts have resulted in the development of nonlinear microbe-explicit models that account for the microbial roles in decomposition and stabilization of SOC (Wieder et al., 2013, 2015; Luo et al., 2016). By representing microbial physiology responses, these models provide better explanations of SOC retention and turnover, especially in a changing environment (Wieder et al., 2013, 2015; Luo et al., 2016).

In conventional ESM, a first-order decay function with fixed coefficients describes the monotonic decrease over time of SOC that has entered a soil pool. However more complex behaviors can arise when the decay coefficients vary with time- and site-dependent microbial activities, or when alternative functional forms, such as Michaelis−Menten kinetics, are incorporated in SOM models (Wang et al., 2014). In fact, microbial models that are based on Michaelis−Menten kinetics to mathematically couple microbial community to C substrate pools better explain priming effects (Kuzyakov, 2010), acclimation (Luo et al., 2001), and soil respiration responses to wet-dry cycles of precipitation (Liu et al., 2002) than conventional models. Application of a microbial-enzyme model, demonstrated that controls of microbial community on the feedback between climate warming and SOC losses explain the observed attenuation of soil C emissions in response to warming (Allison et al., 2010). Further advancement came from the development of a microbial-enzyme-mediated decomposition (MEND) model that simulated the dynamics of physically defined pools of SOC different from the traditional models with fast/show/passive pools (Wang et al., 2013) and a new biogeochemical module that explicitly simulates microbial biomass pools in the Community Land Model (CLM; Wieder et al., 2013). This CLM microbial model explained significantly more variations in the soil C observations compared to traditional models such as the daily century model or the Community Land model version 4 (CLM4cn). The CLM microbial model was further developed into the Microbial-Mineral Carbon Stabilization (MIMICS) model that incorporates the relationship between microbial physiology, substrate chemical quality, and the physical stabilization of SOM (Wieder et al., 2015). In MIMICS, growth physiology of microbial communities is parameterized based on copiotrophic and oligotrophic growth strategies. MIMICS adequately simulated litter decomposition and was better in capturing the responses of experimental warming to SOM pools compared with the conventional model structure (Wieder et al., 2015).

Although studies have shown that microbial-explicit models may improve global C cycle, these nonlinear models can exhibit unrealistic temporal oscillations in response to small perturbations (Wang et al., 2014; Hararuk et al., 2015). However, calibrations of model parameters can improve microbial-based model prediction by reducing the observed oscillatory SOC dynamics (Wang et al., 2014; Hararuk et al., 2015). For example, Wang et al. (2015) quantified uncertainties in parameters and model simulations using the Critical Objective Function Index method based on a global stochastic optimization algorithm, as well as model complexity and observational data availability. Together, these model extrapolations of the incubation study show that long-term soil incubations with experimental data for multiple carbon pools are conducive to estimate both decomposition and microbial parameters. In addition, representing microbial processes in other ways such as by making decomposition a function of substrate chemistry and enzyme-based microbial guilds (Fujita et al., 2014), and embedding soil enzyme dynamics in an ecosystem model (Sistla and Schimel, 2013) may avoid the undesirable oscillations, yet account for microbial roles in SOC decomposition. Oscillations can also result from oversimplification of the spatial structure of soil heterogeneity or microbial communities. Upscaling biophysical models that capture the dynamic interactions of microbial communities (including microbial physiology, evolution, and ecology) from microsites to global scales is essential to support to future field- and global-scale simulations, and enable more confident predictions of feedbacks between environmental change and carbon cycling (Table 3.3).

Table 3.3 Challenges in Explicitly Representing Soil Microbial Process in Earth System Models (ESMs) and Needs to Address These Challenges

Challenge	Needs
Estimating and interpreting model parameters: While "decay" constants in the first-order models can be adjusted as a function of environmental variables, the highly dynamic microbial interactions across scales provides significant challenge in deriving parameter values in microbial-explicit models for facilitating model prediction at global scale	Combining manipulative experimentation (i.e., mesocosm approaches) with computational and functional genomics to generate and test mechanistic knowledge and then incorporate microbial mechanistic behaviors (cellular, community, and ecosystem) in predictive models that can be used to interpolate or extrapolate observed interactions among microbes and their environment
Incorporating nutrient dynamics that feed-back SOC turnover: Plant−microbe interactions that regulate SOC turnover are highly dependent on the availability of other nutrients particularly N and P. However nutrient dynamics in plant-microbial feedbacks are not incorporated in nonlinear models to project SOC turnover at global scale	Application of single-cell genomics, bioinformatics, and metabolic network modeling to provide a genomic/ mechanistic basis for biological feedbacks to terrestrial C cycle
	Develop theory, data, and statistical models to identify mechanisms and demonstrate microbial control over biogeochemical cycles and quantitative linkages between microbial diversity and/or community shifts and ecosystem functioning
Quantify and parameterize microbial-mineral and microbe-faunal interactions constrained within soil heterogeneity: Soil aggregate structure and mineralogy greatly affect the function and composition of soil microbial communities with direct impact on soil C stabilization and decomposition. In addition, interactions between microbes and fauna are the major drivers of SOC dynamics. However, multitude of interactions and feedbacks among different binding agents, environmental factors, microbial and faunal communities, and the natural heterogeneity of the soil as three-dimensional systems is not fully known and hence has not been taken into account in microbial-based models	Use of a metaanalysis to benchmark models by aggregating the response of a number of different climate change experiments across spatial and temporal scales to converge upon an average ecosystem or biome response
	Upscaling an aggregate biophysical model that incorporate the dynamic interaction of microbial communities in hierarchy of soil aggregates scheme to reduce the inconsistencies between theoretical formulations and microbial-explicit models
	Combination of long-term and cross-biome microbial datasets to probe the morphological and molecular makeup, diversity, evolution, and ecology of soil microbial communities for parameterizations that directly interface between molecular, genomic, and physiological datasets scaling the ecological models from the gene to ecosystem level
Rigorous evaluation with observations for benchmarking: First-order conventional models are validated by experimental/observational results, but similar tests for microbial models are rare. Such validations are essential for more mechanistic representation of microbial-mediated processes for improving soil C components in ESMs	Wider engagement across diverse scientific disciplines to advance understanding of soil C dynamics from aggregates to ecosystems

Note that the needs are not related to a particular challenge and instead represent various steps that need to be undertaken to tackle the challenges and incorporate microbial process in ESMs.

In addition, findings on the regulatory role of microbial communities on SOC turnover provides a strong framework for improved predictions on soil C dynamics that could be achieved by adopting a gene-centric approach incorporating the abundance of functional genes into process models (Box 3.2).

BOX 3.2 GENE-CENTRIC APPROACH TO INCORPORATE MICROBIAL COMMUNITIES IN SOIL ORGANIC CARBON (SOC) MODELS

Rapid advances in sequencing technologies have yielded an unprecedented amount of data about the evolutionary relationships and functional traits of microbial communities, however biogeochemical models are trailing in the wake of the environmental genomics revolution—such models rarely incorporate explicit representations of bacteria and archaea, nor are they compatible with nucleic acid or protein sequence data. A major obstacle in the marriage of the two techniques is the differences in the dataset generated by sequencing technologies and the variable required to parametrize Earth system models (ESMs). For example, sequencing data refer to marker and functional genes, whereas ESMs typically simulate chemical concentrations and biomass, grouping organisms according to their function as opposed to genetic identity. Although at present there is a divide between modeling efforts and genomics studies, there is yet much to be gained by integrating these fields—e.g., mechanistic insight into biogeochemical processes, model-based hypothesis development for guiding meta "omic" studies, and improved predictive power. Studies that provide empirical evidence on the microbial regulation of soil C turnover provide a strong framework for improved predictions on soil C dynamics that could be achieved by adopting a gene-centric approach incorporating the abundance of functional genes into process models. By defining genomic signatures for metagenome datasets and applying functional metaanalyses (e.g., metatranscriptomics and metaproteomics), it will be possible to define the rate and mode of microbial community processes related to SOC turnover without the need of time consuming and laborious process level measurements of soil functions. Increasing capacity to rapidly analyze the significant volume of DNA sequencing data can aid the generation of global microbial observatories for effectively monitor the indigenous microbial populations for the defining characteristic of their growth, survival, and trophic lifestyles. Although in its infancy, we postulate that the functional gene approach holds great promise in describing the SOC dynamics in terrestrial ecosystems and their resident microbial communities with the potential to increase the predictive power of models from local to global scales.

3.6 CONCLUDING REMARKS (SEE ALSO BOX 3.3)

Microbial communities play a vital and undisputable role in soil C storage, but microbial control over processes that facilitate soil C storage remains a topic of debate. Numerous basic questions remain to be addressed to understand microbial control over C cycling in terrestrial ecosystems that range from unraveling genome based microbial ecophysiology to the incorporation of microbial mechanistic behaviors (cellular, community, and ecosystems) to predictive models. Recent breakthroughs in sophisticated techniques for metaomics analysis, molecular scale process visualization, and data integration offer new opportunities for investigating the structure and functional properties of soil microbial communities and their interactions with the environment (see also Box 3.4). The biological challenge for these technologies is to simultaneously measure multiple chemical and biological species at multiple spatial and temporal hierarchies within complex, heterogonous cellular and environmental systems. The ability to predict microbial species responses to a changing environment will be critical in realizing the potential of soil microbial community in soil C sequestration.

BOX 3.3 TAKE-HOME MESSAGE

- Soil microbes differ in their survival strategies and ability to influence the terrestrial carbon (C) pool.
- Substrate specificity and diversity of transporters in bacterial genome sequences could provide direct evidence in support of bacterial lifestyles such as oligotrophy or copiotrophy.
- Soil microbial communities have a direct and significant control on C storage and decomposition.
- Soil aggregates and their associated microbial communities have differential relationships to key processes in soil organic carbon dynamics.

- Better management practices (e.g., conservation agriculture; agroforestry) and land-use (forestry) can promote soil microbial communities with positive effects on soil C accrual and retention.
- Statistical modeling provided evidence that phylogenetic and gene-centric data can further improve prediction and should be explicitly considered in future Earth system models (ESMs).
- New ESM models that incorporate microbial data can significantly improve predictions of soil C stocks particularly in the future climate scenarios.
- Improved predictions through these new generation ESMs can aid policy makers to develop better-informed strategies for mitigating and adapting to climate change.

BOX 3.4 TOOLS AND TECHNIQUES FOR LINKING THE STRUCTURE AND FUNCTION OF MICROBIAL COMMUNITY WITH SOIL C TURNOVER

Sequencing technologies: Several high-throughput sequencing platforms have been developed and are widely used, including the Illumina (e.g., HiSeq, MiSeq), Roche 454 GS FLX, SOLiD 5500 series, and Ion Torrent/Ion Proton platforms. The use of *targeted* and *shotgun* high-throughput sequencing of DNA through these platforms provide snapshots of the gene content and genetic diversity of microbial communities. Currently, the majority of the studies use target gene sequencing approach (as by using MiSeq) with phylogenetic (e.g., 16S rRNA) or functional genes (e.g., *nifH, amoA*) to provide information of the structure of total or a specific group within microbial community. To query whole genetic and functional diversity and to identify novel genes, shotgun metagenome sequencing (as with HiSeq) has been widely used. The shotgun sequencing approach provides community level information on both the phylogenetic and functional makeup of microbial communities. Metatranscriptomic sequencing involves random sequencing of expressed microbial community RNA and is used to provide new insights into the functional potential of microbial communities as well as the discovery of novel genes and regulatory elements.

Array based technologies: Various types of DNA microarrays that rely on specific probes have been developed for microbial detection and community analyses, including phylogenetic (e.g., Phylochip) and functional (e.g., Geochip) gene arrays. Phylogenetic gene arrays often target rRNA genes, which are useful to study phylogenetic relationships by identifying specific taxa within microbial communities. Functional gene arrays contain probes targeting genes involved in various biogeochemical cycling processes or metagenomes, and are useful for monitoring the functional composition and structure of microbial communities. A functional gene array, GeoChip (version 5.0) contains about 167,000 oligonucleotide probes covering $\sim$ 395,000 coding sequences from 1590 functional genes related to carbon, nitrogen, sulfur, and phosphorus cycling, energy metabolism, antibiotic resistance, metal homeostasis and resistance, secondary metabolism, organic remediation, stress responses, bacteriophages, and virulence.

Stable isotope probing (SIP): SIP enables direct linkage of microbial metabolic capability to phylogenetic and metagenomic information within a community context by tracking isotopically labeled substances (e.g., ^{13}C or ^{15}N) into phylogenetically and functionally informative biomarkers. Separation and molecular analysis of labeled nucleic acids (DNA-SIP or RNA-SIP), protein (protein-SIP), or pho phospholipid-derived fatty acids (PLFA-SIP) reveals phylogenetic and functional information about the microorganisms responsible for the metabolism of a particular substrate.

(Continued)

BOX 3.4 (CONTINUED)

Ecological modeling: Ecological modeling is the construction and analysis of mathematical models of ecological processes, including both purely biological and combined biophysical models. Models can be analytic or simulation-based and are used to understand complex ecological processes and predict how real ecosystems might change.

Earth system models (ESMs): ESMs include processes, impacts, and complete feedback cycles to integrate the interactions of atmosphere, ocean, land, ice, and biosphere to estimate the state of regional and global climate under a wide variety of conditions. ESMs include physical processes like those in other climate models, but they can also simulate the interaction between the physical climate, the biosphere, and the chemical constituents of the atmosphere, ocean, and terrestrial ecosystems. ESMs can show huge uncertainties in simulating the processes they include, but they are useful tools for extrapolating what we know about the present Earth system to the past and the future.

Soil Carbon models: Models that simulate cycling of soil C to estimate components of soil C budgets, such as: (1) The carbon pool of soil; (2) Changes in the carbon pool of soil over time; and (3) Carbon dioxide emissions from soil as a result of decomposition of organic carbon compounds in soil (heterotrophic soil respiration). The soil C models used in model-based soil C monitoring systems are generally dynamic (i.e., they account for time) rather than static. There are already established dynamic soil carbon models that can be used and have been used as parts of model-based soil carbon monitoring systems, such as: CENTURY; RothC; SOILN; ROMUL; Yasso or Yasso07. These models differ from each other in complexity and requirements of input information. Also, microbial models (e.g., MEND, DENZY, MIMICS) that explicitly integrate mechanistic microbial mechanisms (including microbial decomposition, active microbial biomass, and enzymes) into SOM models have been developed to improve projections on SOC dynamics.

Statistical modeling: A simplified, mathematically-formalized way to approximate reality (i.e., what is generated by the dataset) and to make predictions from this approximation. Statistical models imply dependent (that is to be described, explained, or predicted) and explanatory (independent variables that are used to explain, describe, or predict dependent variables) variables that may be single or multiple, qualitative or quantitative. Details of different statistical modeling approaches such as Random Forest, Structure Equation modeling, variance partitioning analysis, and the multimodel inference approach is provided in Chapter 9, Projecting Soil C Under Future Climate and Land-Use Scenarios (Modeling).

We envisage that a systems ecology approach that incorporates long-term functional genomic and metagenomics cross-biome microbial data, and diverse tools including (meta)genomic assembly, annotation, network inference and modeling, will allow researchers to combine a diverse line of evidence to create robust predictive models that can be used to interpolate or extrapolate observed interactions among microbes and their environment (see also Box 3.4). Collaborations of microbial ecologists, biogeochemists, macroecologists, and big-data analysts/modelers will be required to provide comprehensive understanding of terrestrial microbial communities and specific processes that determine the rate and fate of C dynamics. Such cross disciplinary integrative research has the potential to increase the likelihood of developing management options for successful manipulation of the terrestrial ecosystem for increasing stable C inventories.

ACKNOWLEDGEMENT

PT acknowledges funding by National Institute of Food and Agriculture (NIFA) grant COL00760.

REFERENCES

A'Bear, A.D., Boddy, L., Kandeler, E., Ruess, L., Jones, T.H., 2014. Effects of isopod population density on woodland decomposer microbial community function. Soil Biol. Biochem. 77, 112−120.

Acosta-Martinez, V., Bell, C.W., Morris, B.E.L., Zak, J., Allen, V.G., 2010. Long-term soil microbial community and enzyme activity responses to an integrated cropping-livestock system in a semi-arid region. Agric. Ecosyst. Environ. 137 (3), 231−240.

Allen, A.P., Gillooly, J.F., 2009. Towards an integration of ecological stoichiometry and the metabolic theory of ecology to better understand nutrient cycling. Ecol. Lett. 12 (5), 369−384.

Allison, S.D., 2012. A trait-based approach for modelling microbial litter decomposition. Ecol. Lett. 15 (9), 1058−1070.

Allison, S.D., Wallenstein, M.D., Bradford, M.A., 2010. Soil-carbon response to warming dependent on microbial physiology. Nat. Geosci. 3 (5), 336−340.

Allison, S.D., Lu, Y., Weihe, C., Goulden, M.L., Martiny, A.C., Treseder, K.K., et al., 2013. Microbial abundance and composition influence litter decomposition response to environmental change. Ecology 94 (3), 714−725.

Bach, E.M., Hofmockel, K.S., 2015. A time for every season: soil aggregate turnover stimulates decomposition and reduces carbon loss in grasslands managed for bioenergy. GCB Bioenergy 8, 588−599.

Bailey, V.L., Smith, J.L., Bolton, H., 2002. Fungal-to-bacterial ratios in soils investigated for enhanced C sequestration. Soil Biol. Biochem. 34 (7), 997−1007.

Baldrian, P., 2014. Distribution of extracellular enzymes in soils: spatial heterogeneity and determining factors at various scales. Soil Sci Soc. Am. J. 78 (1), 11−18.

Barcenas-Moreno, G.E.M.A., Gomez-Brandon, M.A.R.I.A., Rousk, J., Bååth, E., 2009. Adaptation of soil microbial communities to temperature: comparison of fungi and bacteria in a laboratory experiment. Global Change Biol. 15 (12), 2950−2957.

Bardgett, R.D., Hobbs, P.J., Frostegård, Å., 1996. Changes in soil fungal: bacterial biomass ratios following reductions in the intensity of management of an upland grassland. Biol. Fertil. Soils 22 (3), 261−264.

Bardgett, R.D., Freeman, C., Ostle, N.J., 2008. Microbial contributions to climate change through carbon cycle feedbacks. ISME J. 2 (8), 805−814.

Barnard, R.L., Osborne, C.A., Firestone, M.K., 2013. Responses of soil bacterial and fungal communities to extreme desiccation and rewetting. ISME J. 7 (11), 2229−2241.

Baumann, F., He, J.S., Schmidt, K., Kuehn, P., Scholten, T., 2009. Pedogenesis, permafrost, and soil moisture as controlling factors for soil nitrogen and carbon contents across the Tibetan Plateau. Global Change Biol. 15 (12), 3001−3017.

Bell, C.W., Fricks, B.E., Rocca, J.D., Steinweg, J.M., McMahon, S.K., Wallenstein, M.D., 2013. High-throughput fluorometric measurement of potential soil extracellular enzyme activities. J. Vis. Exp. 81, 50961.

Berg, B, Laskowski, R., 2006. Litter decomposition; a guide to carbon and nutrient turnover. Adv. Ecol. Res. 38, 20−71.

Billings, S.A., Ballantyne, F., 2013. How interactions between microbial resource demands, soil organic matter stoichiometry, and substrate reactivity determine the direction and magnitude of soil respiratory responses to warming. Global Change Biol. 19 (1), 90−102.

Blagodatskaya, E., Blagodatsky, S., Anderson, T.H., Kuzyakov, Y., 2014. Microbial growth and carbon use efficiency in the rhizosphere and root-free soil. PLoS one 9 (4), e93282.

Bodman, R.W., Rayner, P.J., Jones, R.N., 2016. How do carbon cycle uncertainties affect IPCC temperature projections? Atmos. Sci. Lett. 17, 236−242.

de Boer, W., Folman, L.B., Summerbell, R.C., Boddy, L., 2005. Living in a fungal world: impact of fungi on soil bacterial niche development. FEMS Microbiol. Rev. 29 (4), 795−811.

Bonan, G.B., 2008. Forests and climate change: forcings, feedbacks, and the climate benefits of forests. Science 320 (5882), 1444−1449.

Bouskill, N.J., Lim, H.C., Borglin, S., Salve, R., Wood, T.E., Silver, W.L., et al., 2013. Pre-exposure to drought increases the resistance of tropical forest soil bacterial communities to extended drought. ISME J. 7 (2), 384−394.

Bradford, M.A., Crowther, T.W., 2013. Carbon use efficiency and storage in terrestrial ecosystems. New Phytol. 199 (1), 7−9.

Brandl, J., Andersen, M.R., 2015. Current state of genome-scale modeling in filamentous fungi. Biotechnol. Lett. 37 (6), 1131−1139.

Brant, J.B., Myrold, D.D., Sulzman, E.W., 2006. Root controls on soil microbial community structure in forest soils. Oecologia 148 (4), 650−659.

Brockett, B.F., Prescott, C.E., Grayston, S.J., 2012. Soil moisture is the major factor influencing microbial community structure and enzyme activities across seven biogeoclimatic zones in western Canada. Soil Biol. Biochem. 44 (1), 9−20.

Brussaard, L., De Ruiter, P.C., Brown, G.G., 2007. Soil biodiversity for agricultural sustainability. Agric. Ecosyst. Environ. 121 (3), 233−244.

Burnham, K.P., Anderson, D.R., Huyvaert, K.P., 2011. AIC model selection and multimodel inference in behavioral ecology: some background, observations, and comparisons. Behav. Ecol. Sociobiol. 65 (1), 23−35.

Burns, J.H., Anacker, B.L., Strauss, S.Y., Burke, D.J., 2015. Soil microbial community variation correlates most strongly with plant species identity, followed by soil chemistry, spatial location and plant genus. AoB Plants 7, plv030.

Campbell, B.J., Polson, S.W., Hanson, T.E., Mack, M.C., Schuur, E.A., 2010. The effect of nutrient deposition on bacterial communities in Arctic tundra soil. Environ. Microbiol. 12 (7), 1842−1854.

Carvalhais, N., Forkel, M., Khomik, M., Bellarby, J., Jung, M., Migliavacca, M., et al., 2014. Global covariation of carbon turnover times with climate in terrestrial ecosystems. Nature 514 (7521), 213−217.

Cavigelli, M.A., Robertson, G.P., 2000. The functional significance of denitrifier community composition in a terrestrial ecosystem. Ecology 81 (5), 1402−1414.

Chamberlain, P.M., McNamara, N.P., Chaplow, J., Stott, A.W., Black, H.I., 2006. Translocation of surface litter carbon into soil by Collembola. Soil Biol. Biochem. 38 (9), 2655−2664.

Chen, R., Senbayram, M., Blagodatsky, S., Myachina, O., Dittert, K., Lin, X., et al., 2014. Soil C and N availability determine the priming effect: microbial N mining and stoichiometric decomposition theories. Global Change Biol. 20 (7), 2356−2367.

Coleman, D.C., 2008. From peds to paradoxes: linkages between soil biota and their influences on ecological processes. Soil Biol. Biochem. 40 (2), 271−289.

Collins, S.L., Sinsabaugh, R.L., Crenshaw, C., Green, L., Porras Alfaro, A., Stursova, M., et al., 2008. Pulse dynamics and microbial processes in aridland ecosystems. J. Ecol. 96 (3), 413−420.

Comte, J., Fauteux, L., del Giorgio, P.A., 2013. Links between metabolic plasticity and functional redundancy in freshwater bacterioplankton communities. Front. Microbiol. 4, 112.

Conant, R.T., Ryan, M.G., Ågren, G.I., Birge, H.E., Davidson, E.A., Eliasson, P.E., et al., 2011. Temperature and soil organic matter decomposition rates−synthesis of current knowledge and a way forward. Global Change Biol. 17 (11), 3392−3404.

Contosta, A.R., Frey, S.D., Cooper, A.B., 2015. Soil microbial communities vary as much over time as with chronic warming and nitrogen additions. Soil Biol. Biochem. 88, 19−24.

Creamer, C.A., de Menezes, A.B., Krull, E.S., Sanderman, J., Newton-Walters, R., Farrell, M., 2015. Microbial community structure mediates response of soil C decomposition to litter addition and warming. Soil Biol. Biochem. 80, 175−188.

Cregger, M.A., Schadt, C.W., McDowell, N.G., Pockman, W.T., Classen, A.T., 2012. Response of the soil microbial community to changes in precipitation in a semiarid ecosystem. Appl. Environ. Microbiol. 78 (24), 8587−8594.

Curiel Yuste, J, Baldocchi, D.D., Gershenson, A., Goldstein, A., Misson, L., Wong, S., 2007. Microbial soil respiration and its dependency on carbon inputs, soil temperature and moisture. Global Change Biol. 13 (9), 2018−2035.

Cusack, D.F., Silver, W.L., Torn, M.S., Burton, S.D., Firestone, M.K., 2011. Changes in microbial community characteristics and soil organic matter with nitrogen additions in two tropical forests. Ecology 92 (3), 621−632.

David, J.F., 2014. The role of litter-feeding macroarthropods in decomposition processes: a reappraisal of common views. Soil Biol. Biochem. 76, 109−118.

Davinic, M., Fultz, L.M., Acosta-Martinez, V., Calderón, F.J., Cox, S.B., Dowd, S.E., et al., 2012. Pyrosequencing and mid-infrared spectroscopy reveal distinct aggregate stratification of soil bacterial communities and organic matter composition. Soil Biol. Biochem. 46, 63−72.

De Deyn, G.B., Cornelissen, J.H., Bardgett, R.D., 2008. Plant functional traits and soil carbon sequestration in contrasting biomes. Ecol. Lett. 11 (5), 516−531.

De Graaff, M.A., Classen, A.T., Castro, H.F., Schadt, C.W., 2010. Labile soil carbon inputs mediate the soil microbial community composition and plant residue decomposition rates. New Phytol. 188 (4), 1055−1064.

De Graaff, M.A., Adkins, J., Kardol, P., Throop, H.L., 2015. A meta-analysis of soil biodiversity impacts on the carbon cycle. Soil 1 (1), 257.

de Vries, W.I.M., Reinds, G.J., Gundersen, P.E.R., Sterba, H., 2006. The impact of nitrogen deposition on carbon sequestration in European forests and forest soils. Global Change Biol. 12 (7), 1151−1173.

DeAngelis, K.M., Pold, G., Topçuoğlu, B.D., van Diepen, L.T., Varney, R.M., Blanchard, J.L., et al., 2015. Long-term forest soil warming alters microbial communities in temperate forest soils. Front. Microbiol. 6, 104.

Delgado-Baquerizo, M., Maestre, F.T., Reich, P.B., Jeffries, T.C., Gaitan, J.J., Encinar, D., et al., 2016a. Microbial diversity drives multifunctionality in terrestrial ecosystems. Nat. Commun. 7, 10541.

Delgado Baquerizo, M., Maestre, F.T., Reich, P.B., Trivedi, P., Osanai, Y., Liu, Y.R., et al., 2016b. Carbon content and climate variability drive global soil bacterial diversity patterns. Ecol. Monogr. 86 (3), 373−390.

Dijkstra, F.A., Carrillo, Y., Pendall, E., Morgan, J.A., 2013. Rhizosphere priming: a nutrient perspective. Front. Microbiol. 4, 216.

Dijkstra, P., Salpas, E., Fairbanks, D., Miller, E.B., Hagerty, S.B., van Groenigen, K.J., et al., 2015. High carbon use efficiency in soil microbial communities is related to balanced growth, not storage compound synthesis. Soil Biol. Biochem. 89, 35−43.

Dungait, J.A., Hopkins, D.W., Gregory, A.S., Whitmore, A.P., 2012. Soil organic matter turnover is governed by accessibility not recalcitrance. Global Change Biol. 18 (6), 1781−1796.

Eilers, K.G., Lauber, C.L., Knight, R., Fierer, N., 2010. Shifts in bacterial community structure associated with inputs of low molecular weight carbon compounds to soil. Soil Biol. Biochem. 42 (6), 896−903.

Eisenlord, S.D., Freedman, Z., Zak, D.R., Xue, K., He, Z., Zhou, J., 2013. Microbial mechanisms mediating increased soil C storage under elevated atmospheric N deposition. Appl. Environ. Microbiol. 79 (4), 1191−1199.

Elliott, E.T., 1986. Aggregate structure and carbon, nitrogen, and phosphorus in native and cultivated soils. Soil Sci. Soc. Am. J. 50 (3), 627−633.

Evans, S.E., Wallenstein, M.D., 2012. Soil microbial community response to drying and rewetting stress: does historical precipitation regime matter? Biogeochemistry 109 (1−3), 101−116.

Evans, S.E., Wallenstein, M.D., 2014. Climate change alters ecological strategies of soil bacteria. Ecol. Lett. 17 (2), 155−164.

Fierer, N., Bradford, M.A., Jackson, R.B., 2007. Toward an ecological classification of soil bacteria. Ecology 88 (6), 1354−1364.

Fierer, N., Strickland, M.S., Liptzin, D., Bradford, M.A., Cleveland, C.C., 2009. Global patterns in belowground communities. Ecol. Lett. 12 (11), 1238−1249.

Fierer, N., Leff, J.W., Adams, B.J., Nielsen, U.N., Bates, S.T., Lauber, C.L., et al., 2012. Cross-biome metagenomic analyses of soil microbial communities and their functional attributes. Proc. Natl. Acad. Sci. 109 (52), 21390−21395.

Finn, D., Kopittke, P.M., Dennis, P.G., Dalal, R.C., 2017. Microbial energy and matter transformation in agricultural soils. Soil Biol. Biochem. 11, 176−192.

Follows, M.J., Dutkiewicz, S., Grant, S., Chisholm, S.W., 2007. Emergent biogeography of microbial communities in a model ocean. Science 315 (5820), 1843−1846.

Fontaine, S., Bardoux, G., Benest, D., Verdier, B., Mariotti, A., Abbadie, L., 2004. Mechanisms of the priming effect in a savannah soil amended with cellulose. Soil Sci. Soc. Am. J. 68 (1), 125−131.

Fontaine, S., Henault, C., Aamor, A., Bdioui, N., Bloor, J.M.G., Maire, V., et al., 2011. Fungi mediate long term sequestration of carbon and nitrogen in soil through their priming effect. Soil Biol. Biochem. 43 (1), 86−96.

Franklin, R.B., Mills, A.L., 2009. Importance of spatially structured environmental heterogeneity in controlling microbial community composition at small spatial scales in an agricultural field. Soil Biol. Biochem. 41 (9), 1833−1840.

Frey, S.D., Elliott, E.T., Paustian, K., 1999. Bacterial and fungal abundance and biomass in conventional and no-tillage agroecosystems along two climatic gradients. Soil Biol. Biochem. 31 (4), 573−585.

Frey, S.D., Lee, J., Melillo, J.M., Six, J., 2013. The temperature response of soil microbial efficiency and its feedback to climate. Nat. Clim. Change 3 (4), 395−398.

Friend, A.D., Lucht, W., Rademacher, T.T., Keribin, R., Betts, R., Cadule, P., et al., 2014. Carbon residence time dominates uncertainty in terrestrial vegetation responses to future climate and atmospheric CO_2. Proc. Natl. Acad. Sci. 111 (9), 3280−3285.

Fuchslueger, L., Bahn, M., Hasibeder, R., Kienzl, S., Fritz, K., Schmitt, M., et al., 2016. Drought history affects grassland plant and microbial carbon turnover during and after a subsequent drought event. J. Ecol. 104 (5), 1453−1465.

Fujita, Y., Witte, J.P.M., Bodegom, P.M., 2014. Incorporating microbial ecology concepts into global soil mineralization models to improve predictions of carbon and nitrogen fluxes. Global Biogeochem. Cycles 28 (3), 223−238.

García-Orenes, F., Morugán-Coronado, A., Zornoza, R., Scow, K., 2013. Changes in soil microbial community structure influenced by agricultural management practices in a Mediterranean agro-ecosystem. PLoS One 8 (11), e80522.

Garcia-Pausas, J., Paterson, E., 2011. Microbial community abundance and structure are determinants of soil organic matter mineralisation in the presence of labile carbon. Soil Biol. Biochem. 43 (8), 1705−1713.

Garrett, S.D., 1951. Ecological groups of soil fungi: a survey of substrate relationships. New Phytol. 50 (2), 149−166.

Geisseler, D., Scow, K.M., 2014. Long-term effects of mineral fertilizers on soil microorganisms−a review. Soil Biol. Biochem. 75, 54−63.

German, D.P., Marcelo, K.R., Stone, M.M., Allison, S.D., 2012. The Michaelis−Menten kinetics of soil extracellular enzymes in response to temperature: a cross latitudinal study. Global Change Biol. 18 (4), 1468−1479.

Geyer, K.M., Kyker-Snowman, E., Grandy, A.S., Frey, S.D., 2016. Microbial carbon use efficiency: accounting for population, community, and ecosystem-scale controls over the fate of metabolized organic matter. Biogeochemistry 127 (2−3), 173−188.

Grandy, A.S., Wieder, W.R., Wickings, K., Kyker-Snowman, E., 2016. Beyond microbes: are fauna the next frontier in soil biogeochemical models? Soil Biol. Biochem. 102, 40−44.

Grayston, S.J., Prescott, C.E., 2005. Microbial communities in forest floors under four tree species in coastal British Columbia. Soil Biol. Biochem. 37 (6), 1157–1167.

Griffiths, R.I., Thomson, B.C., James, P., Bell, T., Bailey, M., Whiteley, A.S., 2011. The bacterial biogeography of British soils. Environ. Microbiol. 13 (6), 1642–1654.

Gupta, V.V., Germida, J.J., 2015. Soil aggregation: influence on microbial biomass and implications for biological processes. Soil Biol. Biochem. 80, A3–A9.

Gupta, V.V.S.R., Germida, J.J., 1988. Distribution of microbial biomass and its activity in different soil aggregate size classes as affected by cultivation. Soil Biol. Biochem. 20 (6), 777–786.

Hacker, N., Ebeling, A., Gessler, A., Gleixner, G., González Macé, O., Kroon, H., et al., 2015. Plant diversity shapes microbe rhizosphere effects on P mobilisation from organic matter in soil. Ecol. Lett. 18 (12), 1356–1365.

Hackl, E., Pfeffer, M., Donat, C., Bachmann, G., Zechmeister-Boltenstern, S., 2005. Composition of the microbial communities in the mineral soil under different types of natural forest. Soil Biol. Biochem. 37 (4), 661–671.

Hagerty, S.B., Van Groenigen, K.J., Allison, S.D., Hungate, B.A., Schwartz, E., Koch, G.W., et al., 2014. Accelerated microbial turnover but constant growth efficiency with warming in soil. Nat. Clim. Change. 4 (10), 903–906.

Hanson, C.A., Allison, S.D., Bradford, M.A., Wallenstein, M.D., Treseder, K.K., 2008. Fungal taxa target different carbon sources in forest soil. Ecosystems 11 (7), 1157–1167.

Hartmann, M., Howes, C.G., VanInsberghe, D., Yu, H., Bachar, D., Christen, R., et al., 2012. Significant and persistent impact of timber harvesting on soil microbial communities in Northern coniferous forests. ISME J. 6 (12), 2199–2218.

Hararuk, O., Smith, M.J., Luo, Y., 2015. Microbial models with data-driven parameters predict stronger soil carbon responses to climate change. Global Change Biol 21 (6), 2439–2453.

Hendrix, P.F., Parmelee, R.W., Crossley, D.A., Coleman, D.C., Odum, E.P., Groffman, P.M., 1986. Detritus food webs in conventional and no-tillage agroecosystems. Bioscience 36 (6), 374–380.

Hobbs, P.R., Sayre, K., Gupta, R., 2008. The role of conservation agriculture in sustainable agriculture. Philos. Trans. R. Soc. B: Biol. Sci. 363 (1491), 543–555.

Hoffman, F.M., Randerson, J.T., Arora, V.K., Bao, Q., Cadule, P., Ji, D., et al., 2014. Causes and implications of persistent atmospheric carbon dioxide biases in Earth System Models. J. Geophys. Res. Biogeosci. 119 (2), 141–162.

Högberg, M.N., Högberg, P., Myrold, D.D., 2007. Is microbial community composition in boreal forest soils determined by pH, C-to-N ratio, the trees, or all three? Oecologia 150 (4), 590–601.

Jassey, V.E., Chiapusio, G., Binet, P., Buttler, A., Laggoun Défarge, F., Delarue, F., et al., 2013. Above and belowground linkages in Sphagnum peatland: climate warming affects plant microbial interactions. Global Change Biol. 19 (3), 811–823.

Jenkinson, D.S., Fox, R.H., Rayner, J.H., 1985. Interactions between fertilizer nitrogen and soil nitrogen—the so called 'priming' effect. Eur. J. Soil Sci. 36 (3), 425–444.

Jenkinson, D.S., Adams, D.E., Wild, A., 1991. Model estimates of CO (2) emissions from soil in response to global warming. Nature 351 (6324), 304.

Jiao, N., Zheng, Q., 2011. The microbial carbon pump: from genes to ecosystems. Appl. Environ. Microbiol. 77 (21), 7439–7444.

Jin, H., Sun, O.J., Liu, J., 2010. Changes in soil microbial biomass and community structure with addition of contrasting types of plant litter in a semiarid grassland ecosystem. J. Plant Ecol. 3 (3), 209–217.

Juarez, S., Nunan, N., Duday, A.C., Pouteau, V., Chenu, C., 2013. Soil carbon mineralisation responses to alterations of microbial diversity and soil structure. Biol. Fertil. Soils 49 (7), 939–948.

Kaiser, C., Franklin, O., Richter, A., Dieckmann, U., 2015. Social dynamics within decomposer communities lead to nitrogen retention and organic matter build-up in soils. Nat. Commun. 6, 8960.

Kallenbach, C.M., Frey, S.D., Grandy, A.S., 2016. Direct evidence for microbial-derived soil organic matter formation and its ecophysiological controls. Nat. Commun. 7, 13630.

Karhu, K., Auffret, M.D., Dungait, J.A., Hopkins, D.W., Prosser, J.I., Singh, B.K., et al., 2014. Temperature sensitivity of soil respiration rates enhanced by microbial community response. Nature 513 (7516), 81–84.

Keiluweit, M., Bougoure, J.J., Nico, P.S., Pett-Ridge, J., Weber, P.K., Kleber, M., 2015. Mineral protection of soil carbon counteracted by root exudates. Nat. Clim. Change 5 (6), 588–595.

King, G.M., 2011. Enhancing soil carbon storage for carbon remediation: potential contributions and constraints by microbes. Trends Microbiol. 19 (2), 75–84.

Kjøller, A., Struwe, S., 1992. Functional groups of microfungi and growth strategies during decomposition. Responses of Forest Ecosystems to Environmental Changes. Springer, Netherlands, pp. 755–756.

Kuramae, E.E., Yergeau, E., Wong, L.C., Pijl, A.S., van Veen, J.A., Kowalchuk, G.A., 2012. Soil characteristics more strongly influence soil bacterial communities than land-use type. FEMS Microbiol. Ecol. 79 (1), 12–24.

Kuramae, E.E., Zhou, J.Z., Kowalchuk, G.A., van Veen, J.A., 2014. Soil-borne microbial functional structure across different land uses. Scientific World J. 2014, 216071.

Kuzyakov, Y., 2010. Priming effects: interactions between living and dead organic matter. Soil Biol. Biochem. 42 (9), 1363–1371.

Lange, M., Eisenhauer, N., Sierra, C.A., Bessler, H., Engels, C., Griffiths, R.I., et al., 2015. Plant diversity increases soil microbial activity and soil carbon storage. Nat. Commun. 6, 6707.

Lauber, C.L., Hamady, M., Knight, R., Fierer, N., 2009. Pyrosequencing-based assessment of soil pH as a predictor of soil bacterial community structure at the continental scale. Appl. Environ. Microbiol. 75 (15), 5111–5120.

Lauro, F.M., McDougald, D., Thomas, T., Williams, T.J., Egan, S., Rice, S., et al., 2009. The genomic basis of trophic strategy in marine bacteria. Proc. Natl. Acad. Sci. 106 (37), 15527–15533.

Leff, J.W., Jones, S.E., Prober, S.M., Barberán, A., Borer, E.T., Firn, J.L., et al., 2015. Consistent responses of soil microbial communities to elevated nutrient inputs in grasslands across the globe. Proc. Natl. Acad. Sci. 112 (35), 10967–10972.

Lennon, J.T., Aanderud, Z.T., Lehmkuhl, B.K., Schoolmaster, D.R., 2012. Mapping the niche space of soil microorganisms using taxonomy and traits. Ecology 93 (8), 1867–1879.

Levine, U.Y., Teal, T.K., Robertson, G.P., Schmidt, T.M., 2011. Agriculture's impact on microbial diversity and associated fluxes of carbon dioxide and methane. ISME J. 5 (10), 1683–1691.

Liang, C., Balser, T.C., 2011. Microbial production of recalcitrant organic matter in global soils: implications for productivity and climate policy. Nat. Rev. Microbiol. 9 (1), 75.

Liang, X., Erickson, J.E., Silveira, M.L., Sollenberger, L.E., Rowland, D.L., 2015. Tissue chemistry and morphology affect root decomposition of perennial bioenergy grasses on sandy soil in a sub tropical environment. GCB Bioenergy 8, 1015–1024.

Litchman, E., Klausmeier, C.A., 2008. Trait-based community ecology of phytoplankton. Annu. Rev. Ecol. Evol. Syst. 39, 615–639.

Liu, X., Wan, S., Su, B., Hui, D., Luo, Y., 2002. Response of soil CO_2 efflux to water manipulation in a tallgrass prairie ecosystem. Plant Soil 240 (2), 213–223.

Lloyd, J., Taylor, J.A., 1994. On the temperature dependence of soil respiration. Funct. Ecology 315–323.

Louis, B.P., Maron, P.A., Viaud, V., Leterme, P., Menasseri-Aubry, S., 2016. Soil C and N models that integrate microbial diversity. Environ. Chem. Lett. 14 (3), 331–344.

Lucas, R.W., Casper, B.B., Jackson, J.K., Balser, T.C., 2007. Soil microbial communities and extracellular enzyme activity in the New Jersey Pinelands. Soil Biol. Biochem. 39 (10), 2508–2519.

Luo, Y., Wan, S., Hui, D., Wallace, L.L., 2001. Acclimatization of soil respiration to warming in a tall grass prairie. Nature 413 (6856), 622–625.

Luo, Y., Ahlström, A., Allison, S.D., Batjes, N.H., Brovkin, V., Carvalhais, N., et al., 2016. Toward more realistic projections of soil carbon dynamics by Earth system models. Global Biogeochem. Cycles 30, 40−56.

Malik, A.A., Chowdhury, S., Schlager, V., Oliver, A., Puissant, J., Vazquez, P.G., et al., 2016. Soil fungal: bacterial ratios are linked to altered carbon cycling. Front. Microbiol. 7, 1247.

Manzoni, S., Taylor, P., Richter, A., Porporato, A., Ågren, G.I., 2012. Environmental and stoichiometric controls on microbial carbon use efficiency in soils. New Phytol. 196 (1), 79−91.

Martiny, J.B., Martiny, A.C., Weihe, C., Lu, Y., Berlemont, R., Brodie, E.L., et al., 2017. Microbial legacies alter decomposition in response to simulated global change. ISME J. 11 (2), 490.

Mau, R.L., Liu, C.M., Aziz, M., Schwartz, E., Dijkstra, P., Marks, J.C., et al., 2015. Linking soil bacterial biodiversity and soil carbon stability. ISME J. 9 (6), 1477−1480.

McGill, W.B., Hunt, H.W., Woodmansee, R.G., Reuss, J.O., 1981. Phoenix, a model of the dynamics of carbon and nitrogen in grassland soils. Ecol. Bull. (Sweden) 33, 49−115.

McGuire, K.L., Bent, E., Borneman, J., Majumder, A., Allison, S.D., Treseder, K.K., 2010. Functional diversity in resource use by fungi. Ecology 91 (8), 2324−2332.

Melillo, J.M., Steudler, P.A., Aber, J.D., Newkirk, K., Lux, H., Bowles, F.P., et al., 2002. Soil warming and carbon-cycle feedbacks to the climate system. Science 298 (5601), 2173−2176.

Miltner, A., Bombach, P., Schmidt-Brücken, B., Kästner, M., 2012. SOM genesis: microbial biomass as a significant source. Biogeochemistry 111 (1−3), 41−55.

Monreal, C.M., Kodama, H., 1997. Influence of aggregate architecture and minerals on living habitats and soil organic matter. Can. J. Soil Sci. 77 (3), 367−377.

Moorhead, D.L., Sinsabaugh, R.L., 2006. A theoretical model of litter decay and microbial interaction. Ecol. Monogr. 76 (2), 151−174.

Morrison, E.W., Frey, S.D., Sadowsky, J.J., van Diepen, L.T., Thomas, W.K., Pringle, A., 2016. Fungal Ecol. 23, 48−57.

Neumann, D., Heuer, A., Hemkemeyer, M., Martens, R., Tebbe, C.C., 2013. Response of microbial communities to long-term fertilization depends on their microhabitat. FEMS Microbiol. Ecol. 86 (1), 71−84.

Nie, M., Pendall, E., Bell, C., Wallenstein, M.D., 2014. Soil aggregate size distribution mediates microbial climate change feedbacks. Soil Biol. Biochem. 68, 357−365.

Nielsen, U.N., Ayres, E., Wall, D.H., Bardgett, R.D., 2011. Soil biodiversity and carbon cycling: a review and synthesis of studies examining diversity−function relationships. Eur. J. Soil Sci. 62 (1), 105−116.

Oades, J.M., Waters, A.G., 1991. Aggregate hierarchy in soils. Soil Res. 29 (6), 815−828.

Ollivier, J., Töwe, S., Bannert, A., Hai, B., Kastl, E.M., Meyer, A., et al., 2011. Nitrogen turnover in soil and global change. FEMS Microbiol. Ecol. 78 (1), 3−16.

Pascault, N., Ranjard, L., Kaisermann, A., Bachar, D., Christen, R., Terrat, S., et al., 2013. Stimulation of different functional groups of bacteria by various plant residues as a driver of soil priming effect. Ecosystems 16 (5), 810−822.

Perveen, N., Barot, S., Alvarez, G., Klumpp, K., Martin, R., Rapaport, A., et al., 2014. Priming effect and microbial diversity in ecosystem functioning and response to global change: a modeling approach using the SYMPHONY model. Global Change Biol. 20 (4), 1174−1190.

Philippot, L., Spor, A., Hénault, C., Bru, D., Bizouard, F., Jones, C.M., et al., 2013. Loss in microbial diversity affects nitrogen cycling in soil. ISME J. 7 (8), 1609−1619.

Placella, S.A., Brodie, E.L., Firestone, M.K., 2012. Rainfall-induced carbon dioxide pulses result from sequential resuscitation of phylogenetically clustered microbial groups. Proc. Natl. Acad. Sci. 109 (27), 10931−10936.

Powell, J.R., Welsh, A., Hallin, S., 2015. Microbial functional diversity enhances predictive models linking environmental parameters to ecosystem properties. Ecology 96 (7), 1985−1993.

Qin, S., Hu, C., He, X., Dong, W., Cui, J., Wang, Y., 2010. Soil organic carbon, nutrients and relevant enzyme activities in particle-size fractions under conservational versus traditional agricultural management. Appl. Soil Ecol. 45 (3), 152−159.

Rabbi, S.M., Daniel, H., Lockwood, P.V., Macdonald, C., Pereg, L., Tighe, M., et al., 2016. Physical soil architectural traits are functionally linked to carbon decomposition and bacterial diversity. Sci. Rep. UK 6.

Ramirez, K.S., Lauber, C.L., Knight, R., Bradford, M.A., Fierer, N., 2010. Consistent effects of nitrogen fertilization on soil bacterial communities in contrasting systems. Ecology 91 (12), 3463−3470.

Ramirez, K.S., Craine, J.M., Fierer, N., 2012. Consistent effects of nitrogen amendments on soil microbial communities and processes across biomes. Global Change Biol. 18 (6), 1918−1927.

Reeve, J.R., Schadt, C.W., Carpenter-Boggs, L., Kang, S., Zhou, J., Reganold, J.P., 2010. Effects of soil type and farm management on soil ecological functional genes and microbial activities. ISME J. 4 (9), 1099−1107.

Reichstein, M., Rey, A., Freibauer, A., Tenhunen, J., Valentini, R., Banza, J., et al., 2003. Modeling temporal and large scale spatial variability of soil respiration from soil water availability, temperature and vegetation productivity indices. Global Biogeochem. Cycles 17 (4), 1104.

Reinsch, S., Ambus, P., Thornton, B., Paterson, E., 2013. Impact of future climatic conditions on the potential for soil organic matter priming. Soil Biol. Biochem. 65, 133−140.

Reischke, S., Kumar, M.G., Bååth, E., 2015. Threshold concentration of glucose for bacterial growth in soil. Soil Biol. Biochem. 80, 218−223.

Rillig, M.C., Mardatin, N.F., Leifheit, E.F., Antunes, P.M., 2010. Mycelium of arbuscular mycorrhizal fungi increases soil water repellency and is sufficient to maintain water-stable soil aggregates. Soil Biol. Biochem. 42 (7), 1189−1191.

Rinnan, R., Bååth, E., 2009. Differential utilization of carbon substrates by bacteria and fungi in tundra soil. Appl. Environ. Microbiol. 75 (11), 3611−3620.

Robertson, G.P., Klingensmith, K.M., Klug, M.J., Paul, E.A., Crum, J.R., Ellis, B.G., 1997. Soil resources, microbial activity, and primary production across an agricultural ecosystem. Ecol. Appl. 7 (1), 158−170.

Roller, B.R., Schmidt, T.M., 2015. The physiology and ecological implications of efficient growth. ISME J. 9 (7), 1481−1487.

Romero-Olivares, A.L., Taylor, J.W., Treseder, K.K., 2015. Neurospora discreta as a model to assess adaptation of soil fungi to warming. BMC Evol. Biol. 15 (1), 198.

Rousk, J., Bååth, E., 2011. Growth of saprotrophic fungi and bacteria in soil. FEMS Microbiol. Ecol. 78 (1), 17−30.

Rousk, J., Bååth, E., Brookes, P.C., Lauber, C.L., Lozupone, C., Caporaso, J.G., et al., 2010. Soil bacterial and fungal communities across a pH gradient in an arable soil. ISME J. 4 (10), 1340.

Rousk, J., Brookes, P.C., Bååth, E., 2009. Contrasting soil pH effects on fungal and bacterial growth suggest functional redundancy in carbon mineralization. Appl. Environ. Microbiol. 75 (6), 1589−1596.

Rousk, J., Frey, S.D., Bååth, E., 2012. Temperature adaptation of bacterial communities in experimentally warmed forest soils. Global Change Biol. 18 (10), 3252−3258.

Rousk, J., Hill, P.W., Jones, D.L., 2015. Priming of the decomposition of ageing soil organic matter: concentration dependence and microbial control. Funct. Ecol. 29 (2), 285−296.

Schimel, D.S., 1995. Terrestrial ecosystems and the carbon cycle. Global Change Biol. 1 (1), 77−91.

Schimel, J., Balser, T.C., Wallenstein, M., 2007. Microbial stress response physiology and its implications for ecosystem function. Ecology 88 (6), 1386−1394.

Schimel, J.P., Gulledge, J.A.Y., 1998. Microbial community structure and global trace gases. Global Change Biol. 4 (7), 745−758.

Schimel, J.P., Schaeffer, S.M., 2012. Microbial control over carbon cycling in soil. Front. Microbiol. 3, 348.

Schmidt, M.W., Torn, M.S., Abiven, S., Dittmar, T., Guggenberger, G., Janssens, I.A., et al., 2011. Persistence of soil organic matter as an ecosystem property. Nature 478 (7367), 49−56.

Sessitsch, A., Weilharter, A., Gerzabek, M.H., Kirchmann, H., Kandeler, E., 2001. Microbial population structures in soil particle size fractions of a long-term fertilizer field experiment. Appl. Environ. Microbiol. 67 (9), 4215−4224.

Singh, B.K., Bardgett, R.D., Smith, P., Reay, D.S., 2010. Microorganisms and climate change: terrestrial feedbacks and mitigation options. Nat. Rev. Microbiol. 8 (11), 779−790.

Sinsabaugh, R.L., Manzoni, S., Moorhead, D.L., Richter, A., 2013. Carbon use efficiency of microbial communities: stoichiometry, methodology and modelling. Ecol. Lett. 16 (7), 930−939.

Sipilä, T.P., Yrjälä, K., Alakukku, L., Palojärvi, A., 2012. Cross-site soil microbial communities under tillage regimes: fungistasis and microbial biomarkers. Appl. Environ. Microbiol. 78 (23), 8191−8201.

Sistla, S.A., Schimel, J.P., 2013. Seasonal patterns of microbial extracellular enzyme activities in an arctic tundra soil: identifying direct and indirect effects of long-term summer warming. Soil Biol. Biochem. 66, 119−129.

Six, J., Bossuyt, H., Degryze, S., Denef, K., 2004. A history of research on the link between (micro) aggregates, soil biota, and soil organic matter dynamics. Soil Tillage Res. 79 (1), 7−31.

Six, J., Frey, S.D., Thiet, R.K., Batten, K.M., 2006. Bacterial and fungal contributions to carbon sequestration in agroecosystems. Soil Sci. Soc. Am. J. 70 (2), 555−569.

Soong, J.L., Vandegehuchte, M.L., Horton, A.J., Nielsen, U.N., Denef, K., Shaw, E.A., et al., 2016. Soil microarthropods support ecosystem productivity and soil C accrual: evidence from a litter decomposition study in the tallgrass prairie. Soil Biol. Biochem. 92, 230−238.

Spring, S., Scheuner, C., Göker, M., Klenk, H.P., 2015. A taxonomic framework for emerging groups of ecologically important marine gammaproteobacteria based on the reconstruction of evolutionary relationships using genome-scale data. Front. Microbiol. 6, 281.

Steinweg, J.M., Plante, A.F., Conant, R.T., Paul, E.A., Tanaka, D.L., 2008. Patterns of substrate utilization during long-term incubations at different temperatures. Soil Biol. Biochem. 40 (11), 2722−2728.

Sterner, R.W., Elser, J.J., 2002. Ecological Stoichiometry: The Biology of Elements from Molecules to the Biosphere. Princeton University Press, Princeton, NJ.

Strickland, M.S., Rousk, J., 2010. Considering fungal: bacterial dominance in soils−methods, controls, and ecosystem implications. Soil Biol. Biochem. 42 (9), 1385−1395.

Strickland, M.S., Lauber, C., Fierer, N., Bradford, M.A., 2009. Testing the functional significance of microbial community composition. Ecology 90 (2), 441−451.

Sylvia, D.M., Fuhrmann, J.J., Hartel, P.G., Zuberer, D.A. (Eds.), 2005. Principles and Applications of Soil Microbiology. Pearson Prentice Hall, Upper Saddle River, NJ.

Talbot, J.M., Bruns, T.D., Taylor, J.W., Smith, D.P., Branco, S., Glassman, S.I., et al., 2014. Endemism and functional convergence across the North American soil mycobiome. Proc. Natl. Acad. Sci. 111 (17), 6341−6346.

Tardy, V., Spor, A., Mathieu, O., Lévèque, J., Terrat, S., Plassart, P., et al., 2015. Shifts in microbial diversity through land use intensity as drivers of carbon mineralization in soil. Soil Biol. Biochem. 90, 204−213.

Thakur, M.P., Milcu, A., Manning, P., Niklaus, P.A., Roscher, C., Power, S., et al., 2015. Plant diversity drives soil microbial biomass carbon in grasslands irrespective of global environmental change factors. Global Change Biol. 21 (11), 4076−4085.

Thoms, C., Gattinger, A., Jacob, M., Thomas, F.M., Gleixner, G., 2010. Direct and indirect effects of tree diversity drive soil microbial diversity in temperate deciduous forest. Soil Biol. Biochem. 42 (9), 1558−1565.

Tiemann, L.K., Grandy, A.S., 2015. Mechanisms of soil carbon accrual and storage in bioenergy cropping systems. GCB Bioenergy 7 (2), 161−174.

Tiemann, L.K., Grandy, A.S., Atkinson, E.E., Marin Spiotta, E., McDaniel, M.D., 2015. Crop rotational diversity enhances belowground communities and functions in an agroecosystem. Ecol. Lett. 18 (8), 761–771.

Todd-Brown, K.E., Hopkins, F.M., Kivlin, S.N., Talbot, J.M., Allison, S.D., 2012. A framework for representing microbial decomposition in coupled climate models. Biogeochemistry 109 (1–3), 19–33.

Treseder, K.K., 2008. Nitrogen additions and microbial biomass: a meta analysis of ecosystem studies. Ecol. Lett. 11 (10), 1111–1120.

Treseder, K.K., Lennon, J.T., 2015. Fungal traits that drive ecosystem dynamics on land. Microbiol. Mol. Biol. Rev. 79 (2), 243–262.

Trivedi, P., Anderson, I.C., Singh, B.K., 2013. Microbial modulators of soil carbon storage: integrating genomic and metabolic knowledge for global prediction. Trends Microbiol. 21 (12), 641–651.

Trivedi, P., Rochester, I.J., Trivedi, C., Van Nostrand, J.D., Zhou, J., Karunaratne, S., et al., 2015. Soil aggregate size mediates the impacts of cropping regimes on soil carbon and microbial communities. Soil Biol. Biochem. 91, 169–181.

Trivedi, P., Delgado-Baquerizo, M., Trivedi, C., Hu, H., Anderson, I.C., Jeffries, T.C., et al., 2016. Microbial regulation of the soil carbon cycle: evidence from gene–enzyme relationships. ISME J. 10 (11), 2593–2604.

Trivedi, P., Delgado-Baquerizo, M., Jeffries, T.C., Trivedi, C., Anderson, I.C., Lai, K., et al., 2017. Soil aggregation and associated microbial communities modify the impact of agricultural management on carbon content. Environ. Microbiol. 19 (8), 3070–3086.

Van Gestel, M., Merckx, R., Vlassak, K., 1996. Spatial distribution of microbial biomass in microaggregates of a silty-loam soil and the relation with the resistance of microorganisms to soil drying. Soil Biol. Biochem. 28 (4), 503–510.

Verhoef, H.A., Brussaard, L., 1990. Decomposition and nitrogen mineralization in natural and agroecosystems: the contribution of soil animals. Biogeochemistry 11 (3), 175–211.

Vos, M., Wolf, A.B., Jennings, S.J., Kowalchuk, G.A., 2013. Micro-scale determinants of bacterial diversity in soil. FEMS Microbiol. Rev. 37 (6), 936–954.

Waldrop, M.P., Firestone, M.K., 2006. Seasonal dynamics of microbial community composition and function in oak canopy and open grassland soils. Microb. Ecol. 52 (3), 470–479.

Wallenstein, M.D., Hall, E.K., 2012. A trait-based framework for predicting when and where microbial adaptation to climate change will affect ecosystem functioning. Biogeochemistry 109 (1–3), 35–47.

Wallenstein, M.D., McMahon, S., Schimel, J., 2007. Bacterial and fungal community structure in Arctic tundra tussock and shrub soils. FEMS Microbiol. Ecol. 59 (2), 428–435.

Wang, G., Post, W.M., Mayes, M.A., 2013. Development of microbial enzyme mediated decomposition model parameters through steady state and dynamic analyses. Ecol. Appl. 23 (1), 255–272.

Wang, G., Jagadamma, S., Mayes, M.A., Schadt, C.W., Steinweg, J.M., Gu, L., et al., 2015. Microbial dormancy improves development and experimental validation of ecosystem model. ISME J. 9 (1), 226–237.

Wang, Y.P., Chen, B.C., Wieder, W.R., Leite, M., Medlyn, B.E., Rasmussen, M., et al., 2014. Oscillatory behavior of two nonlinear microbial models of soil carbon decomposition. Biogeosciences 11 (7), 1817.

Wardle, D.A., Bardgett, R.D., Klironomos, J.N., Setälä, H., Van Der Putten, W.H., Wall, D.H., 2004. Ecological linkages between aboveground and belowground biota. Science 304 (5677), 1629–1633.

Wardle, D.A., Jonsson, M., Bansal, S., Bardgett, R.D., Gundale, M.J., Metcalfe, D.B., 2012. Linking vegetation change, carbon sequestration and biodiversity: insights from island ecosystems in a long term natural experiment. J. Ecol. 100 (1), 16–30.

Waring, B.G., Averill, C., Hawkes, C.V., 2013. Differences in fungal and bacterial physiology alter soil carbon and nitrogen cycling: insights from meta analysis and theoretical models. Ecol. Lett. 16 (7), 887–894.

Weand, M.P., Arthur, M.A., Lovett, G.M., McCulley, R.L., Weathers, K.C., 2010. Effects of tree species and N additions on forest floor microbial communities and extracellular enzyme activities. Soil Biol. Biochem. 42 (12), 2161–2173.

Wei, H., Guenet, B., Vicca, S., Nunan, N., AbdElgawad, H., Pouteau, V., et al., 2014. Thermal acclimation of organic matter decomposition in an artificial forest soil is related to shifts in microbial community structure. Soil Biol. Biochem. 71, 1–12.

Wickings, K., Grandy, A.S., 2011. The oribatid mite *Scheloribates moestus* (Acari: Oribatida) alters litter chemistry and nutrient cycling during decomposition. Soil Biol. Biochem. 43 (2), 351–358.

Wickings, K., Grandy, A.S., Reed, S.C., Cleveland, C.C., 2012. The origin of litter chemical complexity during decomposition. Ecol. Lett. 15 (10), 1180–1188.

Wieder, W.R., Bonan, G.B., Allison, S.D., 2013. Global soil carbon projections are improved by modelling microbial processes. Nat. Clim. Change 3 (10), 909–912.

Wieder, W.R., Grandy, A.S., Kallenbach, C.M., Taylor, P.G., Bonan, G.B., 2015. Representing life in the Earth system with soil microbial functional traits in the MIMICS model. Geosci. Model Dev. 8 (6), 1789–1808.

Wild, B., Schnecker, J., Alves, R.J.E., Barsukov, P., Bárta, J., Čapek, P., et al., 2014. Input of easily available organic C and N stimulates microbial decomposition of soil organic matter in arctic permafrost soil. Soil Biol. Biochem. 75, 143–151.

Wood, S.A., Sokol, N., Bell, C.W., Bradford, M.A., Naeem, S., Wallenstein, M.D., et al., 2016. Opposing effects of different soil organic matter fractions on crop yields. Ecol. Appl. 26 (7), 2072–2085.

Xiao, C., Janssens, I.A., Liu, P., Zhou, Z., Sun, O.J., 2007. Irrigation and enhanced soil carbon input effects on below ground carbon cycling in semiarid temperate grasslands. New Phytol. 174 (4), 835–846.

Xu, M., Qi, Y., 2001. Soil surface CO_2 efflux and its spatial and temporal variations in a young ponderosa pine plantation in northern California. Global Change Biol. 7 (6), 667–677.

Xue, K., Wu, L., Deng, Y., He, Z., Van Nostrand, J., Robertson, P.G., et al., 2013. Functional gene differences in soil microbial communities from conventional, low-input, and organic farmlands. Appl. Environ. Microbiol. 79 (4), 1284–1292.

You, Y., Wang, J., Huang, X., Tang, Z., Liu, S., Sun, O.J., 2014. Relating microbial community structure to functioning in forest soil organic carbon transformation and turnover. Ecol. Evol. 4 (5), 633–647.

Yuste, J.C., Barba, J., Fernandez Gonzalez, A.J., Fernandez Lopez, M., Mattana, S., Martinez Vilalta, J., et al., 2012. Changes in soil bacterial community triggered by drought induced gap succession preceded changes in soil C stocks and quality. Ecol. Evol. 2 (12), 3016–3031.

Zhou, J., Xue, K., Xie, J., Deng, Y., Wu, L., Cheng, X., et al., 2012. Microbial mediation of carbon-cycle feedbacks to climate warming. Nat. Clim. Change 2 (2), 106–110.

Zimmerman, A.E., Martiny, A.C., Allison, S.D., 2013. Microdiversity of extracellular enzyme genes among sequenced prokaryotic genomes. ISME J. 7 (6), 1187–1199.

FURTHER READING

Bardgett, R.D., van der Putten, W.H., 2014. Belowground biodiversity and ecosystem functioning. Nature 515 (7528), 505–511.

Lavorel, S., Garnier, E., 2002. Predicting changes in community composition and ecosystem functioning from plant traits: revisiting the Holy Grail. Funct. Ecol. 16 (5), 545–556.

FURTHER READING

LEVERAGING A NEW UNDERSTANDING OF HOW BELOWGROUND FOOD WEBS STABILIZE SOIL ORGANIC MATTER TO PROMOTE ECOLOGICAL INTENSIFICATION OF AGRICULTURE

Stephen A. Wood[1,2] **and Mark A. Bradford**[1]

[1]*Yale University, New Haven, CT, United States* [2]*The Nature Conservancy, Arlington, VA, United States*

4.1 INTRODUCTION

Research has emphasized the important role of soil microbiota in building and stabilizing soil carbon (Lehmann and Kleber, 2015). There is growing evidence that biotic interactions among larger fauna can also play a key role in stabilizing soil carbon (Bradford, 2016). At the same time, there is a growing appreciation globally of the important role soil carbon plays in providing multiple ecosystem services, including those underpinning agriculture (Banwart et al., 2014). Yet little work has synthesized the potential for a new understanding of soil biotic interactions and soil carbon stabilization to inform agricultural management and the promotion of sustainable agriculture.

High-intensity agriculture that relies on large amounts of external inputs is the source of elevated greenhouse gas production as well as contamination of water systems with soluble, mineral nutrients. Because greater intensity of agriculture has led to pressure on global planetary boundaries for greenhouse gases and nutrient enrichment (Rockstrom et al., 2009), there are many calls to intensify agriculture sustainably—to continue crop yield gains while minimizing environmental degradation (Burney et al., 2010; Caron et al., 2014; Foley et al., 2011; Garnett et al., 2013; Godfray et al., 2010; Tilman et al., 2011).

The gaseous and aqueous losses of excess agricultural nutrients—which characterize unsustainable intensification—are mediated by soil properties and processes that determine the leakiness of nutrients into the atmosphere and water. Processes such as denitrification, nitrification, immobilization, and organic matter stabilization can govern whether and the rate by which nutrients are lost

Soil Carbon Storage. DOI: https://doi.org/10.1016/B978-0-12-812766-7.00004-4

from soil into connected systems. These processes are largely governed directly by soil microbiota, with indirect control by other biotic taxa that influence the dynamics of soil microbial communities. Other soil organisms—particularly those directly associated with plant roots—can further help the move toward sustainable agriculture by increasing the efficiency by which plants acquire nutrients and, thus, potentially lowering the need for high amounts of supplemental nutrients. Hence, understanding the drivers and dynamics of soil biotic activity has been identified—mainly by soil ecologists—as a key aspect of sustainable agroecosystem management (Cheeke et al., 2012).

Work outside of the biophysical sciences has challenged whether sustainable intensification is sufficient to achieve true sustainability (Loos et al., 2014). Critiques of the term "sustainable intensification" argue that the concept is misleading because it inadequately addresses several key concepts in sustainability, including: Failure to address the ecological consequences associated with even the most "sustainable" form of agriculture; and, failure to address procedural and distributive injustices and inequalities within the food system that lead to disproportionate access to food (Loos et al., 2014).

These critiques highlight the importance of conceptual clarity and rigor in defining and justifying one's use of the concept of sustainability. Before reviewing concepts of soil food webs and their potential importance to sustainable agriculture, we will highlight an important distinction in types of sustainability—known as weak versus strong sustainability. We will then use that framework to advocate for an approach to agriculture rooted in ecological principles, particularly those of soil food webs.

4.2 NOTIONS OF SUSTAINABILITY

The original, and most common, definitions of sustainability hinge on the criterion of nondeclining human wellbeing through time (Arrow et al., 2012; United Nations, 1987). The extent to which the environment is important to sustainability is then in its contribution to human wellbeing. The ability of the environment to contribute to human wellbeing is referred to as natural capital—the environmental stocks and processes that support functioning ecosystems and contribute to ecosystem services that people rely on (de Groot et al., 2002). Because ecosystem services are important for human wellbeing, degradation of natural capital should not be sustainable because it would lead to violation of the condition of nondeclining human utility.

Yet, natural capital has been persistently degraded while human wellbeing has increased, which some have called the environmentalist's paradox (Raudsepp-Hearne et al., 2010). This paradox highlights an important distinction between two notions of sustainability—weak and strong sustainability (Fig. 4.1). Proponents of weak sustainability assert that all forms of capital—including natural capital—are substitutable for other types of capital, as long as human welfare is nondeclining. Phrased another way, depletion of natural capital is justifiable as long as the rents from depletion of exhaustible resources are invested in reproducible capital that maintains total capital stocks and human consumption and utility (Hartwick, 1977). Applying the notion of weak sustainability to agriculture, nondeclining human utility is the essential criterion of agricultural sustainability, although an economy-wide view would allow for decreased agricultural production as long as human utility was maintained through gains in other

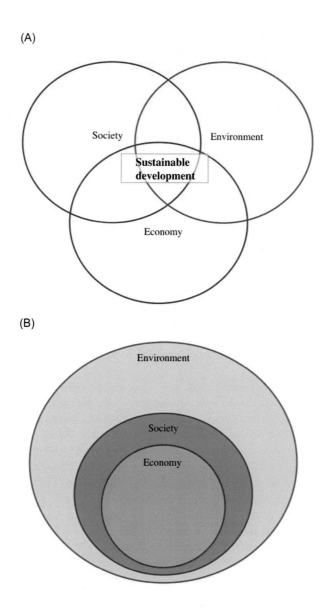

FIGURE 4.1

Common representation of weak versus strong sustainability. Weak sustainability (A) advocates for nondeclining total capital stocks through time, where different types of capital are seen as separate and substitutable, hence not perfectly overlapping circles. Strong sustainability (B) emphasizes that certain forms of natural capital cannot and/or should not be substituted for other forms of capital.

sectors. Maintaining human utility in agriculture—presumably through maintaining crop productivity—allows for the depletion of soil nutrient stocks as long as the earnings associated with that depletion are reinvested in other forms of capital, such as mineral fertilizers, that allow continued productivity through time.

Proponents of strong sustainability, however, challenge the notion that natural and other forms of capital can be realistically substituted (Daly et al., 1995). An advocate for strong sustainability might claim that the value of the stratospheric ozone layer cannot be replaced by sunscreen (Jacobs, 1995). Under strong sustainability, natural capital stocks require protection independent of total capital stocks so that natural capital stocks must always be nondeclining. The justification for the separate protection of natural capital stocks is that certain forms of natural capital are seen to be not readily monetized and, therefore, cannot be compared to reproducible capital stocks to evaluate whether total stocks are in fact nondeclining (Trainor, 2006). Complex ecological processes, such as food web interactions, can play an important role in ecosystem functioning (Dobson, 2009; Hawlena et al., 2012; Strickland et al., 2013), but accurately monetizing such complex interactions is not straightforward to the point of being implausible. As a result, advocates of strong sustainability often propose discussing natural capital in physical, rather than monetary, language (Özkaynak et al., 2004). Because certain types of natural capital cannot be reliably monetized, it is implied that not all forms of natural capital can be reliably substituted for human-made capital. This notion opens the door to *critical natural capital*, which is the subset of natural capital that must be maintained for total capital stocks to be nondeclining. Under this viewpoint, environmental damage is only acceptable if critical natural capital—that which cannot be replaced by human capital—is unaffected.

We raise this distinction between strong and weak sustainability because whether or not soil subsystems—and, in particular, soil biotic interactions—are necessary for agriculture hinges on whether one adopts a weak or strong view of sustainability. As mentioned earlier, an advocate of weak sustainability would argue that the decline in soil resource stocks—such as organic matter—would be justifiable as long as that resource depletion was compensated for by the use of mineral fertilizers to maintain agricultural productivity. We argue for a strong approach to sustainability because of evidence that natural soil fertility is nonsubstitutable. We propose three arguments. First, all viable approaches to soil nutrient management—such as mineral fertilizers—are themselves an exhaustible natural resource stock. Phosphorus in fertilizers is rock-derived, the stores of which are estimated to be depleted within the century (Cordell et al., 2009). Nitrogen, while renewable, requires high energy to convert to plant available forms, which may be less viable in the future depending on the trajectory in the energy sector. Second, there is increasing evidence that mineral fertilizers alone are insufficient to maintain crop yields. In sub-Saharan Africa, there is a growing body of evidence demonstrating that on certain fields, crop yields are nonresponsive to fertilizers even for farms with low starting levels of yield (Tittonell et al., 2008). Though the mechanisms behind these patterns are unresolved, the pattern highlights that soil fertility management cannot be divorced from ecological processes—i.e., natural capital. Third, the ecological processes that generate soil nutrient cycling are complex processes that likely defy valuation. The goal of this chapter is to highlight emerging understanding of soil science—including interactions between agronomy, soil organic matter (SOM), and soil biotic interactions—to substantiate this claim that soil-based natural capital may not be easily quantified, thus justifying a strong sustainability approach.

4.3 THE NEED FOR A MULTIPOOL ORGANIC MATTER VISION FOR AGRICULTURE

In the previous section, we argued that soil ecological processes strongly interact with human technology-based efforts to maintain agricultural productivity, highlighting research on nonresponsive soils in Africa. Though the specific mechanisms for these nonresponsive soils is inconclusive, there is substantial and growing evidence that sufficient amounts of SOM are necessary for the use of mineral fertilizers to succeed (Tittonell et al., 2008; Tittonell and Giller, 2013). SOM increases the exchange capacity of soil, allowing for greater nutrient and water availability to crops through time.

Because of this elevated water and nutrient retention—as well as impacts on soil physical structure—SOM has long been recognized as a critical component of soil (Palm et al., 2007). As such there is broad interest in managing for SOM to increase carbon storage and agricultural productivity (Lal, 2004), and hence a critical need to create specific, quantitative targets for SOM levels that would optimize these outcomes (Oldfield et al., 2015). To date, these targets have remained elusive, in part due to site-level differences in the relationship between organic matter and agronomic outcomes.

One potential reason for this site-level variability is that different types of organic matter may contribute to outcomes in different ways. SOM—and, by association, soil carbon—does not make up a single, homogenous entity in soil, but is instead a gradient of different organic compounds that vary in energy content for microbes, nutrient content, and the likelihood of being stabilized onto mineral surfaces or into aggregates (Lehmann and Kleber, 2015). Evidence has shown that different types of organic matter can contribute to crop yield in different ways, with some relationships being positive and others negative (Cates and Ruark, 2017; Pieralli, 2017; Wood et al., 2016). This evidence is further supported by findings that fast-cycling organic matter, usually measured by soil respiration after incubation, is the measure of SOM that is most strongly related to crop yields (Culman et al., 2013). Faster cycling pools, by virtue of their quicker decomposition, are thought to be highly related to crop yields because decomposition releases nutrients into the soil matrix that can be taken up by plants (Haynes, 2005).

The existence of multiple "pools" of organic matter that cycle at different rates has long been recognized (Fig. 4.2A). Steady-state ecosystem models, such as CENTURY, have had multiple pools of organic matter embedded since their development (Parton et al., 1987). However, only recently has it become fully appreciated that these different pools turn over based on different mechanisms and biological pathways (Fig. 4.2B). Previously, the existence of stable, long-term organic matter was thought to be the byproducts and leftovers of shorter-term decomposition—it is now understood that these pools can be formed independently from one another (Lehmann and Kleber, 2015). Under prior conceptualizations, highly recalcitrant, humic compounds were argued to make up the bulk of stable organic matter. The explanation for the existence of these compounds was that they were either (1) compounds that microbes could not decompose because of high molecular complexity, or (2) long-chain compounds that were stitched together by abiotic processes, creating highly recalcitrant material that was in turn undecomposable by microorganisms (Lehmann and Kleber, 2015). More recent evidence has demonstrated that old soil carbon compounds (as determined by ^{14}C dating methods) show little resemblance to high molecular

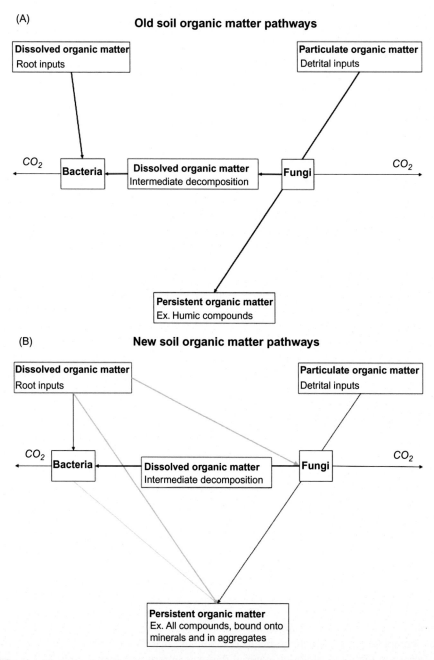

FIGURE 4.2

Soil organic matter stabilization pathways, new (B) versus old (A). New links are shown in *gray*. Early models of SOM stabilization highlighted the importance of recalcitrant substances, like humic acids, to long-term stabilization. New understanding demonstrates that labile inputs become stabilized either directly or by passing through a microbial filter, which could be bacterial or fungal. Stabilization is defined as physical binding onto minerals or isolation into aggregates.

weight, complex compounds. Instead, long-term stable organic matter is now thought to be largely made up of simple compounds—such as root exudates—that are used by microbes whose cellular products and byproducts become stabilized onto reactive mineral surfaces, or caught up in structurally stable soil aggregates. New thinking also highlights that as high molecular weight compounds get further and further decomposed (and hence simpler and smaller in structure), they become more likely to bind to mineral surfaces—in contrast with old ideas that stable organic matter was commonly not highly processed nor decomposed.

Because long-term organic matter is now thought to be made up of simple compounds, both bacteria and fungi should play a direct role in the buildup of slower cycling organic matter pools. Previously, bacteria were thought to contribute mostly to faster cycling pools and fungi to slower cycling pools. The differential role of these two soil biotic kingdoms highlights the potential for soil biotic interactions to influence soil carbon cycling—with an implication for the agronomic benefits people derive from soil.

4.4 EVOLVING UNDERSTANDING OF SOIL FOOD WEBS

Some of the best-known multipool SOM models were initially developed in the 1980s (Coleman and Jenkinson, 1996; Parton et al., 1987), and are now widely applied to inform such activities as rangeland management, national soil carbon accounting, and in projecting carbon cycle-climate feedbacks. These multipool SOM models were based on biologically mediated decomposition, but the processes underpinning them were "implicit" and no organismal pools (other than plants) were represented (Todd-Brown et al., 2013).

In the same decade in which multipool SOM models were developed, a parallel, but largely separate modeling effort, gained prominence. In contrast to the multipool models, in the parallel modeling effort of soil food webs, the carbon and nitrogen pools represented included detritus (i.e., dead organic matter) and the organismal groups that consumed them. The assumption made in these modeling efforts was that the food webs were primarily maintained by a detrital energy channel with a distinct but much less energetically important channel sustained through root inputs (Fig. 4.3A; Hunt et al., 1987; Moore and William Hunt, 1988; De Ruiter et al., 1993). The detrital channel was then further subdivided into saprotrophic bacterial- and fungal-based compartments, meaning that secondary production of these two microbial groups sustained animals at higher trophic levels (Moore et al., 2004). These animals were binned into functional groups, including fungivorous nematodes and bacterial-feeding protists, which were then in turn fed upon by a variety of animals (e.g., predatory mites) at increasingly higher trophic levels (Fig. 4.3A). The initial focus in these food web models was to help understand the stability and diversity of belowground food webs, and to a certain extent the dynamics of nitrogen and its resupply to plants following the decomposition of plant detritus (De Ruiter et al., 1993; Hunt et al., 1987; Moore and William Hunt, 1988).

Although multipool SOM models and detrital-based food web models had different objectives, a focus on building SOM in the interests of sustainability became a common focus of both efforts in the late 1990s. Observations that high SOM contents, slower nutrient cycling, and efficient reuse of carbon and nutrients was associated with dominance of the fungal

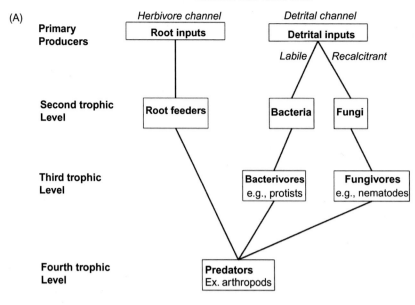

Classic soil food web

(A)

| *Herbivore channel* | *Detrital channel* |

Primary Producers — Root inputs | Detrital inputs

Labile / \ *Recalcitrant*

Second trophic Level — Root feeders | Bacteria | Fungi

Third trophic Level — Bacterivores e.g., protists | Fungivores e.g., nematodes

Fourth trophic Level — Predators Ex. arthropods

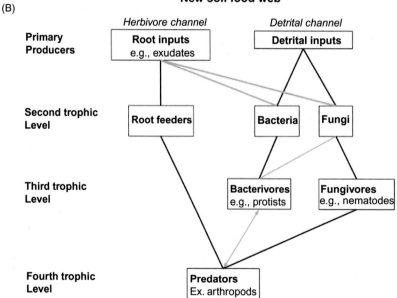

New soil food web

(B)

| *Herbivore channel* | *Detrital channel* |

Primary Producers — Root inputs e.g., exudates | Detrital inputs

Second trophic Level — Root feeders | Bacteria | Fungi

Third trophic Level — Bacterivores e.g., protists | Fungivores e.g., nematodes

Fourth trophic Level — Predators Ex. arthropods

FIGURE 4.3

Soil food web pathways, new (B) versus old (A). New links are shown in *gray*. Early conceptualizations of food webs emphasized root and detrital channels as distinct, with the detrital channel further divided into a bacterial and fungal component. New thinking shows that root inputs directly feed bacteria and fungi, as well as detrital inputs, emphasizing the linked nature of these channels. Multichannel feeding is also now known to be widespread. Not all possible food web interactions are shown. Instead, we highlight the key new understanding that shows how food web thinking has changed.

compartment, led to the common conception that fungal-based dominance was necessary for system sustainability (Bardgett and McAlister, 1999). These ideas were extended to agricultural systems to increase nutrient retention (De Vries and Bardgett, 2012). Such conceptions also lent support to efforts that minimize disturbance in agricultural systems by adoption of conservation tillage, facilitating the establishment of highly connected, slower responding, and so disturbance sensitive, belowground food webs (Morriën et al., 2017)—though mathematical network theory suggests that the presence of many weak links, rather than strong network links, promotes network stability (Neutel, 2002). Fungal-based food webs, as opposed to those based on bacterial energy flows, were (and still are) considered most resilient to disturbance given their high interconnectedness and use of lower quality substrates that decompose slowly. Just as in multipool SOM models, the idea was that if you push the system toward dominance of energy flows through slower-cycling organic matter pools, then SOM will accumulate because decomposition is slowed per unit detritus and nutrients are made available at a rate where uptake is rapid enough to prevent leaching or gaseous losses (Strickland and Rousk, 2010; van der Heijden et al., 2008).

Further evidence, however, blurs distinctions between root and detrital-based energy channels, and also conceptions of distinct bacterial- and fungal-based compartments within the detrital channel (Fig. 4.3B; Strickland and Rousk, 2010; Pollierer et al., 2012). Instead, low molecular weight root inputs (such as exudates that can be rapidly assimilated by bacteria and fungi without extracellular enzymatic degradation) are suggested to fuel more than half of the activity of belowground food webs (Gilbert et al., 2014; Pollierer et al., 2007; van Hees et al., 2005). Rather than suggesting a quantitatively more important root versus detrital channel, these observations challenge the original dichotomy by coupling the root and detrital channel. That is, root and detrital channels are not discrete, with root exudates increasingly considered to be a major basal resource flow to the majority of soil organisms (Bardgett et al., 2014). The likely dominance of root exudates as a food source for symbiotic and free-living bacteria and fungi, is helping to stimulate calls for a reevaluation of how soil food webs are structured and operate (de Vries and Caruso, 2016). The emerging model is that the prevalence of both bacterial and fungal feeding on labile root exudates unites the three classic channels—root, bacterial and fungal. The unification then temporally and spatially couples aboveground and belowground dynamics much more closely because factors that affect exudation rates and patterns (i.e., living plant physiological responses), have immediate consequences for energy flow to the base of the food web of nearly all soil organisms. As such, sustainable intensification efforts focused on building the fungal-based detrital compartment in agricultural soils by maximizing the amount of residues returned to soils, and minimizing physical disturbances such as tillage, may need to be rethought to ensure that the performance of living plants within the system is prioritized for yield and also soil functioning.

Fungi will likely still feature prominently in efforts to sustainably intensify agriculture and it may be that fungal dominance is essential, but for different reasons to those proposed in classic conceptions of compartmentalized soil food webs. Ballhausen and de Boer (2016) proposed that much of the labile carbon exuded by roots is first consumed not by mycorrhiza and bacteria, but rather by saprotrophic fungi. That is, those organisms classically viewed as decomposing low-quality, macromolecular plant inputs, instead feed on the exudates thought to fuel fast-cycling organisms such as bacteria. Fungal saprotrophs appear to have their own equivalent of rhizosphere bacteria, necessitating an expansion of the rhizosphere to consider the sapro-rhizosphere niche, where bacteria function as primary consumers around roots and secondary consumers around fungal hyphae.

The unification of energy channels through basal resource flows is mirrored by a similar unification from the top-down, with consumers originally thought to specialize on distinct channels or compartments instead exhibiting what is termed multichannel omnivory (Wolkovich, 2016). Under this conception, consumers feed across multiple energy channels and such multichannel feeding is now considered the norm in most food webs. Admittedly, the classic energy channel concept does include multichannel feeding, but only at the highest trophic levels (Moore and William Hunt, 1988), whereas empirical data increasingly suggest that multichannel feeding occurs at the lowest trophic level, the primary consumers of bacterial and fungal biomass. For example, soil protists, which play a central role in classical soil food webs as the primary bacterivores and dominant agents in the remineralization of nitrogen, are key multichannel feeders. They appear to unite the bacterial and fungal energy compartments through bacterivory combined with obligate and facultative mycophagy (Geisen, 2016). In some instances, new observations of protists even reveal that they feed on the nematodes still commonly considered their predators (Geisen, 2016). Overall then, observations over the past decade are forcing a revisioning of how soil food webs are structured and function belowground (Bradford, 2016). The consequence of this revisioning for energy and nutrient dynamics has, to date, received little attention. Yet to build confidence in sustainable intensification efforts, these new ideas must be reconciled with the rationale for management practices intended to build SOM and promote yields.

4.5 THE RELEVANCE OF EMERGING SOIL FOOD WEB CONCEPTS FOR SOIL ORGANIC MATTER CYCLING

Emerging conceptualizations of SOM stabilization focus on microbial biomass and secondary products as the filter through which plant carbon inputs pass before entry into slowly cycling SOM pools that commonly comprise the bulk of total soil carbon stocks (Grandy and Neff, 2008; Miltner et al., 2011). It then stands to reason that carbon inputs to soil in forms that microbes can grow on efficiently should be transferred to stable SOM because a greater fraction of the substrate goes to microbial production as opposed to loss from soil as CO_2 (Bradford et al., 2013). These new ideas in soil carbon stabilization, when coupled with emerging concepts in soil food webs, highlight the critical role that root exudates likely play as precursors to SOM stabilization. They also highlight the "microbial paradox" that the new models of SOM stabilization raise— that microbes serve both as decomposers and as the *formers* of SOM. Notions that labile root exudes cause a priming effect whereby microbial activity is increased, with consequent increases in the decomposition of SOM and reductions in stocks, is inconsistent with the new idea that higher microbial activity should lead to greater SOM formation. It may simply be that the focus on carbon mineralization in the priming effects field, as opposed to the response of the total SOM stock, led to the development of priming effects theory that has no use for understanding SOM stock responses (Bradford et al., 2008).

What does seem certain under the emerging SOM concepts, is that the carbon atoms in exudates such as sugars and amino acids that microbes grow on most efficiently, fuel production of saprotrophic bacteria and fungi to a much greater extent than structural plant-carbon inputs such as lignin. Notably, lignins are an inefficient substrate for growth, with arguments that their irregular chemical structure and the need for extracellular degradation, means that there has to be substantive investment in lignin degradation with potentially little net energy gain (Cotrufo et al., 2013; Lehmann and Kleber, 2015). However, microbes also appear to use some low molecular weight compounds, such as organic acids (e.g., citric and oxalic acid), extremely inefficiently and likely cannot grow on them as sole carbon sources (Frey et al., 2013). These organic acids, instead, might be critical in the acquisition of other macronutrients, such as phosphorus (Keiluweit et al., 2015), making them indirectly important in microbial biomass production and hence soil carbon stabilization.

Emerging conceptualizations of SOM stabilization and soil food webs therefore both emphasize the likely fundamental importance of labile root exudates to the flows and stabilization of carbon in soils. The new food web conceptualizations tie these labile inputs to the faster cycling bacterial and now also the slower-cycling fungal pools, helping to resolve why observations of fungal dominance appear to go hand-in-hand with building stable SOM stocks. One issue remaining, however, is that each trophic transfer incurs a net loss of energy from the system. If exudates flow first through fungal saprotrophs and then bacteria—per the sapro-rhizosphere concept (Ballhausen and de Boer, 2016)—then less carbon may be retained in soils when labile exudates flow through fungal saprotrophs, and not directly to rhizosphere bacteria.

Integrating emerging soil food web concepts with those for SOM cycling, therefore, appears to demand that we understand how soil food web structure and resource supply affects the efficiency with which substrate carbon is used and then passed through trophic networks or to stable SOM pools. As highlighted by Rousk (2016), recent SOM models that explicitly represent microbial processes tend to assign a key mechanistic role to the size of the microbial biomass (Allison et al., 2010; Hagerty et al., 2014; Sulman et al., 2014; Tang and Riley, 2014; Wieder et al., 2014). Whereas they also reveal microbial growth efficiencies as a key control on SOM stocks and turnover rates, Rousk (2016) instead suggests that the models be based on organism growth rates (and hence turnover), given that growth relates most strongly to the biogeochemical process rates food web models are often used to predict (Hagerty et al., 2014). Doing so necessitates that new SOM cycling concepts are placed in the framework of the population biology and ecological interactions of soil microbes (Buchkowski et al., 2017). The role of multichannel omnivory as a top-down control on bacterial and fungal turnover may thus be a key, but as yet unappreciated, regulator of SOM stabilization rates. However, despite many decades of soil food web models, soil animals remain absent from both traditional and the rapidly proliferating microbial-explicit SOM models (Grandy et al., 2016; Soong and Nielsen, 2016). Representation of faunal processes in conceptual SOM models will likely lead to rapid advances in our understanding of how animals shape soil processes, mirroring similar gains achieved through the representation of microbial physiology and interactions. Rapid advances might also be made through representation of how nontrophic interactions, chemical communication, and mycorrhizae shape interactions in belowground food webs and the efficiencies of resource flows (Averill et al., 2014; DeAngelis, 2016; Hawlena et al., 2012; Phillips et al., 2013).

Yet, translating the growing number of empirical observations and associated conceptual advances into SOM model structures will demand substantive resources and may improve confidence in management recommendations, but not necessarily the uncertainty in projected outcomes (Bradford et al., 2016).

4.6   POTENTIAL GAINS AND PITFALLS OF IMPLEMENTING NEW THINKING ABOUT SOIL ORGANIC MATTER CYCLING FOR AGRICULTURE

As agricultural management increasingly emphasizes maintaining SOM, soil biodiversity and biological activity, there has been a drive to champion managements that return organic materials to the soil. The idea of these managements is that the nutrient needs of plants can, in part, be met by these inputs because soil organisms liberate nutrients from the organic material through decomposition, transforming them into plant-available inorganic forms. Hence, high soil biological activity is required to facilitate effective nutrient cycling, with managements emphasizing the return of plant residues varying from high- to low-quality for decomposition, and hence high- to low-nutrient release rates. Such motivations have helped form the idea of soil health, because only living organisms can be described from a health perspective. Tied up with this concept is the idea—borne from traditional soil food web models—that belowground communities which are fungal-based will be most diverse and cycle nutrients most tightly. Research looking at the structure and successional dynamics of soil food webs under different managements (de Vries et al., 2012)—and prior agriculture abandoned to return to grassland and forest (Morriën et al., 2017)—appears to support this viewpoint. There is also evidence that nitrogen addition—a widespread agricultural strategy—can significantly modify below-ground food webs (Box 4.1). Emerging concepts in soil food webs would then emphasize the value of maintaining a large and living root biomass in the soil, to provide the labile compounds that most efficiently fuel microbial productivity and hence the soil animals that rely on this basal resource. However, the traditional view of soil food webs (which instead would emphasize the return of low quality, structurally complex, aboveground plant residues to build SOM) is still communicated as part of the scientific rationale for promoting soil biology by agencies such as the US Department of Agriculture. Nevertheless, these same agencies now stress in their practice recommendations, the importance of the living root instead of aboveground inputs for maintaining active and diverse soil communities, and similarly to promote this soil biology the need to minimize soil disturbance, to maintain plant diversity and to maintain a residue layer to moderate soil microclimate. Therefore, even if their scientific concepts are out-of-date, their advice for practice seems commensurate with emerging concepts for SOM stabilization and food webs.

Efforts to draw on soil ecology principles to sustainably intensify agriculture have largely been founded in older conceptualizations of how soil biology links to SOM cycling and agronomic outcomes. For instance, Palm et al. (2001) created an organic resources database that describes the properties of organic inputs as a function of their lability. Their argument is that crop production requires a balance of fast-cycling organic matter to provide nutrients to crops in the short-term and slow-cycling organic matter to build soil structure and water holding capacity as well as to provide a slower, but steady supply of nutrients. This notion of organic matter

BOX 4.1 ROLE OF BIOTIC INTERACTION IN ORGANIC MATTER DECOMPOSITION UNDER NITROGEN ADDITIONS

In agricultural systems, nitrogen addition can dramatically disrupt biotic communities that regulate ecosystem processes. Most research on the impacts of nitrogen addition in agricultural settings has focused exclusively on microorganisms, because of their key direct role in nutrient cycling processes (Wood et al., 2015a,b). As highlighted in this chapter, other biotic interactions are essential to regulating microbial activity and their control of ecosystem processes. Crowther et al. (2015) studied how manipulation of soil food web structure in the context of nitrogen addition—as well as warming—could structure microbial functional activity and process rates. The authors' aim was to test whether top-down control of cord-forming fungi by isopod grazers can mediate the direct effects of external drivers, such as nitrogen addition. They hypothesized that nitrogen addition would stimulate fungal biomass and functional activity, but that the presence of fungal grazing would counteract these increases. They found that fungal biomass was greatest under nitrogen deposition where isopod grazers were excluded. Similarly, decomposition rates and enzyme potential were elevated where fungal cords were present without isopods (Fig. 4.4). Nitrogen deposition—with and without warming—stimulated enzymatic activity as well (Fig. 4.4). Because microbial byproducts and microbial biomass are now known to be precursors to stable SOM (Lehmann and Kleber, 2015), these microbial-animal interaction responses to exogenous nitrogen addition are likely to play an important regulatory role in the buildup of SOM in the context of sustainable agriculture.

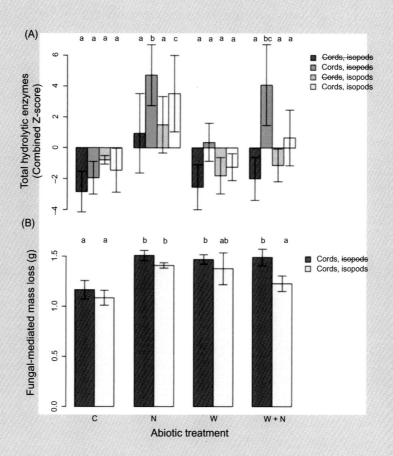

FIGURE 4.4

Response of enzyme potential and organic matter decomposition to removal of fungal cords and isopods under control (C), elevated nitrogen (N), warming (W), and warming + nitrogen addition (W + N).

From Crowther, T.W., Thomas, S.M., Maynard, D.S., Baldrian, P., Covey, K., Frey, S.D., et al., 2015. Biotic interactions mediate soil microbial feedbacks to climate change. Proc. Natl. Acad. Sci. 112, 7033–7038.

cycling is rooted in an older notion of chemical recalcitrance as governing stabilization. Newer ideas that organic matter reflects a continuous gradient of both potential energy and likelihood to physically interact with mineral surfaces are not represented in older thinking about how to manage SOM for agroecosystems.

What might a newer conceptualization of soil management for sustainable agriculture look like? Under new ideas of organic matter cycling, an updated framework would likely still need to promote different types of organic inputs because early-stage decomposition is still governed by plant nutrient chemistry (Cleveland et al., 2013; Cornwell et al., 2008; Hector et al., 2003; Papa et al., 2014). This means that labile recent litter inputs will decompose faster and provide more nutrients to plants, as conceptualized by Palm et al. (2001). However, instead of promoting recalcitrant inputs to build long-term stable SOM, a new framework would emphasize organic inputs as a function of their likelihood to stabilize. Because stable compounds are known to be low molecular weight in origin, one might propose increasing inputs of low molecular weight compounds to soil as a means to build up SOM. These compounds are often produced by plant roots in the form of carbohydrates, sugars, organic acids, and other compounds. This could lead to the following principle of organic matter management for ecologically intensive agriculture: *Maximize the input of nutrient-rich aboveground inputs for short-term nutrient needs; for longer-term organic matter, maximize root exudation—particularly through fine roots that actively exude and cover a large area of soil—both spatially and temporally.*

Because plants regularly produce and exude organic inputs to soil through their root system, the amount of inputs available for carbon stabilization should be, in part, a function of the amount of time an arable soil has plant cover. Thus, implementing the earlier principle could be accomplished by increasing the amount of time a soil is under plant cover. This thinking is reflected in several specific methods of ecologically intensive agriculture, such as cover cropping (Poeplau and Don, 2015), perennial agriculture (Crews et al., 2016), permaculture, regenerative agriculture, agroforestry (Isaac et al., 2005; Lorenz and Lal, 2014; Wood et al., 2016), and holistic grazing.

Yet, many of the proposed techniques are controversial (Briske et al., 2008, 2013, 2014a,b; Teague, 2014) and their effectiveness in building SOM is not necessarily based in empirical evidence. In fact, despite the principle proposed earlier, there may be counteracting ecological forces that could lead to lower agronomic productivity as a function of new understanding of SOM cycling. For instance, there is evidence that the production of certain root exudates—particularly, simple organic acids—can liberate phosphorus to the biologically active nutrient pool (Keiluweit et al., 2015). Yet these simple acids are inefficiently used by microorganisms for growth, meaning that these root-exudate carbon sources could drive reductions in soil microbial biomass (Buchkowski et al., 2017). As a result, SOM formation might decrease with high inputs of these organic acids, eventually decreasing the fertility of the soil. Thus, simply promoting root exudation may be an inefficient way to build up SOM, requiring instead a focus on the classes of exudates that are promoted in order to improve agronomic outcomes.

4.7 CONCLUSIONS

Approaches to soil management in agriculture that draw on a paradigm of weak sustainability largely rely on advocacy for efficient use of external nutrient sources (Hoben et al., 2011; Vitousek et al., 2009)

BOX 4.2 TAKE HOME MESSAGES

1. Weak sustainability is when natural capital can be substituted for other forms of capital; strong sustainability is when certain forms of natural capital cannot be substituted for by human-produced capital.
2. Soil ecological processes—such as food web interactions—are critical to stabilizing and decomposing soil organic matter.
3. Because these ecological interactions are complex, we advocate for a strong sustainability approach to soil management for agriculture that emphasizes promoting these ecological interactions to build ecologically sustainable agriculture.
4. Emerging evidence suggests that root inputs are a primary source of energy for both bacteria and fungi, which then process inputs into forms that can be stabilized into long-term stabilized soil organic matter.
5. Agricultural management approaches that aim to build up soil organic matter over long periods of time should focus on promoting root inputs to soil food webs.

---

or biotic inocula and products to increase nutrient efficiency (Baas et al., 2016). Targeted approaches that increase nutrient use efficiency have an important place in sustainable intensification of agriculture. Yet, even in the presence of sufficient macronutrients, crop production and soil fertility may be deficient (Tittonell et al., 2008), emphasizing the need to understand broader ecological dynamics in soil.

In this chapter, we advocate for an approach to soil management for agriculture rooted in strong sustainability (Box 4.2). We claim that the soil processes responsible for generating fertility and building up key nutrient stocks for other target outcomes—such as soil carbon for climate objectives—depend on complex ecological interactions that are difficult to substitute for human or physical capital. We draw on emerging ideas in soil food web science to demonstrate that biotic interactions linking macrofauna through to microbiota can play an essential role in governing stocks and cycling of SOM. These new concepts lend support to agricultural management approaches that promote continued input of labile organic substrates as a means to build SOM. We advocate for the following principle for soil management in agriculture: *Maximize the input of nutrient-rich organic inputs for short-term nutrient needs; for longer-term organic matter, maximize active fine roots, both spatially and temporally.* This principle, rooted in a new understanding of soil food webs and SOM dynamics, aims to promote both soil fertility and soil carbon sequestration. Though some work has claimed that these two outcomes are necessarily related to total organic matter (Lal, 2004), more recent work has demonstrated that these two goals may be at odds (Cates and Ruark, 2017; Wood et al., 2016). Our chapter—and principle—are aimed at establishing win-wins in soil management based on cutting-edge concepts in soil fertility management.

REFERENCES

Allison, S.D., Wallenstein, M.D., Bradford, M.A., 2010. Soil-carbon response to warming dependent on microbial physiology. Nat. Geosci. 3, 336–340.

Arrow, K.J., Dasgupta, P., Goulder, L.H., Mumford, K.J., Oleson, K., 2012. Sustainability and the measurement of wealth. Environ. Dev. Econ. 17, 317–353.

Averill, C., Turner, B.L., Finzi, A.C., 2014. Mycorrhiza-mediated competition between plants and decomposers drives soil carbon storage. Nature 505, 543–545.

Baas, P., Bell, C., Mancini, L.M., Lee, M.N., Conant, R.T., Wallenstein, M.D., 2016. Phosphorus mobilizing consortium Mammoth P™ enhances plant growth. PeerJ 4, e2121.

Ballhausen, M.-B., de Boer, W., 2016. The sapro-rhizosphere: carbon flow from saprotrophic fungi into fungus-feeding bacteria. Soil Biol. Biochem. 102, 14–17.

Banwart, S., Black, H., Cai, Z., Gicheru, P., Joosten, H., Victoria, R., et al., 2014. Benefits of soil carbon: report on the outcomes of an international scientific committee on problems of the environment rapid assessment workshop. Carbon Manage. 5, 185–192.

Bardgett, R.D., McAlister, E., 1999. The measurement of soil fungal:bacterial biomass ratios as an indicator of ecosystem self-regulation in temperate meadow grasslands. Biol. Fertil. Soils 29, 282–290.

Bardgett, R.D., Mommer, L., De Vries, F.T., 2014. Going underground: root traits as drivers of ecosystem processes. Trends Ecol. Evol. 29, 692–699.

Bradford, M.A., 2016. Re-visioning soil food webs. Soil Biol. Biochem. 102, 1–3.

Bradford, M.A., Fierer, N., Reynolds, J.F., 2008. Soil carbon stocks in experimental mesocosms are dependent on the rate of labile carbon, nitrogen and phosphorus inputs to soils. Funct. Ecol. 22, 964–974.

Bradford, M.A., Keiser, A.D., Davies, C.A., Mersmann, C.A., Strickland, M.S., 2013. Empirical evidence that soil carbon formation from plant inputs is positively related to microbial growth. Biogeochemistry 113, 271–281.

Bradford, M.A., Wieder, W.R., Bonan, G.B., Fierer, N., Raymond, P.A., Crowther, T.W., 2016. Managing uncertainty in soil carbon feedbacks to climate change. Nat. Clim. Change 6, 751–758.

Briske, D.D., Derner, J.D., Brown, J.R., Fuhlendorf, S.D., Teague, W.R., Havstad, K.M., et al., 2008. Rotational grazing on rangelands: reconciliation of perception and experimental evidence. Rangel. Ecol. Manage. 61, 3–17.

Briske, D.D., Bestelmeyer, B.T., Brown, J.R., Fuhlendorf, S.D., Briske, B.D.D., Polley, H.W., 2013. The savory method can not green deserts or reverse climate change. Rangelands 35, 72–74.

Briske, D.D., Ash, A.J., Derner, J.D., Huntsinger, L., 2014a. Commentary: A critical assessment of the policy endorsement for holistic management. Agric. Syst. 125, 50–53.

Briske, D.D., Bestelmeyer, B.T., Brown, J.R., Briske, B.D.D., 2014b. Savory's unsubstantiated claims should not be confused with multipaddock grazing. Rangelands 36, 39–42.

Buchkowski, R.W., Bradford, M.A., Grandy, A.S., Schmitz, O.J., Wieder, W.R., 2017. Applying population and community ecology theory to advance understanding of belowground biogeochemistry. Ecol. Lett. 20, 231–245.

Burney, J.A., Davis, S.J., Lobell, D.B., 2010. Greenhouse gas mitigation by agricultural intensification. PNAS 107, 12052–12057.

Caron, P., Biénabe, E., Hainzelin, E., 2014. Making transition towards ecological intensification of agriculture a reality: the gaps in and the role of scientific knowledge. Curr. Opin. Environ. Sustain. 8, 44–52.

Cates, A.M., Ruark, M.D., 2017. Soil aggregate and particulate C and N under corn rotations: responses to management and correlations with yield. Plant Soil 415, 521–533.

Cheeke, T.E., Colemand, D.C., Wall, D.H., 2012. Microbial Ecology in Sustainable Agroecosystems. CRC Press, Boca Raton, FL.

Cleveland, C.C., Reed, S.C., Keller, A.B., Nemergut, D.R., O'Neill, S.P., Ostertag, R., et al., 2013. Litter quality versus soil microbial community controls over decomposition: a quantitative analysis. Oecologia 174, 283–294.

Coleman, K., Jenkinson, D.S., 1996. RothC-26.3 – a model for the turnover of carbon in soil. Evaluation of Soil Organic Matter Models. Springer Berlin Heidelberg, Berlin, Heidelberg, pp. 237–246.

Cordell, D., Drangert, J.-O., White, S., 2009. The story of phosphorus: global food security and food for thought. Global Environ. Change 19, 292–305.

Cornwell, W.K., Cornelissen, J.H.C., Amatangelo, K., Dorrepaal, E., Eviner, V.T., Godoy, O., et al., 2008. Plant species traits are the predominant control on litter decomposition rates within biomes worldwide. Ecol. Lett. 11, 1065–1071.

Cotrufo, M.F., Wallenstein, M.D., Boot, C.M., Denef, K., Paul, E., 2013. The Microbial Efficiency-Matrix Stabilization (MEMS) framework integrates plant litter decomposition with soil organic matter stabilization: do labile plant inputs form stable soil organic matter? Global Change Biol. 19, 988−995.

Crews, T.E., Blesh, J., Culman, S.W., Hayes, R.C., Jensen, E.S., Mack, M.C., et al., 2016. Going where no grains have gone before: from early to mid-succession. Agric. Ecosyst. Environ. 223, 223−238.

Crowther, T.W., Thomas, S.M., Maynard, D.S., Baldrian, P., Covey, K., Frey, S.D., et al., 2015. Biotic interactions mediate soil microbial feedbacks to climate change. Proc. Natl. Acad. Sci. 112, 7033−7038.

Culman, S.W., Snapp, S.S., Green, J.M., Gentry, L.E., 2013. Short- and long-term labile soil carbon and nitrogen dynamics reflect management and predict corn agronomic performance. Agron. J. 105, 493−502.

Daly, H., Jacobs, M., Skolimowski, H., 1995. Discussion of Beckerman's critique of sustainable development. Environ. Values 4, 49−70.

DeAngelis, K.M., 2016. Chemical communication connects soil food webs. Soil Biol. Biochem. 102, 48−51.

De Ruiter, P.C., Van Veen, J.A., Moore, J.C., Brussaard, L., Hunt, H.W., 1993. Calculation of nitrogen mineralization in soil food webs. Plant Soil 157, 263−273.

De Vries, F.T., Bardgett, R.D., 2012. Plant-microbial linkages and ecosystem nitrogen retention: lessons for sustainable agriculture. Front. Ecol. Environ. 10, 425−432.

de Vries, F.T., Caruso, T., 2016. Eating from the same plate? Revisiting the role of labile carbon inputs in the soil food web. Soil Biol. Biochem. 102, 4−9.

de Vries, F.T., Manning, P., Tallowin, J.R.B., Mortimer, S.R., Pilgrim, E.S., Harrison, K.A., et al., 2012. Abiotic drivers and plant traits explain landscape-scale patterns in soil microbial communities. Ecol. Lett. 15, 1230−1239.

Dobson, A., 2009. Food-web structure and ecosystem services: insights from the Serengeti. Philos. Trans. R. Soc. Lond. B Biol. Sci. 364, 1665−1682.

Foley, J.A., Ramankutty, N., Brauman, K.A., Cassidy, E.S., Gerber, J.S., Johnston, M., et al., 2011. Solutions for a cultivated planet. Nature 478, 337−342.

Frey, S.D., Lee, J., Melillo, J.M., Six, J., 2013. The temperature response of soil microbial efficiency and its feedback to climate. Nat. Clim. Change 3, 1−4.

Garnett, T., Appleby, M.C., Balmford, A., Bateman, I.J., Benton, T.G., Bloomer, P., et al., 2013. Sustainable intensification in agriculture: premises and policies. Science 341, 33−34.

Geisen, S., 2016. The bacterial-fungal energy channel concept challenged by enormous functional versatility of soil protists. Soil Biol. Biochem. 102, 22−25.

Gilbert, K.J., Fahey, T.J., Maerz, J.C., Sherman, R.E., Bohlen, P., Dombroskie, J.J., et al., 2014. Exploring carbon flow through the root channel in a temperate forest soil food web. Soil Biol. Biochem. 76, 45−52.

Godfray, H.C.J., Beddington, J.R., Crute, I.R., Haddad, L., Lawrence, D., Muir, J.F., et al., 2010. Food security: the challenge of feeding 9 billion people. Science 327, 812−818.

Grandy, A.S., Neff, J.C., 2008. Molecular C dynamics downstream: the biochemical decomposition sequence and its impact on soil organic matter structure and function. Sci. Total Environ. 404, 297−307.

Grandy, A.S., Wieder, W.R., Wickings, K., Kyker-Snowman, E., 2016. Beyond microbes: are fauna the next frontier in soil biogeochemical models? Soil Biol. Biochem. 102, 40−44.

de Groot, R.S., Wilson, M.A., Boumans, R.M., 2002. A typology for the classification, description and valuation of ecosystem functions, goods and services. Ecol. Econ. 41, 393−408.

Hagerty, S.B., van Groenigen, K.J., Allison, S.D., Hungate, B.A., Schwartz, E., Koch, G.W., et al., 2014. Accelerated microbial turnover but constant growth efficiency with warming in soil. Nat. Clim. Change 4, 903−906.

Hartwick, J.M., 1977. Intergenerational equity and the investing of rents from exhaustible resources. Am. Econ. Rev. 67, 972−974.

Hawlena, D., Strickland, M.S., Bradford, M.A., Schmitz, O.J., 2012. Fear of predation slows plant-litter decomposition. Science 336, 1434−1438.

Haynes, R.J., 2005. Labile organic matter fractions as central components of the quality of agricultural soils: an overview. Adv. Agron. 85, 221−268.

Hector, A., Beale, A., Minns, A., Otway, S., Lawton, J., 2003. Consequences of the reduction of plant diversity for litter decomposition: effects through litter quality and microenvironment. Oikos 90, 357−371.

van Hees, P.A.W., Jones, D.L., Finlay, R., Godbold, D.L., Lundström, U.S., 2005. The carbon we do not see— the impact of low molecular weight compounds on carbon dynamics and respiration in forest soils: a review. Soil Biol. Biochem. 37, 1−13.

van der Heijden, M.G.A., Bardgett, R.D., van Straalen, N.M., 2008. The unseen majority: soil microbes as drivers of plant diversity and productivity in terrestrial ecosystems. Ecol. Lett. 11, 296−310.

Hoben, J.P., Gehl, R.J., Millar, N., Grace, P.R., Robertson, G.P., 2011. Nonlinear nitrous oxide (N_2O) response to nitrogen fertilizer in on-farm corn crops of the US Midwest. Global Change Biol. 17, 1140−1152.

Hunt, H.W., Coleman, D.C., Ingham, E.R., Ingham, R.E., Elliott, E.T., Moore, J.C., et al., 1987. The detrital food web in a shortgrass prairie. Biol. Fertil. Soils 3, 57−68.

Isaac, M.E., Gordon, A.M., Thevathasan, N., Oppong, S.K., Quashie-Sam, J., 2005. Temporal changes in soil carbon and nitrogen in west African multistrata agroforestry systems: a chronosequence of pools and fluxes. Agroforestry Syst. 65, 23−31.

Jacobs, M., 1995. Sustainable development, capital substitution and economic humility: a response to Beckerman. Environ. Values 4, 57−68.

Keiluweit, M., Bougoure, J.J., Nico, P.S., Pett-Ridge, J., Weber, P.K., Kleber, M., 2015. Mineral protection of soil carbon counteracted by root exudates. Nat. Clim. Change 5, 588−595.

Lal, R., 2004. Soil carbon sequestration impacts on global climate change and food security. Science 304, 1623−1627.

Lehmann, J., Kleber, M., 2015. The contentious nature of soil organic matter. Nature 528, 60−68.

Loos, J., Abson, D.J., Chappell, M.J., Hanspach, J., Mikulcak, F., Tichit, M., et al., 2014. Putting meaning back into "sustainable intensification". Front. Ecol. Environ. 12, 356−361.

Lorenz, K., Lal, R., 2014. Soil organic carbon sequestration in agroforestry systems. A review. Agron. Sustain. Dev. 34, 443−454.

Miltner, A., Bombach, P., Schmidt-Brücken, B., Kästner, M., 2011. SOM genesis: microbial biomass as a significant source. Biogeochemistry 111, 41−55.

Moore, J.C., William Hunt, H., 1988. Resource compartmentation and the stability of real ecosystems. Nature 333, 261−263.

Moore, J.C., Berlow, E.L., Coleman, D.C., Ruiter, P.C., Dong, Q., Hastings, A., et al., 2004. Detritus, trophic dynamics and biodiversity. Ecol. Lett. 7, 584−600.

Morriën, E., Hannula, S.E., Snoek, L.B., Helmsing, N.R., Zweers, H., de Hollander, M., et al., 2017. Soil networks become more connected and take up more carbon as nature restoration progresses. Nat. Commun. 8, 14349.

Neutel, A.-M., 2002. Stability in real food webs: weak links in long loops. Science 296, 1120−1123.

Oldfield, E.E., Wood, S.A., Palm, C.A., Bradford, M.A., 2015. How much SOM is needed for sustainable agriculture? Front. Ecol. Environ. 13, 527.

Özkaynak, B., Devine, P., Rigby, D., 2004. Operationalising strong sustainability: definitions, methodologies and outcomes. Environ. Values 13, 279−303.

Palm, C.A., Gachengo, C.N., Delve, R.J., Cadisch, G., Giller, K.E., 2001. Organic inputs for soil fertility management in tropical agroecosystems: application of an organic resource database. Agric. Ecosyst. Environ. 83, 27−42.

Palm, C., Sanchez, P., Ahamed, S., Awiti, A., 2007. Soils: a contemporary perspective. Annu. Rev. Environ. Resour. 32, 99−129.

Papa, G., Scaglia, B., Schievano, A., Adani, F., 2014. Nanoscale structure of organic matter could explain litter decomposition. Biogeochemistry 117, 313–324.

Parton, W.J., Schimel, D.S., Cole, C.V., Ojima, D.S., 1987. Analysis of factors controlling soil organic matter levels in Great Plains Grasslands. Soil Sci. Soc. Am. J. 51, 1173.

Phillips, R.P., Brzostek, E., Midgley, M.G., 2013. The mycorrhizal-associated nutrient economy: a new framework for predicting carbon-nutrient couplings in temperate forests. New Phytol. 199, 41–51.

Pieralli, S., 2017. Introducing a new non-monotonic economic measure of soil quality. Soil Tillage Res. 169, 92–98.

Poeplau, C., Don, A., 2015. Carbon sequestration in agricultural soils via cultivation of cover crops – a meta-analysis. Agric. Ecosyst. Environ. 200, 33–41.

Pollierer, M.M., Langel, R., Körner, C., Maraun, M., Scheu, S., 2007. The underestimated importance of belowground carbon input for forest soil animal food webs. Ecol. Lett. 10, 729–736.

Pollierer, M.M., Dyckmans, J., Scheu, S., Haubert, D., 2012. Carbon flux through fungi and bacteria into the forest soil animal food web as indicated by compound-specific 13C fatty acid analysis. Funct. Ecol. 26, 978–990.

Raudsepp-Hearne, C., Peterson, G.D., Tengö, M., Bennett, E.M., Holland, T., Benessaiah, K., et al., 2010. Untangling the environmentalist's paradox: why is human well-being increasing as ecosystem services degrade? Bioscience 60, 576–589.

Rockstrom, J., Steffen, W., Noone, K., Persson, A., Chapin, F.S., Lambin, E.F., et al., 2009. A safe operating space for humanity. Nature 461, 472–475.

Rousk, J., 2016. Biomass or growth? How to measure soil food webs to understand structure and function. Soil Biol. Biochem. 102, 45–47.

Soong, J.L., Nielsen, U.N., 2016. The role of microarthropods in emerging models of soil organic matter. Soil Biol. Biochem. 1–3.

Strickland, M.S., Rousk, J., 2010. Considering fungal:bacterial dominance in soils – methods, controls, and ecosystem implications. Soil Biol. Biochem. 42, 1385–1395.

Strickland, M.S., Hawlena, D., Reese, A., Bradford, M.A., Schmitz, O.J., 2013. Trophic cascade alters ecosystem carbon exchange. Proc. Natl. Acad. Sci. U.S.A. 11–14.

Sulman, B.N., Phillips, R.P., Oishi, A.C., Shevliakova, E., Pacala, S.W., 2014. Microbe-driven turnover offsets mineral-mediated storage of soil carbon under elevated CO_2. Nat. Clim. Change 4, 2–5.

Tang, J., Riley, W.J., 2014. Weaker soil carbon–climate feedbacks resulting from microbial and abiotic interactions. Nat. Clim. Change 5, 56–60.

Teague, B.R., 2014. Deficiencies in the Briske et al. Rebuttal of the Savory Method. Rangelands 36, 37–38.

Tilman, D., Balzer, C., Hill, J., Befort, B.L., 2011. Global food demand and the sustainable intensification of agriculture. Proc. Natl. Acad. Sci. U.S.A. 108, 20260–20264.

Tittonell, P., Giller, K.E., 2013. When yield gaps are poverty traps: the paradigm of ecological intensification in African smallholder agriculture. Field Crop Res. 143, 76–90.

Tittonell, P., Vanlauwe, B., Corbeels, M., Giller, K.E., 2008. Yield gaps, nutrient use efficiencies and response to fertilisers by maize across heterogeneous smallholder farms of western Kenya. Plant Soil 313, 19–37.

Todd-Brown, K.E.O., Randerson, J.T., Post, W.M., Hoffman, F.M., Tarnocai, C., Schuur, E.A.G., et al., 2013. Causes of variation in soil carbon simulations from CMIP5 Earth system models and comparison with observations. Biogeosciences 10, 1717–1736.

Trainor, S.F., 2006. Realms of value: conflicting natural resource values and incommensurability. Environ. Values 15, 3–29.

United Nations, 1987. Our Common Future – Brundtland Report.

Vitousek, P., Naylor, R., Crews, T., David, M., Drinkwater, L., Holland, E., et al., 2009. Nutrient imbalances in agricultural development. Science 324, 1519–1520.

Wieder, W.R., Grandy, A.S., Kallenbach, C.M., Bonan, G.B., 2014. Integrating microbial physiology and physiochemical principles in soils with the MIcrobial-MIneral Carbon Stabilization (MIMICS) model. Biogeosci. Discuss. 11, 1147−1185.

Wolkovich, E.M., 2016. Reticulated channels in soil food webs. Soil Biol. Biochem. 102, 18−21.

Wood, S.A., Almaraz, M., Bradford, M.A., McGuire, K.L., Naeem, S., Neill, C., et al., 2015a. Farm management, not soil microbial diversity, controls nutrient loss from smallholder tropical agriculture. Front. Microbiol. 6, 1−10.

Wood, S.A., Bradford, M.A., Gilbert, J.A., McGuire, K.L., Palm, C.A., Tully, K.L., et al., 2015b. Agricultural intensification and the functional capacity of soil microbes on smallholder African farms. J. Appl. Ecol. 52, 744−752.

Wood, S.A., Sokol, N., Bell, C.W., Bradford, M.A., Naeem, S., Wallenstein, M.D., et al., 2016. Opposing effects of different soil organic matter fractions on crop yields. Ecol. Appl. 0, 1−14.

CLIMATE, GEOGRAPHY, AND SOIL ABIOTIC PROPERTIES AS MODULATORS OF SOIL CARBON STORAGE

Manuel Delgado-Baquerizo[1], Senani B. Karunaratne[2,3], Pankaj Trivedi[4] and Brajesh K. Singh[2]

[1]*University of Colorado, Boulder, CO, United States* [2]*Western Sydney University, Penrith, NSW, Australia* [3]*Department of the Environment and Energy, Parkes, ACT, Australia* [4]*Colorado State University, Fort Collins, CO, United States*

5.1 INTRODUCTION

Soils store three times more carbon (C) than either the atmosphere or terrestrial vegetation (Schmidt et al., 2011)—the total global soil organic carbon (SOC) stocks are estimated as 2053 Gt (Bernoux and Chevallier, 2014). As previous chapters illustrate, carbon cycling is one of the most important ecosystem processes playing a critical role in supporting key ecosystem services such as climate regulation (e.g., CO_2 and CH_4 fluxes), soil fertility, and food and fiber production (Tiessen et al., 1994). SOC is a proxy for organic material that controls the long-term capital stock and availability of nutrients in terrestrial ecosystems. Soil C storage is, therefore, critical for multiple ecosystem services. Because of this, predictions of soil C balance in terrestrial ecosystems have become a global priority during the past decades with the development of Earth System Models (ESMs) as the primary tool for predicting climate impacts on soil C from local to global scale.

SOC storage is highly vulnerable to on-going global climate changes and land use intensification such as deforestation, use of mineral fertilizers, and grazing by livestock (Schlesinger, 1996; Nazaries et al., 2015; Eldridge and Delgado-Baquerizo, 2016) (see Chapter 8: Impact of Global Changes on Soil C Storage-Possible Mechanisms and Modeling Approaches). Understanding the direct and indirect effects of climate and human impacts on soil C storage is therefore critical to accurately assess the future response of ecosystem functioning to global change. To date, the magnitude of C shifts in response to global environmental change (climate change and land management) is uncertain, and to quantify C losses in response to global change, we strongly rely on our capacity to predict changes in soil C using ESMs. Therefore, identifying the major predictors of soil C and improving our understanding on the drivers of soil C storage are a fundamental requirement for formulating appropriate monitoring, management, and conservation policies as well as for predicting how the soil C balance will respond under changing environments.

Soil C is highly variable across the landscape. In fact, even within a given paddock there is a high variation of soil C, where the spatial pattern of soil C is driven largely by abiotic and biotic

Soil Carbon Storage. DOI: https://doi.org/10.1016/B978-0-12-812766-7.00005-6

factors at the microsite scale (Schlesinger et al., 1990; Schlesinger, 1996). Because of this, predictions of soil C are not straight-forward and are largely dependent on the scale at which soil C needs to be predicted. Scientists have recognized the importance of scaling issues to predict ecological patterns for more than five decades as explained by Levin (1992). In this paper, the author explained that the prediction of ecological processes (e.g., soil C sequestration) required the interfacing of phenomena that occur at different spatial scales. The reason for this is that ecosystems naturally show characteristic variability on a range of spatial ecological scales. In the case of soil C stocks, predictions can be made from very dissimilar spatial scales ranging from soil aggregates to the global scale. Local predictions of soil C can be of interest from the point of view of identifying hotspots of nutrient availability and C storage within a particular ecosystem, while prediction of soil C at the global scale is essential for the proper understanding of climate regulation and conservation-related proposals. In this chapter, we briefly summarize the current knowledge on the major environmental drivers of soil C across different spatial scales and provide support with new analysis of data from global to micro scales. Our main aim is to use existing data to identify differences in dominant drivers and mechanisms of soil C storage from global to soil aggregate scales and propose new concepts where appropriate. To do this, we use both de novo analyses using a variety of statistical modeling (Box 5.1) and report examples from the literature.

BOX 5.1 STATISTICAL AND SPATIAL MODELING

- *Structural equation modeling (SEM) (Mechanistic modeling).* SEM can particularly achieve a mechanistic understanding on the network of linkages among key drivers of soil C in natural ecosystems. This technique allows the partitioning of causal influences among multiple variables, and separation of the direct and indirect effects of model predictors (Grace, 2006). SEM is specially recommended to identify the main predictors of environmental response variables in correlative studies. SEM can also be used to improve our mechanistic understanding on the drivers of soil carbon from experimental approaches. SEM allow us to deal with both parametric and nonparametric data.
- *Multimodel inference (Predictive modeling).* Unlike the classical approach to fitting models, based on traditional hypothesis testing, multimodel inference uses information theory and ordinary least squares regression to assess the probability that a given model is the most appropriate description of the observed data. This type of modeling is specially recommended to identify key predictors which explain a unique portion of the variation in a response variable, not explained by other predictors. Models are ranked according to the second-order Akaike information criterion (AICc). Differences <2 in AICc between alternative models indicate that they are approximately equivalent in explanatory power (Burnham and Anderson, 2002). Models with Differences >2 and >7 in AICc between alternative models indicate that two models are substantially different or totally different models, respectively. Multimodel inference is recommended when dealing with observational data collected over large spatial scales and environmental gradients, however, it can be also used in experimental approaches. A nonparametric alternative comes from distance-based multimodel inference.
- *Spatial modeling and mapping of soil C*: Soil carbon is highly variable across the landscape. Therefore, in spatial modeling and mapping of soil carbon, other than the spatial component, incorporation of the deterministic component is essential. This deterministic component is included in terms of environmental covariates. For this, quantitative models are developed based on fundamentals of pedology as explained by Jenny (1941). Hybrid mapping techniques based on *scorpan* models (McBratney et al., 2003) are widely used for digital mapping of soil C:

$$S = f(s, c, o, r, p, a, n) + e$$

where: S is a soil property of interest (e.g., soil carbon), f is a prediction method or function which incorporates covariates related to soil (s), climate (c), organisms (o), relief (r), parent material (p), time (a), space (n), and e is the spatially correlated errors. The function (f) can be statistical and mathematical approaches which are used to explain either linear or nonlinear relationships with targeted soil property.

5.2 A BRIEF SUMMARY OF KEY CLIMATIC, SPATIAL, AND ABIOTIC MODULATORS OF SOIL C STORAGE

5.2.1 LOCATION: LATITUDE, LONGITUDE, ALTITUDE, AND SOIL DEPTH

Soil C stocks vary strongly across different ecosystem types and Earth regions (Fig. 5.1). Global soil C stocks follow a right-skew distribution (Fig. 5.1A), meaning that the majority of SOC is stored in a few locations, while most other places on Earth store a relative low amount of carbon in their soils. Geographical location per se cannot be considered a major driver of soil C storage, however, because its close relationship with other biophysical drivers of soil C storage such as climate or vegetation; geographical location can still be a critical predictor of the distribution of soil C—especially at the global and regional scales. Moreover, the importance of geographical location as a predictor of soil C storage is expected to be strongly reduced from global to local scales.

Studies suggest that, in general, soil total C decreases from northern high latitude areas to the Southern Hemisphere (Xu et al., 2013; Wieder et al., 2013). The highest soil total C found in northern high latitudes is likely the result of organic matter accumulation due to slow decomposition in Taiga and Tundra ecosystems, despite extreme environmental conditions strongly limiting net primary production (NPP) and low plant and animal diversity in these environments (Ernakovich et al., 2014). Using data from the WoSIS dataset (http://www.isric.org/explore/wosis) ($n = 4381$) and land use maps from the European Space Agency (http://due.esrin.esa.int/page_globcover.php), we found that, for the top 10 cm of soil, the biggest reservoirs of global soil C stocks are located in boreal (>66 degrees of latitude), cold (continental climate), and temperate forests (Fig. 5.1B)—pointing out the importance of these biomes for the storage of C and climate change regulation (Wieder et al., 2013). Note that the top 10 cm soil C data in this dataset is highly correlated to soil C across different soil depths (i.e., 0–20, 0–50 and 0–100 cm; see Table S2 in Delgado-Baquerizo et al., 2017). Therefore, major patterns reported here are expected to be valid across the top 100 cm. Similar trends to those reported here have been found by other authors using a meta-analysis approach (Xu et al., 2013). In particular, these authors highlighted that tundra and natural wetlands are the largest reservoirs of soil C on Earth. Carbon storage in wetlands can be especially important for particular regions of Earth. For example, according to Chapman et al. (2009), Scottish peat bogs hold around 1620 Mt of Carbon, which represents 56% of the total carbon in all Scottish soils. In these ecosystems, soil C is accumulated during millennia as a consequence of the low temperatures, anoxic conditions and low pH which strongly limit processes such as soil respiration, still allowing slow litter decomposition rates by fungi and acidophilic bacteria, but leading to slow organic matter accumulation. These ecosystems are highly vulnerable to climate change and land management practices such as draining, burning, overgrazing, afforestation, and extraction and therefore prompt enormous losses of C to the atmosphere. Also in agreement with Xu et al. (2013), we found that arid regions, often located in mid-latitudes, have the lowest amount of C stored in soil for a natural ecosystem (Fig. 5.1B). Given this result, increases in the extension of drylands—which are very low productive ecosystems—are predicted to rise by 23% by the end of this century (Huang et al., 2016) with climate change which might result in a reduction in global C storage in soil (Schlesinger, 1996; Delgado-Baquerizo et al., 2013).

Similar to geographical location, altitude is not a major driver of soil C storage, but can still be a useful predictor of this variable given its close relationship with other "factual" drivers of soil C

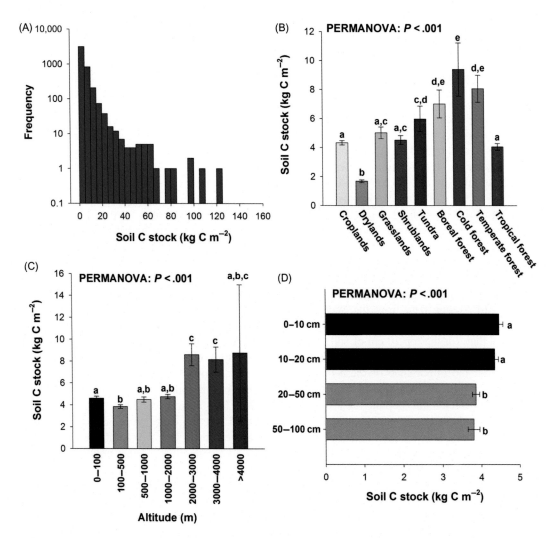

FIGURE 5.1

Global soil C stocks (top 10 cm). Panel (A) represents a histogram with the distribution of data from soil C stocks across the globe. Panel (B) represents soil C stocks (mean ± SE) across different ecosystem types. Panel (C) represents a bar graph soil C stocks (mean ± SE) across different altitudinal ranges. Panel (D) represents a bar graph soil C stocks (mean ± SE) across different soil depths.

storage—such as climate, vegetation, or soil properties. Across altitudinal patterns, the highest carbon contents are expected to occur at the highest altitudes as a consequence of organic matter accumulation due to slow decomposition under strongly decreased temperatures near mountain tops. Using data from the WoSIS dataset (http://www.isric.org/explore/wosis) ($n = 4381$), we found that, for the top 10 cm of soil, the largest reservoirs of global soil C stocks are located in regions >2000 m above sea level (Fig. 5.1C)—highlighting the importance of high mountain ecosystems for C sequestration and climate regulation.

Another major global pattern is the strong reduction in soil C stocks with depth. For example, Jobbágy and Jackson (2001), conducted a global metaanalysis and found that soil C stocks strongly decreased across the top 100 cm of soil for different ecosystem types (e.g., grasslands, shrublands, and forests) and Earth biomes (e.g., deserts, croplands, tundra, boreal, temperate, and tropical forests). Here, using the WoSIS dataset, we found a similar trend for soil C stocks with depth at the global scale (Fig. 5.1D). The highest amount of soil C stocks in the top 20 cm of soil is likely to be related to the highest litter decomposition rates in this soil layer—which is often covered by litter from aboveground communities (Banwart, 2011; Jobbágy and Jackson, 2001). These results suggest that top soil is key for nutrient cycling, but is also highly vulnerable to degradation and erosion (Jobbágy and Jackson, 2001; Banwart, 2011).

5.2.2 CLIMATE AND SOIL PROPERTIES

Major climatic predictors of SOC storage include temperature, precipitation, and evapotranspiration. Climate is considered a true driver of soil C cycling and can control soil C storage both via direct (e.g., emission) and indirect effects (e.g., plant productivity and decomposition processes) on C cycling. Regions of Earth with high mean annual precipitation, often see high plant productivity and litter decomposition rates (García-Palacios et al., 2013a), ultimately leading to high C-fixation rates from the atmosphere and soil organic production. On the other hand, regions with high evapotranspiration, such as those from dryland ecosystems, often lead to reduced plant productivity, ultimately limiting C cycling and C storage (Delgado-Baquerizo et al., 2013). This is supported by the extensive body of literature which suggests that compared with mesic environments arid regions have less plant biomass (Maestre et al., 2012; Delgado-Baquerizo et al., 2013), leading to lower concentrations of soil C and N (biologically controlled) (Delgado-Baquerizo et al., 2013). For temperature, regions with high maximum temperatures are expected to store less C compared to other regions with milder temperatures. Thus, it is likely that in regions with high maximum temperatures the balance between plant production and soil respiration (Crowther et al., 2016) is negative, suppressing C sequestration (see Chapter 8: Impact of Global Changes on Soil C Storage-Possible Mechanisms and Modeling Approaches for details).

Soil properties such as texture, pH or nutrient pools often covary with soil organic matter (SOM) in terrestrial ecosystems and, therefore, can provide useful information for the prediction of C sequestration (see Schlesinger, 1996 and Paul, 2007 for detailed explanations on these variables). In brief, the amount of SOM regulates decomposition and mineralization rates regulating the availability of nutrients in soil. Given the direct link between soil C stocks and nutrient cycling, nutrient availability could be potentially used to predict the amount of C storage in terrestrial environments. Other soil properties such as texture, and clay in particular, are major factors controlling C storage in terrestrial ecosystems (Schlesinger, 1996; Paul, 2007) as SOM often gets bound to, and protected by, the mineral surface. In relation to soil pH, acidification can increase with increasing organic matter (e.g., humic and fulvic acids) concentration (Adams et al., 2012). On the contrary, human activities such as agriculture and pastures often effect soil pH. For example, the use of inorganic fertilizers often decreases soil pH, however, soil lime is used to reverse this effect. Because of this, soil pH can potentially be used as a predictor of the amount of soil C.

5.2.3 NET PRIMARY PRODUCTIVITY AND PLANT TRAITS

Solar energy is fixed by plants into chemical energy in terrestrial ecosystems—e.g., providing the entrance of new carbon to the soil in the form of root exudates and leaf litter. Heterotrophs perform the decomposition of litter which ultimately are incorporated into the soil in the form of organic matter (Hooper et al., 2000; Wardle et al., 2004). Plant productivity is therefore a critical driver of soil C sequestration and is considered an important predictor for C stocks. Over and above total net primary productivity, the intrinsic attributes of the species in a particular plant community, or plant traits, can strongly influence soil C sequestration directly (e.g., via the amount of C in litter or litter production) or indirectly via altering key ecosystem processes such as decomposition rates (e.g., via regulating litter quality). For example, the specific leaf area (SLA) index is strongly related to litter quality and decomposability, both of which influence the amount of C incorporated into the soil (Cornelissen, 1999; García-Palacios et al., 2013b; also see Chapter 2: Plant Communities as Modulators of Soil Carbon Storage).

5.2.4 HUMAN IMPACTS

Land use intensification (e.g., grazing) and cropping are well-known to promote strong reductions in soil C storage (Trivedi et al., 2016; Eldridge and Delgado-Baquerizo, 2016). For example, a metaanalysis carried out by Guo and Gifford (2002) reported that land use changes from pasture and forest to croplands lead to reductions in soil C by 59% and 42%, respectively. They further reported that land-use conversion from secondary forests to pasture and cropland to pasture resulted in increases in soil C by approximately 53% and 19%, respectively. Similarly, Don et al. (2011) reported that conversion of primary forest into cropland led to reductions of soil C by 25%. Using the WoSIS dataset and land use maps from the European Space Agency (http://due.esrin.esa.int/page_globcover.php), we found that croplands strongly reduced soil C stocks, for the top 10 cm, compared to natural ecosystems (Fig. 5.1B). Predicted conversion from forest to cropping following increases in the global population by 36% over the next 40 years to support food demand (Charles et al., 2010) and increases in the extent of arid regions by 23% (Huang et al., 2016) with climate change, will result in a significant loss in the global soil C stock (Schlesinger, 1996).

Interestingly, studies over the past two decades suggest that the effect of cropland on soil C sequestration needs to be evaluated in conjunction with climate, as it is the interaction of these two factors which ultimately drive the amount of C sequestered or lost from ecosystem conversion from natural to cropland. For example, Rabbi et al. (2015) found that aridity index (precipitation/evapotranspiration) is the largest predictor of soil C storage across drylands from Eastern Australia, while changes in land use accounted for a reduced portion of the variation in soil C sequestration. These results suggest that efforts to promote C sequestration via changes in land use intensification (e.g., reductions in grazing) or cropping may not significantly influence C sequestration in water limited ecosystem such as drylands.

In the context of agricultural production systems, Hutchinson et al. (2007), identified the following agronomic and management practices that affect soil C levels including converting cropland to grassland, forage in crop rotation systems, reduced tillage, reduction of bare fallow (increased cropping frequency), nutrient additions via fertilizers, interaction of fertilization and cropping frequency, type of crop in rotation, and weather conditions. Hutchinson et al. (2007) further reported that from the different agronomic and management practices listed above, reduced tillage is considered as the most efficient practice for sequestering carbon in agricultural production systems (see Chapter 8: Impact of Global Changes on Soil C Storage-Possible Mechanisms and Modeling Approaches for details on impact of management practices on soil C).

5.3 DOMINANT MECHANISMS THAT EXPLAIN THE IMPACT OF MODULATORS ON SOIL C STORAGE

The main predictors of soil C are expected to strongly vary across scales (Levin, 1992). Therefore, proper identification of the major drivers of soil C across scales needs to be taken into account. Little explicit attention has been given to scale for the prediction of soil C stocks—an empirical comparison of the well-established predictors of soil C across scale is largely lacking. Herein, we used both new analysis of data from global to microaggregate levels using a variety of statistical modeling approaches (Box 5.1) and report examples from the literature to highlight the role of multiple predictors of soil C stocks across different spatial scales. To do so, we used three published datasets: (1) Global and regional scales—we used information from the top 10 cm C stock data calculated from the WoSIS dataset. The WoSIS dataset used here include more than 4000 locations from all continents except Antarctica, and all major biomes and vegetation types. (2) Microsite/microhabitat scale—we used the C concentrations from the top 5 cm of soils reported in Delgado-Baquerizo et al. (2016). This dataset includes twenty forest ecosystems from New South Wales, Australia (20 plots) including mesic and dryland ecosystems and four microsites (trees, N-fixing shrubs, grasses and bare soil). (3) Soil aggregates scale—we used a dataset on soil aggregation (three soil aggregate sizes) from Trivedi et al. (2017), including information on soil C concentration across three locations from Australia with multiple land management strategies.

5.3.1 GLOBAL SCALE

Global predictions of global C stocks often include parameters such as geolocation (i.e., latitude, longitude), altitude, climate (temperature and precipitation), soil properties (texture and pH), plant productivity, and human impact. Note that the human impact index used here (Sanderson et al., 2002) is a measure of direct human impact based on eight measures of human presence, namely: Population density km^{-2}, score of railroads, score of major roads, score of navigable rivers, score of coastlines, score of nighttime stable lights, values of urban polygons, and cover categories (urban areas, irrigated agriculture, rain-fed agriculture, other cover types including forests, tundra, and deserts). Using a multimodel approach (Box 5.1), we are able to confirm that all these predictors have the capability to explain a unique portion of the variation in soil C stocks at global scale (Fig. 5.2A and B). Thus, the removal of any of these groups of predictors from our best model substantially reduced the model fit ($\Delta AICc > 6.15$; Fig. 5.2A; Burnham and Anderson, 2002; Burnham et al., 2011). Climate followed by soil properties and plant productivity were found to be the major drivers of soil C stocks at global scale (Fig. 5.1C) (Jobbágy and Jackson, 2001; Wieder et al., 2013). Thus, removal of climate or soil properties resulted in a much worse model versus the best model with increases in AICc of 444 and 205, respectively.

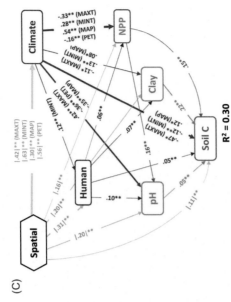

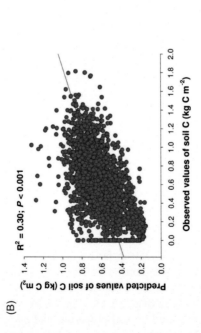

FIGURE 5.2

Modeling for soil C stocks (top 10 cm) at the global scale (WoSIS dataset). Panels (A) and (B) represents multimodel inferences aiming to identify major predictors of soil C stocks. Panel (C) represents a SEM aiming to identify the direct and indirect effects of selected predictors (from the best model from multimodel inference) on soil C stocks. Alt, altitude; HII, human impact index; Lat, latitude; Lon, longitude; MAP, mean annual precipitation; $MAXT$, maximum temperature in the warmest quarter (http://www.worldclim.org/); $MINT$, minimum temperature in the coldest quarter; NPP, net primary productivity (NDVI index; https://neo.sci.gsfc.nasa.gov) (WCS 2005).

To achieve a better understanding of the mechanisms regulating soil C stocks at the global scale, we used Structural Equation Modeling (SEM) (Box 5.1), which allows us to consider the multiple direct and indirect effects of selected predictors from a multimodel approach on soil C stocks. Our SEM provided evidence that, from a mechanistic point of view, soil C stocks are strongly limited by extreme temperature conditions and annual means. In precipitation in particular, we found that increases in maximum temperature has a strongly significant and negative direct effect on C stocks on Earth (Fig. 5.2C). Interestingly, besides the direct effects found, maximum temperature also had an indirect negative effect on soil C via reductions in plant productivity (Fig. 5.2C). Mean annual precipitation is the other key climatic driver of soil C at the global scale. In fact, our SEM detected strong positive and significant direct and indirect (via NPP and percent of clay) effects on soil C stocks (Fig. 5.2C). Texture was the major soil property controlling soil C stocks at the global scale. Locations with high mean annual precipitation over millennia can promote the formation of clay in soil via rock weathering—a process that happen in parallel to a reduction in soil pH during pedogenesis (Schlesinger, 1996) (see Fig. 5.2C).

5.3.2 REGIONAL SCALE

Similar predictors of soil C stocks were selected in our multimodel approach across the main climatic regions on Earth (Fig. 5.3). Climate and soil properties are again the major drivers of soil C stocks. Interestingly, following increases in AICc index when excluding particular groups of predictors (Fig. 5.3), soil properties become more important than climate for predictions of soil C stocks in arid and tropical regions, while climate remains the most important driver in temperate regions. Our SEM provides further evidence for a stronger direct effect of percent of clay on soil C stocks in arid and tropical compared to temperate regions (Fig. 5.3). This result suggests that in arid and tropical regions, where C stocks are lower than in temperate regions (Fig. 5.1B), most soil C stocks may be strongly attached to the smallest particles of soil. Supporting this idea, increases in percent of sand versus clay have been found to negatively impact the amount of total soil C and N in soil in drylands at the global scale (Delgado-Baquerizo et al., 2013).

Our analyses provided some specific insights when comparing climatic predictors of soil C stocks across main regions of Earth (Fig. 5.3). For example, potential evapotranspiration (PET) rather than mean annual precipitation is found in the best predictive model for arid regions compared to tropical or temperate regions (Fig. 5.3). Biological activity is strongly limited by the lack of water availability in drylands in arid regions (Whitford, 2002). In agreement with this, our SEM indicated that increases in PET has a strong negative effect on soil C stocks, which is indirectly driven by reductions in plant productivity. Interestingly, while maximum temperatures are still a key predictor of soil C stocks in arid and temperate regions, our SEM suggest that, for tropical regions, the importance of this predictor is strongly reduced—likely due to the fact that temperature ranges are smaller in the tropics compared to high- and mid-latitudinal regions. In tropical regions, mean annual precipitation is likely to be the major climatic driver of soil C stocks. The positive effect of mean annual precipitation on soil C is likely to be indirectly driven via positive effects on litter decomposition rates (García-Palacios et al., 2013a), net primary productivity, organic matter production, and via abiotic and biotic weathering (via plant production) that promote the dominance of clay particles in soil (Fig. 5.3C).

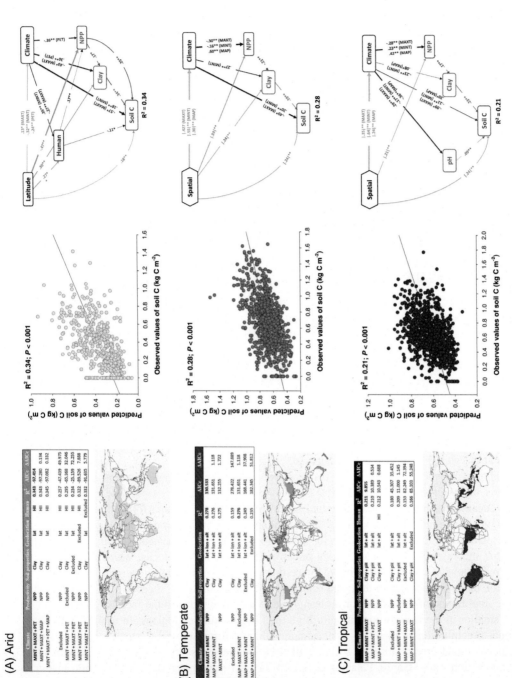

FIGURE 5.3

Modeling for soil C stocks (top 10 cm) at the global scale and across different bioclimatic regions (WoSIS dataset). Panels within each climate represents (1) multimodel inferences aiming to identify major predictors of soil C stocks and (2) SEMs aiming to identify the direct and indirect effects of selected predictors (from the best model from multimodel inference) on soil C stocks. *Alt*, altitude; *HII*, human impact index; *Lat*, latitude; *Lon*, longitude; *MAP*, mean annual precipitation; *MAXT*, maximum temperature in the warmest quarter; *MINT*, minimum temperature in the coldest quarter; *NPP*, net primary productivity (NDVI index; https://neo.sci.gsfc.nasa.gov); *PET*, potential evapotranspiration (http://www. worldclim.org/) (WCS 2005).

5.3.3 **PLOT, MICROSITE AND MICROHABITAT SCALE**

In highly heterogeneous ecosystems, soil C stocks are also expected to be greatly variable across plots particularly with different microsites or microhabitats given by particular biotic (e.g., identity of plant species and functional groups) and abiotic features (e.g., bare ground areas, litter patches and rocks) within a particular location. Although different microsites are present in all biomes of Earth, they are especially important in drylands (arid, semiarid and dry-sub humid ecosystems)—the largest biome on Earth (45% of land surface) which supports over 38% of the global human population (Maestre et al., 2016). In fact, a major feature of dryland ecosystems is their spatial heterogeneity (Maestre et al., 2016). Drylands are characterized by a sparse coverage of plants, which are separated by open areas devoid of perennial vegetation (Fig. 5.4A). Plant patches include a wide variety of vegetation types such as grasses, nitrogen (N)-fixing shrubs, and trees, while open areas are often covered by biocrusts—soil communities dominated by mosses, lichens, and cyanobacteria. Different microsites are expected to support different C sequestration in these regions. For example, at the global scale, biocrusts have been reported to promote soil C concentration (top 4 cm) compared to bare ground (Delgado-Baquerizo et al., 2016). The microsites found within a particular location can potentially regulate the capacity of a given ecosystem to store C via regulating important ecosystem processes such us photosynthesis, N-fixation (e.g., Leguminosae family) and P mineralization (plant species with cluster roots from Proteaceae family in Australian soils) (Lambers and Plaxton, 2015). Different microsites can also modify microclimate, infiltration and soil water availability dynamics (Cerda, 1997; Maestre et al., 2003) with consequences for ecosystem processes such as litter decomposition and mineralization (García-Palacios et al., 2013a; Robertson and Groffman, 2007). Other than via regulating ecosystem processes, microsite—especially those dominated by plants and biocrusts—can regulate C storage indirectly via key plant traits such as plant height and SLA index (Cornelissen, 1999). Including microsites in models to predict C stocks can be especially important when considering climate change and land use intensification scenarios. For example, predicted increases in aridity for the late 21st century in most drylands (Huang et al., 2016) will negatively impact upon vascular vegetation cover in drylands (Maestre et al., 2012; Delgado-Baquerizo et al., 2013), and this may increase the proportion of bare ground areas (Delgado-Baquerizo et al., 2016). Increases in aridity are also expected to shift plant composition in drylands (Maestre et al., 2016). Changes in the relative abundance of microsites could potentially alter the capacity of an ecosystem to store C, however, the role of microsites for C sequestration is still relatively unexplored and poorly understood.

Using a plot scale dataset with a clear microsite differentiation (vegetation vs bare soils) from New South Wales, Australia (20 plots) including four microsites (trees, N-fixing shrubs, grasses and bare soil) from drylands and more mesic ecosystems, we found that, on average, soil C concentration (top 10 cm) is higher under trees than under other plant microsites and much higher (more than two times) than under bare soils lacking vegetation (Fig. 5.4B). Using a multimodel inference approach (Box 5.1), we found that soil properties (inorganic N and starch degradation) are the main predictors of soil C concentration at this scale followed by plant traits (SLA index), microsite (Eucalyptus trees) and climate (Fig. 5.4C). All of these groups of predictors explained a unique portion of the variation in soil C concentrations (Fig. 5.4C). Our capacity to predict soil C at the microsite scale doubled those from the global and regional scale including similar type of redictors (Figs. 5.2B, 5.3D, and 5.4D). Other predictors in our model also included human impact,

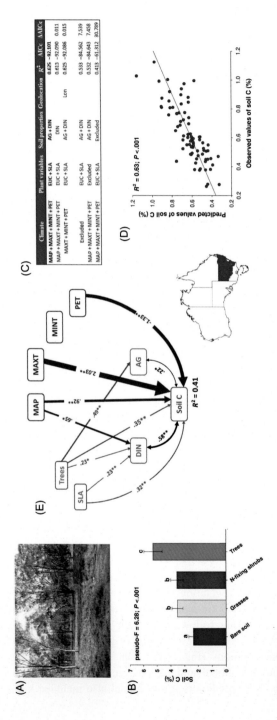

FIGURE 5.4

Modeling for soil C stocks (top 5 cm) at the microsite scale. Panel (A) includes a picture of the studied ecosystem. Panel (B) represents soil C stocks (mean ± SE) across different ecosystem microsites. Panels (C) and (D) represent multimodel inferences aiming to identify major predictors of soil C stocks. Panel (E) represents a SEM aiming to identify the direct and indirect effects of selected predictors (from the best model from multimodel inference) on soil C stocks. *AG*, Alpha-glucosidase; *DIN*, dissolved inorganic nitrogen; *EUC*, eucalyptus tree microsite (Trees); *Lon*, longitude; *MAP*, mean annual precipitation; *MAXT*, maximum temperature in the warmest quarter; *MINT*, minimum temperature in the coldest quarter; *PET*, potential evapotranspiration (http://www.worldclim.org/); *SLA*, surface lead area index.

plant production, total plant cover, and diversity or geolocation, which were not selected in the best model (Fig. 5.4C).

From a mechanistic point of view, our SEM indicated that plant traits, such as SLA index, has a direct positive and significant effect on soil C concentration. Litter with high SLA index often lead to high litter decomposition rates promoting SOM production (Cornelissen, 1999; García-Palacios et al., 2013b). The SLA index was also positively related to inorganic N in soil—both litter decomposition and mineralization are always coupled in terrestrial ecosystems (Schimel and Bennett, 2004; Robertson and Groffman, 2007; Delgado-Baquerizo et al., 2015). Similarly, tree microsite showed positive and significant direct effects on the concentration of soil C and DIN. Tree microsites are expected to produce a higher amount of litter because of their larger biomass compared to grasses and shrubs, resulting in the highest concentrations of soil C and DIN for this microsite. As reported at the global scale for arid regions (Fig. 5.3A), PET also reduced the amount of soil C across this environmental gradient from Australia, most likely indirectly via reductions in NPP. Interestingly, NPP was not selected as an important soil C predictor by our multimodel approach, suggesting that from a predictive point of view, PET absorbed the predictive power of NPP at this spatial scale. Moreover, precipitation—the major limiting factor for biological activity in drylands—had a positive direct effect on the concentration of soil C. Altogether, our analyses suggest that microsite and plant traits could be useful predictors of soil C at the regional scale in highly heterogeneous ecosystems as they explained unique portions of the variation of soil C and provided useful mechanistic information for the understanding of soil C sequestration at this scale.

5.3.4 **SOIL AGGREGATE SCALE**

At the soil aggregate scale (a few centimeters) and within a particular microsite or microhabitat (e.g., beneath plant canopies), the storage of soil C is highly influenced by the process of soil aggregation (Gupta and Germida, 1988; Six et al., 2004, 2006). The highest amount of soil C is often reported to be stocked in the smallest soil aggregates (microaggregates) (Gupta and Germida, 1988; Six et al., 2004, 2006). Using a dataset on soil aggregation including three different sized aggregates (mega-($>250 \mu m$), macro-($250-50 \mu m$), and micro-($<50 \mu m$)] from several locations across Australia and multiple land managements, we found that microaggregates store up to three/four more times C than the largest aggregates (megaaggregates) (Fig. 5.5A). The different size of soil aggregates is highly responsive to different management practices such as tilling, residual retention, or crop rotation (Six et al., 2006; Tiemann et al., 2015; Trivedi et al., 2017). Because of this, understanding the role of soil aggregation and human management in controlling soil C is especially important at this spatial scale (Six et al., 2006; Trivedi et al., 2017).

Using the multimodel approach (Box 5.1) we found that factors such as soil aggregation size, soil properties (starch degradation and pH), human management, and mean annual precipitation are key predictors of soil C concentration at this scale—all of these groups of predictors explain a unique portion of the variation in soil C concentration (Fig. 5.5B). Soil aggregation size and soil properties were the major drivers of soil C at this scale—removal of these groups of predictors from the best model resulted in a much worse model (Fig. 5.5B). Other predictors such as plant production, maximum and minimum temperatures, PET, human impacts, and geolocation were not selected in our best model (Fig. 5.5B). Interestingly, our capacity to predict soil C at this scale was almost four times larger than at the global and regional scales.

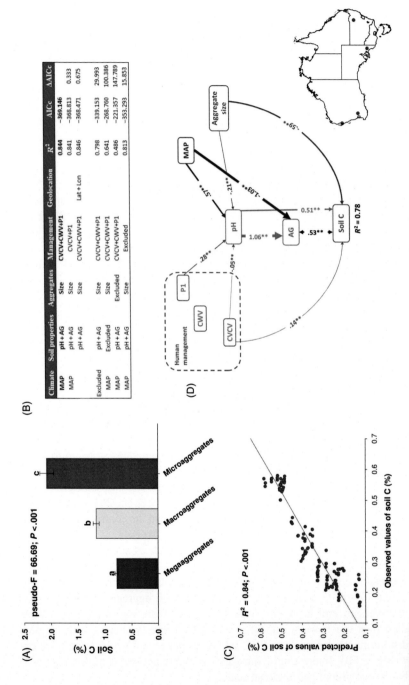

FIGURE 5.5

Modeling for soil C stocks (top 5 cm) at the aggregate scale. Panel (A) represents soil C stocks (mean ± SE) across different aggregate sizes (mega-(>250 μm); macro-(250–50 μm); and micro-(<50 μm) aggregates). Panels (B) and (C) represent multimodel inferences aiming to identify major predictors of soil C stocks. Panel (D) represents a SEM aiming to identify the direct and indirect effects of selected predictors (from the best model from multimodel inference) on soil C stocks. *AG*, Alpha-glucosidase; *CVCV*, cotton-vetch rotation; *CWV*, cotton-wheat-vetch; *Lon*, longitude; *MAP*, mean annual precipitation (http://www.worldclim.org/); *P1*, continuous cereal; no-tillage disc seeder (Barley-Oat-Wheat); *Size*, aggregate size.

Our SEM shed some light on the mechanisms driving soil C concentration across different soil aggregates (Fig. 5.5D). As expected, aggregate size had a direct negative and significant effect on soil C concentration. The smallest aggregates are considered to be the repository of the most stable C pool in soils due to physicochemical properties (Trivedi et al., 2015). The highest amount of C in microaggregates may be related to strong interactions between polyvalent metals and organic ligands at this scale, and also to organic debris surrounded by clay particles that result in strong protected structures that store soil C for long periods of time (Lehmann et al., 2007). Human management and climate also played indirect/direct roles in controlling the amount of soil C at this spatial scale. For example, particular crop management such as cotton-vetch rotation had direct positive effects on the amount of C retained in soil. In addition, mean annual precipitation had an indirect positive effect on soil C concentration via starch degradation—a proxy of decomposition. This result suggests that locations with a higher precipitation might promote higher litter decomposition rates and promote higher organic matter production rates (Fig. 5.5D) (see Chapter 6: Soil Nutrients and Soil Carbon Storage: Modulatorsand Mechanisms).

Mega and macroaggregates protect plant-derived and microbial-derived SOM, contain high amounts of fungal biomass and are enriched with labile C and N originating predominately from plant residues. On the other hand, microaggregates are formed by microbially induced bonding of clay particles, polyvalent metals and organometal complexes, and are characterized by lower concentrations of new and labile carbon (lower C:N ratio) and increased amounts of physically protected and biochemically more recalcitrant C compared to macroaggregates (Trivedi et al., 2017). SOM turnover in different sized soil aggregates is influenced by a number of variables including management practices, organic matter loading, mechanical disturbance, temperature and moisture, along with functional groups of soil microbes (Six et al., 2006; Tiemann et al., 2015; Trivedi et al., 2015). For example, a greater quantity and quality of residues entering soils in high diversity rotations enhances microbial activity with (1) positive impacts on megaaggregate formation and stabilization, and (2) concomitant increases in microbial by-products that accelerates microaggregate formation, resulting in increasing stocks of stable SOC (Tiemann et al., 2015; Trivedi et al., 2015).

5.3.5 SOIL C PREDICTIONS ACROSS DIFFERENT SPATIAL SCALES

Soil C stocks can be predicted up to a certain point using largely available information on common predictors such as geolocation, climate, soil properties, plant production, and human impact from local to global scales. However, the relative importance of these predictors and our capacity to predict soil C largely change across scales (Fig. 5.6). The analyses conducted here suggest that the classical paradigm from Levin (1992). "The problem of pattern and scale in ecology" largely apply to the predictions of soil C stocks across different spatial scales. For instance, our capacity to predict soil C at the global scale is much more limited than that one at the microsite and microaggregates scales. This result suggests that while predictions at the microsite/microaggregate scales—essential for understanding changes in nutrient cycling and soil function under changing scenarios—can be accurately done with common predictors, predictions of soil C stocks at the global scale—critical for the understanding of climatic regulation—are challenged by the used of common predictors. In this respect, future studies need to continue searching for predictors of soil C explaining a unique portion of the variation of soil C stocks at the global scale that can be used in ESMs. Our results suggest that the relative importance of common predictors of soil C stocks

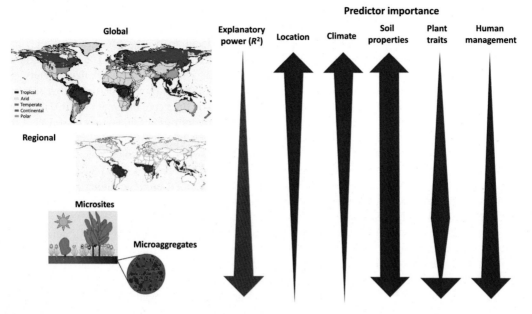

FIGURE 5.6

Conceptual figure summarizing main results derived from Figs. 5.2 to 5.5.

largely change across different spatial scales. Thus, while climate and geolocation are key drivers of soil C stocks at the global and regional scale, the importance of these drivers is reduced at the microsite and microaggregate scales. At these scales, other factors such as human management, microsite, or plant traits needs to be considered to accurately predict soil C stocks. Interestingly, only soil properties are universal predictors of soil C across spatial scales, being the major predictor of soil C stocks from local to global scales (Fig. 5.6). Altogether, the analyses conducted in this chapter suggest that any prediction of soil C stocks needs to consider the limitations in common predictors of soil C given by different spatial scales. This knowledge is critical for developing proper strategies of data collection for the prediction of soil C stocks across multiple scales.

5.4 PREDICTIVE MODELING OF SOIL C

5.4.1 WHY SOIL C MAPS?

There is an increasing interest in mapping soil C across in multiple scales and with depth. The recent launch of the "4 per 1000" initiative at the COP21 in Paris aims to increase global SOC stocks by 4 per 1000 (0.4%) per year in order to cope with the rising levels of global greenhouse gas emissions due to anthropogenic activities. Together with improved agronomic practices and land management practices to achieve this target, it is essential to have a clear understanding about the spatial distribution of soil C across the landscape and with depth. Therefore, C maps will provide first-hand information on land areas across the globe that have the potential to increase soil C using appropriate policies and procedures.

Digital mapping of soil C is carried out across multiple scales from global scale to farm scale or even at paddock scale. Some examples for global scale mapping of soil C includes work carried out by Stockmann et al. (2015), Hengl et al. (2014) and Hengl et al. (2017). Examples of studies that have mapped soil C at continental scale includes work carried out by Hengl et al. (2015) for Africa and Viscarra Rossel et al. (2014) for Australia. There is also an increasing trend in mapping soil C at country the level, for example work carried out by Adhikari et al. (2014) for Denmark, Akpa et al. (2016) for Nigeria, and Poggio and Gimona (2014) for Scotland. Miklos et al. (2010) and Malone et al. (2017) provide some examples of farm scale mapping of soil C.

5.4.2 HARMONIZATION OF SOIL DATA COLLECTED USING DIFFERENT SOIL HORIZONS OR WITH VARIABLE DEPTHS

Data collected using soil horizons are generally required to be harmonized into interested depth intervals prior to modeling. Harmonization of soil C data will standardize it and allow one to carry out direct comparison of data collected across the landscape. A widely used method to harmonize soil datasets is through fitting equal-area quadratic splines (Bishop et al., 1999). In addition, use of depth functions which includes negative exponential (Minasny et al., 2006), exponential function (Mishra et al., 2009), interpolation with continuous depth function, and log−log model for two data points (Viscarra Rossel et al., 2014) are also used for data preparation prior to mapping soil C. Recent work by Orton et al. (2016) suggests a novel approach having a single model without any prior harmonization of the datasets. This approach is promising for future modeling and mapping of soil C especially for countries where data is limited and only available from multiple surveys with variety of depth intervals.

5.4.3 MAPPING SOIL C: MODEL FITTING PROCESSES AND MODEL VALIDATION

The mapping of soil C contents/concentrations/stocks is generally carried out using a hybrid approach commonly known as regression-kriging (Hengl et al., 2015; Viscarra Rossel et al., 2014; Adhikari et al., 2014; Malone et al., 2009). More recently, there is a trend to use machine learning and data mining algorithms for the regression component of the hybrid approach. Use of machine learning and data mining algorithms such as Random Forest, CUBIST, Boosted Regression Trees, and Neural Networks enable one to incorporate a large number of environmental covariates to represent the scorpan model (Box 5.1). Furthermore, use of these analytical methods will avoid limitations such as multicollinearity similarly to when trying to fit traditional statistical data models. Additionally, these machine learning and data mining algorithms handle nonlinear relationships between measured soil C and environmental covariates more effectively.

Most studies split data at the beginning as model calibration (training) and model validation. Due to a limited number of soil C data most studies carried out interval validation using methods such as leave one out cross validation (LOOCV) and k-fold validation. Independent data collected using a design-based sampling scheme (probabilistic) will provide the true accuracy of the derived soil C maps (Brus et al., 2011; Bishop et al., 2015). Nevertheless, such applications are limited due to associated costs related to field data collection.

5.4.4 ENVIRONMENTAL COVARIATES FOR SOIL C MODELING AND MAPPING

Global studies of mapping soil C utilizes GIS datasets (e.g., land use/land cover maps, soil maps), freely available satellite data and their derived products such as MODIS EVI (available at different spatial resolutions, 1 km, 500 m and 250 m) and SRTM DEM ($\sim$90 m, available from http://srtm.csi. cgiar.org/SELECTION/inputCoord.asp), DEM-derived primary and secondary terrain attributes which are included in the model fitting process to represent the *scorpan* model (Hengl et al., 2014, 2017; Stockmann et al., 2015). Nevertheless, at the farm scale, other than those environmental covariates, multi sensor platform data—e.g., EM 31, gamma radiometric datasets—are also included as environmental covariates to present as drivers of soil C (Miklos et al., 2010; Malone et al., 2017).

The importance of these environmental covariates as drivers of soil C varies with the scale of the study and with the depth interval from the surface of the soil we are interested in order to carry out the mapping. For example, Hengl et al. (2017) in their global scale study reported that observed depth from the surface, mean monthly temperature for January and mean monthly temperature for February as the top three most important environmental covariates (drivers of soil C) useful for mapping soil C. In another study by Viscarra Rossel et al. (2014) on mapping top soil C (0−0.3 m) across Australia reported that depending on bioclimatic zone, environmental covariates used for model "conditions" and subsequent "linear model development" in their CUBIST model varies— details of the CUBIST model is given below in the case study mapping of soil C. For example, in the Savanna regions of Australia, the top three model development "conditions" were rainfall, Prescott index (a common climatic index; Prescott, 1950), and minimum temperature while in the cooler temperature regions it was solar radiation and PET. This provides evidence that in mapping soil C, depending on the region or bioclimatic region, their spatial drivers vary across the landscape.

Some studies of mapping soil C also focused on mapping with multiple depth intervals. In such situations, the importance of environmental covariates used as drivers of soil C are found to vary with the multiple depth intervals considered (Adhikari et al., 2014; Akpa et al., 2016; Gray et al. 2015). For example Gray et al. (2015) carried out mapping of soil C across multiple depth intervals, namely 0−5, 5−15, 15−30, 30−60, and 60−100 cm, and found that the respective model prediction capabilities in terms of Lin's concordance correlations decrease with depth. They further reported that relative importance of temperature and land use/land cover decreases with depth as drivers of soil C while importance of parent material increases.

5.4.5 PRACTICAL APPLICATIONS OF SOIL CARBON MAPS

There are many practical applications of digitally mapped soil C maps. In Australia, when producing national inventory reports related to land use, land use change and forestry sector, The Full Carbon Accounting Model process-based model utilizes digitally mapped soil C and its fraction maps to initialize the process-based model and report estimates of C stocks and its changes. This information is used for reporting greenhouse gas emissions under international treaties such as the Kyoto protocol and the United Nations Framework Convention on Climate Change

(Department of the Environment and Energy, 2016). Gray et al. (2016) demonstrated that digital soil C mapping techniques can be used to derive a preclearance carbon map for eastern Australia using soil samples collected across undisturbed natural vegetation. Furthermore, Karunaratne et al. (2014a) showed derived soil C maps for two time periods can be used to assess the change in soil C over the time and map the areas where statistically significant change occurred between two survey periods. De Gruijter et al. (2016) demonstrated the practical application of soil C maps in carbon auditing at the farm scale. These types of approaches are important for monitoring soil resources which are an important component to maintain soil health at a viable level.

5.4.6 MAPPING SOIL CARBON ACROSS THE LANDSCAPE AND WITH DEPTH: A CATCHMENT SCALE CASE STUDY FROM NORTHERN NEW SOUTH WALES, AUSTRALIA

Here, we used a catchment scale dataset collected from Northern New South Wales (NSW) in Australia as an example, including the above knowledge on soil C modeling, for soil C mapping at the regional scale. We acquired soil profiles collected by the Office of Environment and Heritage of NSW between 1999 and 2001 that are scattered across the Cox's Creek catchment (1358 km^2) in Namoi Valley, a major cropping region in NSW. For this particular analysis, we selected soil profiles that reported a maximum depth up to 1 m which resulted in 59 soil profiles.

In this dataset, soil C contents are reported by soil horizons. Therefore, we used equal-area quadratic splines (Bishop et al., 1999) to harmonize the soil C data in to four depth intervals namely 0−10, 10−20, 20−50 and 50−100 cm. This harmonization process allows us to compare the soil C estimates across the profiles and incorporate them into modeling processes. We prepared a variety of environmental covariates to represent factors in the *scorpan* model outlined by McBratney et al. (2003). These environmental covariates include DEM derived from the NASA Shuttle Radar Topography Mission (SRTM) (http://srtm.csi.cgiar.org/), other terrain attributes calculated from DEM namely wetness index and multiresolution valley bottom flatness index (MRVBF). The DEM and its derivatives represents the "relief" component in the *scorpan* model. Landsat 7 ETM + image acquired on September 4, 2000 from the USGS (http://glovis.usgs.gov/) was included in modeling, namely Landsat image bands 1 [0.45−0.52 μm], 2 [0.52−0.60 μm], 3 [0.63−0.69 μm], 4 [0.75−0.90 μm], 5 [1.55−1.75 μm], 7 [2.09−2.35 μm], band ratios (5/7, 3/2, 3/7), and normalized difference vegetation index (NDVI = [band 4 − band 3]/[band 4 + band 3]) as environmental covariates. These Landsat derived environmental covariates represent the "organisms" component in the *scorpan* model. In order to present the "soil" components in the *scorpan* model we included digitally mapped clay minerals, namely kaolinite, illite, and smectite (Viscarra Rossel, 2011) with the silica index (Gray et al., 2015). As the "climate" component of the *scorpan* model, we included mean annual precipitation and mean annual temperature available through worldclim (www.worldclim.org). Prior to model fitting processes all these spatial datasets were converted to a common spatial resolution of 90 m through resampling. Once the environmental covariates were extracted to soil profile locations, the model development was carried out using the CUBIST data mining algorithm.

Initially we tried to develop four predictive models based on harmonized depth intervals, but model predictive power was low. Therefore, similar to Akpa et al. (2014), we stacked four harmonized depth interval datasets (which resulted in 236 observations—i.e., 59 soil profiles*four harmonized depth intervals), created a single dataset, and a single model was fitted. We added a vector of mean depth value for the four harmonized depth intervals and included it as a predictor variable for each observation (i.e., 5, 15, 35, and 75 for 0−10, 10−20, 20−50, and 50−100 cm harmonized depth intervals). Finally, to account for the soil sampling depth in the prediction grid (the produce maps), mean depth value for each depth interval (e.g., 5 for 0−10 cm depth) was populated into the prediction grid depending on the considered depth interval prior to carrying out spatial predictions.

The CUBIST model forms a piecewise linear decision tree and partitions the response data (in this case soil C) into subsets within which their characteristics are similar with respect to the environmental covariates (predictors) (Viscarra Rossel et al., 2014). The model uses a series of rules derived using "if and else" statements to define the data partitions, and those rules are arranged in a hierarchy. "Conditions" within the model are developed either using one environmental covariate or by using a variety of environmental covariates. The fitted CUBIST model resulted in eight rules that explained the relationships between soil C and environmental covariates. For example, the first rule says that if the soil depth is greater than ($>$) 15 cm then "soil carbon $= -0.033 - 0.0068$ depth $- 0.0088$ silica $+ 3.6$ illite $+ 0.5$ smectite $+ 0.35$ kaolinite $- 0.0025$ ls_3 $+ 0.003$ ls_2 $+ 0.0012$ ls_7 $+ 0.0004$ DEM," while in the second rules is defined as: If MRVBF less than or equal to ($< =$) 3.85 and average mean temperature greater than ($>$) 17.9°C, then "soil carbon $= 416.023 - 0.3027$ mean annual precipitation $- 12.18$ mean annual temperature $- 0.0215$ depth $+ 0.2$ smectite $+ 0.16$ kaolinite. This shows how the simple multiple regression model was developed in CUBIST for each rule derived using "if" and "else" statements. The spatially fitted CUBIST regression model explained ∼89% of the total variability of the soil C in the dataset. Therefore, we did not model the obtained residuals—i.e., difference between measured soil C and model predicted soil C values—for its spatial autocorrelation through geostatistical analysis—variogram analysis (see typical spatial modeling framework given in Fig. 5.7).

Results revealed that in the CUBIST model, following environmental covariates namely, depth (69%), MRVBF (56%), Landsat band ratio 3/7 (34%), Landsat band 7, (16%), mean annual temperature (16%), mean annual precipitation (11%) and wetness index (9%), were identified as important environmental covariates for the modeling of soil C across the catchment. For the model development, out of 20 environmental covariates included in the initial model fitting processes, depth (100%), smectite (91%), kaolinite (91%), Landsat band 7 (76%), Landsat band 3 (76%), mean annual temperature (50%), Landsat band 3 (50%), mean annual precipitation (49%), wetness index (43%), DEM (42%), illite (42%), silica (42%), Landsat band 3 (34%), Landsat band ratio 3/7 (40%), MRVBF (24%) were included in regression models. Values inside the parenthesis indicate the model usage of those environmental covariates.

It is well-documented that soil C content declines with depth in most ecosystems. Therefore, having depth included in a predictor has significantly improved the prediction capability of the CUBIST model. Similar results were also observed by Hengl et al. (2017). Higher model usage of selected environmental covariates ($> = 50$) namely, smectite (91%), kaolinite (91%), Landsat band 7 (76%), Landsat band 3 (76%), mean annual temperature (50%) and Landsat band 3 (50%) suggested the importance of inclusion of soil attributes, climate, and biota in modeling soil C across the catchment scale through the *scorpan* model explained in Box 5.1.

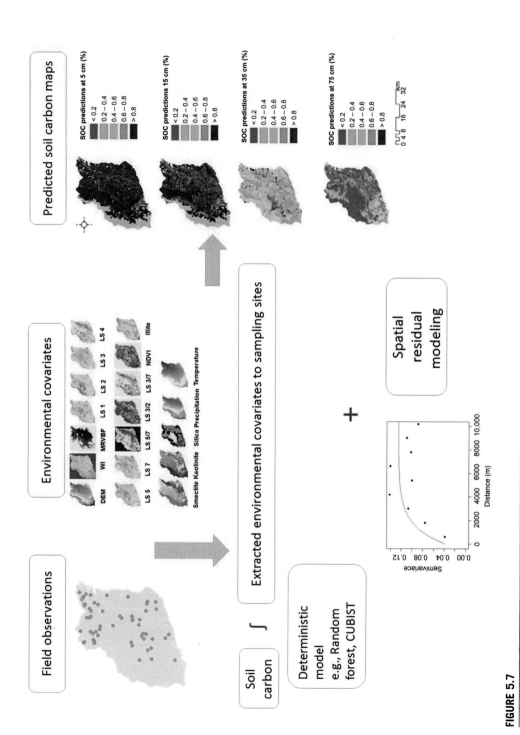

FIGURE 5.7

General framework for spatial modeling of soil carbon. In the example provided in this chapter modeling of residuals were ignored.

Validation of fitted models is an important step in spatial modeling of soil C as these maps are used as inputs for process-based models, regional planning and for policy directions (Karunaratne et al., 2014a,b; Ratnayake et al., 2016) (see Section 5.4.5). The LOOCV results indicated a model accuracy (Root Mean Squared Error—RMSE) of 0.41% (ideally close to zero) while model bias (Mean Error—ME) was -0.002 (ideally close to zero). Additionally, Lin's concordance value (correlation calculated to see how well the measured data and validation data fitted along the 1:1 line) was indicated as 0.80 (ideally 1). These results suggested that reasonably accurate soil C spatial prediction models can be developed even when using sparse datasets scattered across the landscape scale. Fig. 5.7 depicts the digitally mapped soil carbon maps for the catchment at four depth intervals.

5.5 NEW INSIGHTS, THE INCLUSION OF WHICH MIGHT IMPROVE PREDICTIONS OF SOIL C STORAGE

As explained earlier, predictions of soil C stocks at the global scale are limited to the use of common predictors such as geolocation, climate, soil properties, plant production, and human impact. Because of this, future endeavors aiming to improve soil C predictions need to identify new groups of predictors that are capable of explaining unique portions of the variation in soil C stocks. Two of the most promising areas in the search for new predictors of soil C stocks include microbial ecology (Allison et al., 2010) and paleoclimatology (Delgado-Baquerizo et al., 2017).

5.5.1 MICROBIAL ECOLOGY

The role of microbial ecology in predicting soil C stocks is addressed with detail in Chapter 3, Microbial Modulators and Mechanisms of Soil Carbon Storage. Studies suggest that inclusion of soil microbial-derived information, including microbial biomass (Wieder et al., 2013) and microbially-driven processes (enzyme activities and nutrient cycling; Allison et al., 2010; Trivedi et al., 2016) can improve the prediction of soil C stocks at regional and global scales. These models included microbial communities using a "Black-Box" approach (Schimel and Schaeffer, 2012)—i.e., we do not know who is in there, but their activity relates to soil C stocks. Much less is known however, on whether the inclusion of omics-derived information on microbial community composition, structure (fungi-to-bacteria ratio) and functional profiles might also help explain unique portions of the variations in soil C stocks. The abundance of functional genes has recently been demonstrated to be strongly linked to soil C concentrations and functions (Trivedi et al., 2016), suggesting that this information may lead to improvements in the predictions of soil C stocks. In addition, microbial community composition per se may also explain a unique portion of the variation of soil C via the information provided by particular microbial life-strategists. For example, oligotrophic groups such as Actinobacteria, Acidobacteria and Deltaproteobacteria (Fierer et al., 2007; Bastian et al., 2009; Trivedi et al., 2013) are expected to promote low respiration rates and to prefer soils with low levels of carbon where they are more competitive. On the contrary, copiotrophic organisms such as Gamma-proteobacteria, Alpha-proteobacteria and

Bacteroidetes (Fierer et al., 2007; Trivedi et al., 2013) seem to prefer environments that are rich in carbon and often promote labile carbon decomposition and high respiration rates in soil. This information can be used to target particular groups of microorganisms—microbial indicators of soil C—to improve predictions of C stocks under future scenarios. The use of this type of data is especially timely as recent studies have provided evidence that (1) major phyla/classes are globally distributed and common across samples (e.g., Ramirez et al., 2012), and (2) the use of high bacterial taxonomic ranks (phyla and classes) is strongly linked to ecosystem functioning (Philippot et al., 2010; Trivedi et al., 2013). In addition to dominant microbial taxa, soil bacteria, and fungi are well-known to have different capacities to decompose organic matter, to perform soil respiration and to incorporate C into the soil (Philippot et al., 2010; Trivedi et al., 2013). Because of this, incorporating both information from bacteria and fungi, as well as from their relative abundance (fungi-to-bacteria ratio), may further improve the prediction of soil C stocks. "Next generation" studies on C modeling will need to assess the utility of this data and to identify the most valuable—from an economic and practicality point of view—microbial predictors of soil C stocks.

5.5.2 PALEOCLIMATOLOGY

One of the most significant developments in the field of soil science involves the increased recognition that most soils are polygenetic—i.e., archival products of pedogenic processes that vary widely over time (Richter and Yaalon, 2012). Thus, we know from chronosequences that soil C sequestration is a long-term process that occurs over thousands of years during pedogenesis (at least as long as 10,000 years) (Schlesinger et al., 1990). Because of this, C projections into the future are conditional on the past, strongly reflecting site history across centennial to millennial time scales. A study in 2017 suggested that paleoclimate information predicts a unique portion of the variation in soil C stocks and the inclusion of long-term climate legacy information may further improve current predictions of soil C at the global scale (Delgado-Baquerizo et al., 2017). Interestingly, this study further suggests that conversion from natural to agricultural ecosystems largely reduces our capacity to predict soil C stocks using paleoclimatic data. Thus, paleoclimate explained five-times more variance in soil C than current climate in natural compared to agricultural systems. As the extension of the agriculture system and its associated disturbance (i.e., reductions in soil C stocks; Fig. 5.1B) continues to increase, the predictive power and utility of paleoclimate as a global predictor of soil C stocks is likely to reduce. Next generation models aiming to predict soil C stocks will need to cross-validate the utility of this data to improve predictions of soil C under different climatic and land use intensification scenarios.

5.6 CONCLUSIONS

In this chapter, we provide evidence that prediction of soil C stocks needs to consider the limitations in common predictors of soil C given by different spatial scales. In particular, we provide an integrative and mechanistic view of the multiple environmental predictors regulating C stocks from global to local scales. For example, geolocation (e.g., latitude, longitude, altitude and soil depth)

BOX 5.2 TAKE HOME MESSAGES

1. Abiotic predictors of soil C vary at different spatial scales. For example, climate and geolocation are major drivers of SOC at global scale, but human activities, microbial activity, microhabitat, and plant traits are better predictors of soil C at plot and microsite scales.
2. Soil properties (e.g., texture, pH) are universal predictors of soil C at all spatial scales.
3. Biome specific variables can further improve predictive capacity. For example, our analysis suggests that different climatic predictors module SOC in temperature, tropical and arid ecosystems.
4. New predictors (e.g., microbial community, paleoclimates) need to be considered explicitly to improve prediction and its accuracy.
5. Most of the variation in soil C storage remains unexplained at all spatial scales suggesting better mechanistic knowledge needs to be generated including interactions between biotic and abiotic predictors.
6. Soil C can be accurately mapped based on the use of dominant predictors which can potentially be used for better farm management practices, in site monitoring, and policy decisions.

---

climate and net primary productivity are key predictors at global scales. At the smallest scales, human management, microhabitat, and plant traits need to be considered to accurately predict soil C stocks. Only soil properties (texture and/or pH) were identified as universal predictors of soil C across different spatial scales. We then used those modulators to map soil C contents at catchment scale in order to demonstrate the utility of such knowledge for monitoring and managing soil C. Accurate soil C maps have the potential to provide crucial information for better farm management and policy decisions in order to sustain or even enhance soil C for the multiple ecosystem services they provide (Box 5.2).

In addition, we also discuss the utility of paleoclimate and microbial attributes for explaining a unique portion in the variation in soil C stock. In future, utilization of these new findings by existing and new simulation models can potentially improve prediction of soil C at global and regional scales. However, it is also evident that a large proportion of soil C variation remains unexplained. This ultimately suggests that new insightful predictors of soil C stocks needs to be found in the future. Obtaining a better mechanistic understanding of, and between, interacting abiotic and biotic variables with management practices and climate change will be needed to provide a conclusive picture for options to enhance soil C storage.

ACKNOWLEDGMENTS

We acknowledge the use of data from ISRIC's WoSIS, which are made available subject to the ISRIC Data Policy (http://www.isric.org/explore/wosis). We thank the Office of Environment and Heritage (OEH) of New South Wales Government for making available the Cox's Creek dataset. Additionally, we extend our thanks to Dr. Jonathan Gray (OEH) for providing access to some digital layers and for his comments on spatial modeling of soil C. M.D-B. and acknowledge support from the Marie Sklodowska-Curie Actions of the Horizon 2020 Framework Programme H2020-MSCA-IF-2016 under REA grant agreement n° 702057. B.K.S. was supported by the ARC projects DP13010484 and DP170104634.

DISCLAIMER

S.K. wants to clarify that the views and opinions expressed in this chapter do not necessarily reflect those of the Australian Government or the Minister for the Department of the Environment and Energy.

REFERENCES

Adams, C.R., Bamford, K.M., Early, M.P., 2012. Principles of Horticulture. Routledge, Abingdon.

Adhikari, K., Hartemink, A.E., Minasny, B., Kheir, R.B., Greve, M.B., Greve, M.H., 2014. Digital mapping of soil organic carbon contents and stocks in Denmark. PLoS One 9 (8), e105519.

Akpa, S.I., Odeh, I.O., Bishop, T.F., Hartemink, A.E., 2014. Digital mapping of soil particle-size fractions for Nigeria. Soil Sci. Soc. Am. J. 78 (6), 1953–1966.

Akpa, S.I., Odeh, I.O., Bishop, T.F., Hartemink, A.E., Amapu, I.Y., 2016. Total soil organic carbon and carbon sequestration potential in Nigeria. Geoderma 271, 202–215.

Allison, S.D., Wallenstein, M.D., Bradford, M.A., 2010. Soil-carbon response to warming dependent on microbial physiology. Nat. Geosci. 3, 336–340.

Banwart, S., 2011. Save our soils. Nature 474, 151–152.

Bastian, F., Bouziri, L., Nicolardot, B., Ranjard, L., 2009. Impact of wheat straw decomposition on successional patterns of soil microbial community structure. Soil Biol. Biochem. 41, 262–275.

Bernoux, M., Chevallier, T., 2014. Carbon in Dryland Soils. Multiple Essential Functions. Dossier thématiques du CSFD (Comité Scientifique Français de la Désertification). N°10. Download the English or the French version on: <http://www.csf-desertification.eu/dossier/item/carbon-in-dryland-soils-dossier>.

Bishop, T.F.A., McBratney, A.B., Laslett, G.M., 1999. Modelling soil attribute depth functions with equal-area quadratic smoothing splines. Geoderma 91 (1), 27–45.

Bishop, T.F.A., Horta, A., Karunaratne, S.B., 2015. Validation of digital soil maps at different spatial supports. Geoderma 241, 238–249.

Brus, D.J., Kempen, B., Heuvelink, G.B.M., 2011. Sampling for validation of digital soil maps. Eur. J. Soil Sci. 62 (3), 394–407.

Burnham, K.P., Anderson, D.R., 2002. Model Selection and Multimodel Inference. A Practical Information-Theoretical Approach. Springer, Heidelberg.

Burnham, K.P., Anderson, D.R., Huyvaert, K.P., 2011. AICc model selection in the ecological and behavioral sciences, some background, observations and comparisons. Behav. Ecol. Sociobiol. 65, 23–35.

Cerda, A., 1997. The effect of patchy distribution of *Stipa tenacissima* L. on runoff and erosion. J. Arid Environ. 36, 37–51.

Chapman, S., Bell, J., Donnelly, D., Lilly, A., 2009. Carbon stocks in Scottish peatlands. Soil Use Manage. 25, 105–112.

Charles, H., et al., 2010. Food security, the challenge of feeding 9 billion people. Science 327, 812–818.

Cornelissen, J.H.C., 1999. A triangular relationship between leaf size and seed size among woody species: allometry, ontogeny, ecology and taxonomy. Oecologia 118, 248–255.

Crowther, T.W., et al., 2016. Nature 540, 104–108.

De Gruijter, J.J., McBratney, A.B., Minasny, B., Wheeler, I., Malone, B.P., Stockmann, U., 2016. Farm-scale soil carbon auditing. Geoderma 265, 120–130.

Delgado-Baquerizo, M., et al., 2013. Decoupling of soil nutrient cycles as a function of aridity in global drylands. Nature 502, 672–676.

Delgado-Baquerizo, M., García-Palacios, P., Milla, R., Gallardo, A., Maestre, F.T., 2015. Soil characteristics determine soil carbon and nitrogen availability during leaf litter decomposition regardless of litter quality. Soil Biol. Biochem. 81, 134−142.

Delgado-Baquerizo, M., Maestre, F.T., Eldridge, D.J., Singh, B.K., 2016. Microsite differentiation drives the abundance of soil ammonia oxidizing bacteria along aridity gradients. Front. Microbiol. 7, 505. Available from: https://doi.org/10.3389/fmicb.2016.00505.

Delgado-Baquerizo, M., et al., 2017. Climate legacies drive global soil carbon stocks in terrestrial ecosystems. Sci. Adv. 3, e1602008.

Department of the Environment and Energy, 2016. National Inventory Report 2014 (revised) Volume 2, Commonwealth of Australia 2016.

Don, A., Schumacher, J., Freibauer, A., 2011. Impact of tropical land-use change on soil organic carbon stocks − a meta-analysis. Global Change Biol. 17, 1658−1670.

Eldridge, D.J., Delgado-Baquerizo, M., 2016. Continental-scale impacts of livestock grazing on ecosystem supporting and regulating services. Land Degrad. Dev. Available from: https://doi.org/10.1002/ldr.2668.

Ernakovich, J.G., Hopping, K.A., Berdanier, A.B., Simpson, R.T., Kachergis, E.J., Steltzer, H., et al., 2014. Predicted responses of arctic and alpine ecosystems to altered seasonality under climate change. Glob. Change Biol. 20, 3256−3269.

Fierer, N., Bradford, M.A., Jackson, R.B., 2007. Toward an ecological classification of soil bacteria. Ecology 88, 1354−1364.

García-Palacios, P., Maestre, F.T., Kattge, J., Wall, D.H., 2013a. Climate and litter quality differently modulate the effects of soil fauna on litter decomposition across biomes. Ecol. Lett. 16, 1045−1053.

García-Palacios, P., Milla, R., Delgado-Baquerizo, M., Martín-Robles, N., Álvaro-Sánchez, M., Wall, D.H., 2013b. Side effects of plant domestication: ecosystem impacts of changes in litter quality. New Phytol. 198, 504−513.

Grace, J.B., 2006. Structural Equation Modeling Natural Systems. Cambridge University Press, New York, NY.

Gray, J.M., Bishop, T.F., Wilson, B.R., 2015. Factors controlling soil organic carbon stocks with depth in eastern Australia. Soil Sci. Soc. Am. J. 79, 1741−1751.

Gray, J.M., Bishop, T.F., Smith, P.L., 2016. Digital mapping of pre-European soil carbon stocks and decline since clearing over New South Wales, Australia. Soil Res. 54 (1), 49−63.

Guo, L.B., Gifford, R.M., 2002. Soil carbon stocks and land use change: a meta analysis. Global Change Biol. 8, 345−360.

Gupta, V.V.S.R., Germida, J.J., 1988. Distribution of microbial biomass and its activity in different soil aggregate size classes as affected by cultivation. Soil Biol. Biochem. 20, 777−786.

Hengl, T., de Jesus, J.M., MacMillan, R.A., Batjes, N.H., Heuvelink, G.B.M., Ribeiro, E., et al., 2014. SoilGrids1km-Global soil information based on automated mapping. PLoS One 9 (8).

Hengl, T., Heuvelink, G.B., Kempen, B., Leenaars, J.G., Walsh, M.G., Shepherd, K.D., et al., 2015. Mapping soil properties of Africa at 250 m resolution: random forests significantly improve current predictions. PLoS One 10, e0125814.

Hengl, T., de Jesus, J.M., Heuvelink, G.B., Gonzalez, M.R., Kilibarda, M., Blagotić, A., et al., 2017. SoilGrids250m: global gridded soil information based on machine learning. PLoS One 12 (2), e0169748.

Hooper, D.U., et al., 2000. Interactions between aboveground and belowground biodiversity in terrestrial ecosystems: patterns, mechanisms, and feedbacks. BioScience 50, 1049−1061.

Huang, J., et al., 2016. Accelerated dryland expansion under climate change. Nat. Clim. Change 6, 166−171.

Hutchinson, J.J., Campbell, C.A., Desjardins, R.L., 2007. Some perspectives on carbon sequestration in agriculture. Agric. Forest Meteorol. 142, 288−302.

Jenny, H., 1941. Factors of Soil Formation: A System of Quantitative Pedology, McGraw-Hill, New York, NY, 281 pp.

Jobbágy, E.G., Jackson, R.B., 2001. The distribution of soil nutrients with depth: Global patterns and the imprint of plants. Biogeochemistry 53, 51−77.

Karunaratne, S.B., Bishop, T.F.A., Odeh, I.O.A., Baldock, J.A., Marchant, B.P., 2014a. Estimating change in soil organic carbon using legacy data as the baseline: issues, approaches and lessons to learn. Soil Res. 52, 349−365.

Karunaratne, S.B., Bishop, T.F.A., Baldock, J.A., Odeh, I.O.A., 2014b. Catchment scale mapping of measureable soil organic carbon fractions. Geoderma 219, 14−23.

Lambers, H., Plaxton, W.C., 2015. Phosphorus: back to the roots. In: Plaxton, W.C., Lambers, H. (Eds.), Annual Plant Reviews, Volume 48, Phosphorus Metabolism in Plants. John Wiley & Sons, Inc, Hoboken, NJ, pp. 3−22.

Lehmann, J., Kinyangi, J., Solomon, D., 2007. Organic matter stabilization in soil microaggregates: implications from spatial heterogeneity of organic carbon contents and carbon forms. Biogeochemistry 85, 45. Available from: https://doi.org/10.1007/s10533-007-9105-3.

Levin, S.A., 1992. The problem of pattern and scale in ecology: The Robert H. MacArthur Award Lecture. Ecology 73, 1943−1967.

Maestre, F.T., Bautista, S., Cortina, J., 2003. Positive, negative and net effects in grass-shrub interactions in Mediterranean semiarid grasslands. Ecology 84, 3186−3197.

Maestre, F.T., et al., 2012. Plant species richness and ecosystem multifunctionality in global drylands. Science 335, 214−218.

Maestre, F.T., et al., 2016. Structure and functioning of dryland ecosystems in a changing world. Annu. Rev. Ecol. Evol. Syst. 47, 215−237.

Malone, B.P., McBratney, A.B., Minasny, B., Laslett, G.M., 2009. Mapping continuous depth functions of soil carbon storage and available water capacity. Geoderma 154 (1), 138−152.

Malone, B.P., Styc, Q., Minasny, B., McBratney, A.B., 2017. Digital soil mapping of soil carbon at the farm scale: a spatial downscaling approach in consideration of measured and uncertain data. Geoderma 290, 91−99.

McBratney, A.B., Santos, M.M., Minasny, B., 2003. On digital soil mapping. Geoderma 117 (1), 3−52.

Miklos, M., Short, M.G., McBratney, A.B., Minasny, B., 2010. Mapping and comparing the distribution of soil carbon under cropping and grazing management practices in Narrabri, north-west New South Wales. Soil Res. 48 (3), 248−257.

Minasny, B., McBratney, A.B., Mendonça-Santos, M.L., Odeh, I.O.A., Guyon, B., 2006. Prediction and digital mapping of soil carbon storage in the Lower Namoi Valley. Soil Res. 44 (3), 233−244.

Mishra, U., Lal, R., Slater, B., Calhoun, F., Liu, D., Van Meirvenne, M., 2009. Predicting soil organic carbon stock using profile depth distribution functions and ordinary kriging. Soil Sci. Soc. Am. J. 73 (2), 614−621.

Nazaries, L., Tottey, W., Robinson, L., Khachane, A., Al-Soud, W.A., Sørensen, S., et al., 2015. Shifts in the microbial community structure explain the response of soil respiration to land-use change but not to climate warming. Soil Biol. Biochem. 89, 123−134.

Orton, T.G., Pringle, M.J., Bishop, T.F.A., 2016. A one-step approach for modelling and mapping soil properties based on profile data sampled over varying depth intervals. Geoderma 262, 174−186.

Paul, A.E., 2007. Soil Microbiology, Ecology, and Biochemistry. Academic Press, Amsterdam.

Philippot, L., et al., 2010. The ecological coherence of high bacterial taxonomic ranks. Nat. Rev. Microbiol. 8, 523−529.

Prescott, J.A., 1950. A climatic index for the leaching factor in soil formation. J. Soil Sci. 1, 9−19.

Poggio, L., Gimona, A., 2014. National scale 3D modelling of soil organic carbon stocks with uncertainty propagation—an example from Scotland. Geoderma 232, 284–299.

Rabbi, S.M.F., Daniel, H., Lockwood, P.V., Macdonald, C., Pereg, L., Tighe, M., et al., 2015. Physical soil architectural traits are functionally linked to carbon decomposition and bacterial diversity. Sci Rep-UK 6.

Ramirez, K.S., Craine, J.M., Fierer, N., 2012. Consistent effects of nitrogen amendments on soil microbial communities and processes across biomes. Global Change Biol. 18, 1918–1927.

Ratnayake, R.R., Karunaratne, S.B., Lessels, J.S., Yogenthiran, N., Rajapaksha, R.P.S.K., Gnanavelrajah, N., 2016. Digital soil mapping of organic carbon concentration in paddy growing soils of Northern Sri Lanka. Geoderma Regional 7 (2), 167–176.

Richter, D.B., Yaalon, D.H., 2012. "The changing model of soil" Revisited. Soil Sci. Soc. Am. J. 76, 766–778.

Robertson, G.P., Groffman, P., 2007. Nitrogen transformations. Microbiology Biochemistry Ecology. Springer, New York, NY.

Sanderson, E.W., Jaiteh, M., Levy, M.A., Redford, K.H., Wannebo, A.V., Woolmer, G., 2002. The human footprint and the last of the wild: the human footprint is a global map of human influence on the land surface, which suggests that human beings are stewards of nature, whether we like it or not. BioScience 52, 891–904.

Schimel, J.P., Bennett, J., 2004. Nitrogen mineralization, challenges of a changing paradigm. Ecology 85, 591–602.

Schimel, J.P., Schaeffer, S.M., 2012. Microbial control over carbon cycling in soil. Front. Microbiol. 3, 348. Available from: https://doi.org/10.3389/fmicb.2012.00348.

Schlesinger, W.H., 1996. Biogeochemistry, an Analysis of Global Change. Academic Press, San Diego, CA.

Schlesinger, W.H., Reynolds, J.F., Cunningham, G.L., Huenneke, L.F., Jarrell, W.M., Virginia, R.A., et al., 1990. Biological feedbacks in global desertification. Science 247, 1043–1048.

Schmidt, M.W., Torn, M.S., Abiven, S., Dittmar, T., Guggenberger, G., Janssens, I.A., et al., 2011. Persistence of soil organic matter as an ecosystem property. Nature 478, 49–56.

Six, J., Bossuyt, H., Degryze, S., Denef, K., 2004. A history of research on the link between (micro) aggregates, soil biota, and soil organic matter dynamics. Soil Tillage Res. 79, 7–31.

Six, J., Frey, S.D., Thiet, R.K., Batten, K.M., 2006. Bacterial and fungal contributions to C-sequestration in agroecosystems. Soil Sci. Soc. Am. J. 70, 555–569.

Stockmann, U., Padarian, J., McBratney, A., Minasny, B., de Brogniez, D., Montanarella, L., et al., 2015. Global soil organic carbon assessment. Global Food Security 6, 9–16.

Tiemann, L.K., Grandy, A.S., Atkinson, E.E., Marin-Spiotta, E., McDaniel, M.D., 2015. Crop rotational diversity enhances belowground communities and functions in an agroecosystem. Ecol. Lett. 18, 761–771.

Tiessen, H., Cuevas, E., Chacon, P., 1994. The role of soil organic matter in sustaining soil fertility. Nature 371, 783–785.

Trivedi, P., Anderson, I.C., Singh, B.K., 2013. Microbial modulators of soil carbon storage: integrating genomic and metabolic knowledge for global prediction. Trends Microbiol. 21, 641–651.

Trivedi, P., Rochester, I.J., Trivedi, C., Van Nostrand, J.D., Zhou, J., Karunaratne, S., et al., 2015. Soil aggregate size mediates the impacts of cropping regimes on soil carbon and microbial communities. Soil Biol. Biochem. 91, 169–181.

Trivedi, P., Delgado-Baquerizo, M., Trivedi, C., Hu, H., Anderson, I.C., Jeffries, T.C., et al., 2016. Microbial regulation of the soil carbon cycle: evidence from gene–enzyme relationships. ISME J. 10, 2593–2604.

Trivedi, P., et al., 2017. Soil aggregation and associated microbial communities modify the impact of agricultural management on carbon content. Environ. Microbiol. 19, 3070–3086.

Viscarra Rossel, R.A., 2011. Fine-resolution multiscale mapping of clay minerals in Australian soils measured with near infrared spectra. J. Geophys. Res.: Earth Surf. 116 (F4), 1–15.

Viscarra Rossel, R.A., Webster, R., Bui, E.N., Baldock, J.A., 2014. Baseline map of organic carbon in Australian soil to support national carbon accounting and monitoring under climate change. Global Change Biol. 20, 2953–2970.

Wardle, D.A., Bardgett, R.D., Klironomos, J.N., Setälä, H., Van Der Putten, W.H., Wall, D.H., 2004. Ecological linkages between aboveground and belowground biota. Science 304, 1629–1633.

Whitford, W.G., 2002. Ecology of Desert Systems. Academic Press, San Diego, CA.

Wieder, W.R., Bonan, G.B., Allison, S.D., 2013. Global soil carbon projections are improved by modelling microbial processes. Nat. Clim. Change 3, 909–912.

WCS, 2005. Wildlife Conservation Society, and Center for International Earth Science Information Network. Columbia University. 2005. Last of the Wild Project, Version 2, 2005 (LWP-2): Global Human Influence Index (HII) Dataset (Geographic).NASA Socioeconomic Data and Applications Center (SEDAC), Palisades, NY. Available from: https://doi.org/10.7927/H4BP00QC.

Xu, X., Thornton, P.E., Post, W.M., 2013. A global analysis of soil microbial biomass carbon, nitrogen and phosphorus in terrestrial ecosystems. Global Ecol. Biogeogr. 22, 737–749.

FURTHER READING

Karhu, K., Auffret, M.D., Dungait, J.A.J., Hopkins, D.W., Prosser, J.I., Singh, B.K., et al., 2014. Temperature sensitivity of soil respiration rates enhanced by microbial community response. Nature 513, 81–84.

Schlesinger, W.H., 1990. Evidence from chronosequence studies for a low carbon-storage potential of soils. Nature 348, 232–234.

United Nations Environment Programme, 1992. World Atlas of Desertification. UNEP, Edward Arnold, London, UK.

SOIL NUTRIENTS AND SOIL CARBON STORAGE: MODULATORS AND MECHANISMS

6

Catriona A. Macdonald[1], Manuel Delgado-Baquerizo[2], David S. Reay[3], Lettice C. Hicks[3] and Brajesh K. Singh[1]

[1]*Western Sydney University, Penrith, NSW, Australia* [2]*University of Colorado, Boulder, CO, United States* [3]*University of Edinburgh, Edinburgh, United Kingdom*

6.1 INTRODUCTION

The biogeochemical cycles of carbon (C), nitrogen (N), and phosphorus (P) are interconnected via key processes such as photosynthesis, decomposition, and respiration from local to global scales. The strong interlink between C:N:P cycles are the result of the conserved elemental stoichiometry of plants and microorganisms—the major biotic drivers of these biogeochemical cycles (Finzi et al., 2011; Peñuelas et al., 2012). The balance between photosynthesis and the decomposition process is fundamental to the global C cycle and determines soil C stocks. These processes play a critical role in locking up carbon dioxide (CO_2) through accumulation of C in above- and belowground biomass and soils. Importantly, photosynthesis and decomposition rates, and, ultimately the capacity of soil to store C, are tightly linked to macronutrients (e.g., N, P, and sulfur), and, to a lesser extent micronutrients (e.g., potassium and calcium). Global N and P cycles have already experienced significant anthropogenic disturbance (e.g., deposition, fertilization, and mining), and changing nutrient availabilities has been identified as a key uncertainty in predicting future ecosystem C sequestration (e.g., Fernández-Martínez et al., 2014). For example, as a consequence of accelerating use of fossil fuels, fertilizers, and anthropogenic N_2 fixation, atmospheric CO_2 and reactive N concentrations have increased. This has created an imbalance between N and P cycles within terrestrial ecosystems affecting their C-sequestration potential (Peñuelas et al., 2012). Increasing, N:P ratios have driven some ecosystems toward P limitation (Vitousek et al., 2010) and predicted that climate warming is expected to further alter biogeochemical cycles and constrain net primary production, and thus limit C-sequestration potential (Finzi et al., 2011).

Changes in soil organic carbon (SOC) are driven by the balance between C inputs from rhizodeposition and litter, and outputs from decomposition—i.e., belowground heterotrophic respiration resulting from the mineralization of soil organic matter; SOM—and leaching (Woodward et al., 2009). The intimate linkage between the cycling of C between above- and belowground ecosystem components and nutrient cycles occurs via the stoichiometric relationships that allow cellular maintenance and growth

(Finzi et al., 2011; Gruber and Galloway, 2008; Sterner and Elser, 2002). The soil microbial biomass, through which most nutrients are recycled, has a relatively low and constrained C:N:P ratio—between 60:7:1 and 42:6:1 (Cleveland and Liptzin, 2007; Xu et al., 2013)—compared to the soil they reside in—186:13:1—and litter substrates they decompose—3144:45:1 (Cleveland and Liptzin, 2007). As a result of this high N and P requirement of microbial communities, nutrients become immobilized within the microbial biomass and are not available for plant uptake until a critical elemental threshold is reached (Zechmeister-Boltenstern et al., 2015). As such, nutrient limitation of the microbial community can control the rate at which fresh litter is decomposed, and at which rate nutrients are mineralized from SOM. For example, litter and SOM C:N ratio is one of the major determinants of decomposition by indirectly affecting microbial-mediated N immobilization and mineralization rates (Robertson and Groffman, 2007). Ultimately, this strongly impacts the extent to which C is either sequestered or lost from soils. Similarly, direct impacts of nutrients on SOM chemistry, stabilization, microbiota, and extracellular enzymes are known (Paul, 2007) and such shifts in chemistry and biology can also alter the magnitude and direction of SOC storage or loss.

Because of the strong link between soil nutrients and C stocks, information on nutrient availability could potentially be used to improve predictions on C storage. Moreover, if we are to improve our capacity to enhance soil C storage, we need to fill key gaps in our understanding of the mechanisms that drive the interactions between C and nutrient cycles—and on the impact that environment (e.g., temperature, water, atmospheric CO_2 concentration, vegetation structure) and management (e.g., land-use and fertilizer inputs) have on these interactions. First-generation models—i.e., coupled climate-C cycle models (Hungate et al., 2003)—did not include these interactions, but attempts have been made to consider the coupling of C and N (and to a lesser extent P) cycling in second-generation models (coupled C-nutrient models)—e.g., Goll et al. (2012), Meyerholt and Zaehle (2015), Yang et al. (2014), Zaehle et al. (2014). However, models are still constrained by a lack of mechanistic knowledge on the impact of nutrients on SOC dynamics under different ecosystems, climatic scenarios, and management practices. Improving our understanding of the mechanisms that modulate the interactions between C and nutrients is, therefore, essential to reduce uncertainties in SOC projections.

In this chapter, we discuss N, P, and other nutrients as key modulators of soil C storage, consider the main mechanisms driving these modulators, the impact that environment and land-use change is likely to have, and the expected consequences for soil C storage. We evaluate proposed mechanisms and, where relevant, propose modifications. To do this, we used both de novo analyses (Boxes 6.1 and 6.2) and reported examples from the literature.

6.2 KEY ELEMENTAL MODULATORS AND DOMINANT MECHANISMS OF SOIL C STORAGE

6.2.1 NITROGEN

Generally, SOC levels are higher in cool versus warm climates, higher in wet/poorly drained soils than drier climates, and higher in fine versus course textured soils (Lal, 2001; Delgado-Baquerizo et al., Chapter 5: Climate, Geography, and Soil Abiotic Properties as Modulators of Soil C Storage) (see Box 6.1). Such climatic and geographical differences have a strong influence on the rate at which nutrients are mineralized from SOM and become available for plant growth. Soil C and N

BOX 6.1 CASE STUDIES ON THE RELATIONSHIP BETWEEN SOIL C AND N

To investigate the relationship between soil C and N at different scales (microaggregate, local, regional and global), we conducted analyses on three datasets including the BIOCOM project global dryland dataset (Delgado-Baquerizo et al., 2013; Maestre et al., 2012), Scotland NSIS (Delgado-Baquerizo et al., 2017), and a recent microsite dataset (Trivedi et al., 2017). We derived soil C:N ratios in each ecosystem within each dataset and explored the relationship between C and N stocks using regression analysis.

Our analysis suggests a strong linkage between soil C and N at the microaggregate, local, regional, and local scales (Fig. 6.1) while providing interesting insights that aligns with previous findings on the relationship between soil C and N. In dryland biomes, the highest levels of soil C are found in humid regions and least in arid regions (Fig. 6.1A). This may, in part, reflect different degrees of degradation in soils with different soil moistures and microbial activity. Soil C: N ratios followed the same trend (Fig. 6.1A). Similarly, at the regional scale obtained from Scotland (a boreal temperate climate zone), the highest levels of soil C were found in bogs (wet) while arable and grassland soils contained the least soil C (Fig. 6.1B). Soil C:N ratios followed the same trend (Fig. 6.1B). These findings support the observation that reducing moisture availability can induce a decoupling between C and N (Evans and Burke, 2013) and lead to lower soil C:N. In our local dataset, soil C:N ratios under bare soil were lower compared to other sites (Fig. 6.1C), which is in line with other findings which found that in desert soils where vegetation was largely absent, soil N becomes enriched relative to C (Xu et al., 2013). Interestingly, our data suggest that the microaggregate fraction had the highest C:N ratio (Fig. 6.1D) which is considered the most recalcitrant pool of soil C, which may represent soil C that has undergone a higher degree of microbial processing.

Overall, our findings suggest that soil which has high N also has high C, supporting other met analyses (Hartman and Richardson, 2013; Xu et al., 2013)—and this trend is largely held across vastly different spatial scales.

---

stocks are tightly coupled and, typically, soil N increases with soil C (Hartman and Richardson, 2013). The reason for this is that most soil C and N are strongly bounded to the organic matter (OM) in soil. However, as a result of a subtle decline in the relative amount of N with increasing soil C, the allometric slope of the relationship between soil C and N at the global scale is not isometric (i.e., 1) but rather 0.88 (Hartman and Richardson, 2013). This deviation is likely driven by differences in the quality of SOM which varies with time and across environmental gradients as a result of age and differing C:N of different plants. Globally, soil C:N ratios average 16:1, and range from 31:1 in boreal forest soils, 13:1 in grassland soils, 19:1 in temperate soils, 16:1 in tropical forest soils, to 24:1 in cold tundra soils (Xu et al., 2013). Nutrient-poor desert soils and nutrient-rich croplands have a considerably lower soil C:N of 10:1 and 12:1 respectively (Xu et al., 2013) (Fig. 6.2). Nitrogen availability is considered to be limiting in many ecosystems, including drylands (Schlesinger, 1996), boreal forests (Jarvis and Linder, 2000), and tundra (Shaver and Chapin, 1980)—as evidenced by numerous observations where NPP increases under N deposition and fertilization (LeBauer and Treseder, 2008; Lu et al., 2011a; Vitousek and Howarth, 1991; Xia and Wan, 2008). Greater N availability usually results in greater aboveground biomass (especially in N-limited ecosystems), but its impact on SOC dynamics remains less well understood. As soil C accounts for two-thirds of the terrestrial C pool, any positive or negative change in total C stock due to availability or addition of N may significantly alter global soil C storage, with consequences for global climate regulation (Huang et al., 2011).

A number of studies have found increasing N availability increases SOC storage, both under mineral N supplementation and N captured via enhanced biological fixation (Johnson and Curtis, 2001; Janssens et al., 2010). However, in some cases, increasing N availability has only modest or

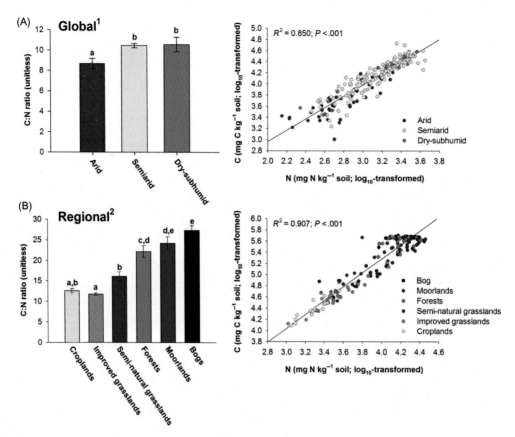

FIGURE 6.1

Mean soil ± SE C:N stoichiometry and the relationship between soil C and N for our (A) Global dataset encompassing different aridity zones (data from De Vries, F.T., Shade, A., 2013. Controls on soil microbial community stability under climate change. Front. Microbiol. 4, 265 (De Vries and Shade, 2013)) and belongs to the BIOCOM project (Maestre et al., 2012), (B) regional Scottish dataset encompassing different vegetation types (data from Delgado-Baquerizo, M., Reich, P.B., Khachane, A.N., Campbell, C.D., Thomas, N., Freitag, T.E., et al., 2017. It is elemental: soil nutrient stoichiometry drives bacterial diversity. Environ. Microbiol. 19, 1176–1188 (Delgado-Baquerizo et al., 2017)), (C) microsite dataset (data from Delgado-Baquerizo, M., Maestre, F.T., Eldridge, D.J., Singh, B.K., 2016a. Microsite differentiation drives the abundance of soil ammonia oxidizing bacteria along aridity gradients. Front. Microbiol. 7, 505 (Delgado-Baquerizo et al., 2016a)), and (D) microaggregate dataset (data from Trivedi, P., Delgao-Baquerizo, M., Jeffries, T.C., Trivedi, C., Anderson, I.C., Lai, K., et al., 2017. Soil aggregation and associated microbial communities modify the impact of agricultural management of carbon content. Environ. Microbiol. doi:10.1111/1462-2920.13779 (Trivedi et al., 2017)). Bars with different letters are significantly different. *Solid lines* on the right hand graphs represent the fitted linear regression. Details of datasets and experiments are available from the references listed above.

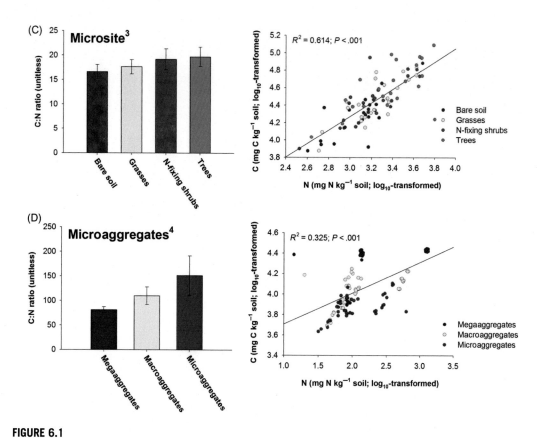

FIGURE 6.1

(Continued).

no impact on SOC, particularly when other nutrient(s) are the major limiting factor (Vitousek, 2004; Vitousek et al., 2010), although significant changes in the C:N ratio of SOM have been observed (Jandl et al., 2003; Johnson and Curtis, 2001). At other sites, N fertilization has doubled SOC stock within 20 years (Franklin et al., 2003). In general, N-rich ecosystems promote SOC storage when other factors (e.g., precipitation, temperature) are not constrained, as there is sufficient available N to support OM turnover (see e.g., Franklin et al., 2003). Conversely, N-poor ecosystems strongly limit SOC storage as a result of either: (1) Immobilization of N within the microbial biomass leading to less assimilation of detrital material into SOM, or (2) the need to scavenge N from SOM to facilitate OM turnover, which results in C losses via respiration or leaching. For example, it has been demonstrated that a stimulatory effect on SOC decomposition appears to be restricted to low N deposition areas (less than 5 kg N ha^{-1} per year) (Knorr et al., 2005), or N-limited ecosystems (Lu et al., 2011b). It should however be noted that, at the global scale, the overall increase in SOC due to N addition was modest at 2.2% (Lu et al., 2011b) and ecosystem dependant whereby forests and grasslands appear to show minimal increase in SOC response to N additions (Lu et al., 2011b), while agricultural soils and non-N limited temperate forest soils typically respond positively to N additions (Lu et al., 2011b; Janssens et al., 2010).

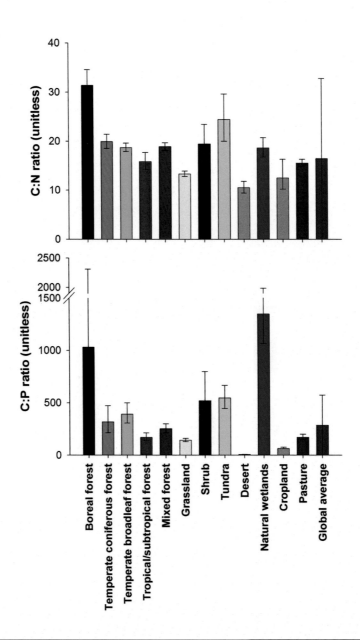

FIGURE 6.2

Mean soil (A) C:N and (B) C:P stoichiometry for major biomes (adapted from Xu, X., Thornton, P.E., Post, W. M., 2013. A global analysis of soil microbial biomass carbon, nitrogen and phosphorus in terrestrial ecosystems. Global Ecol. Biogeogr. 22, 737−749). Error bars are upper and lower 95% confidence boundaries. Bars sharing the same letter are not significantly different from each other ($P > 0.05$).

BOX 6.2 CASE STUDIES ON THE RELATIONSHIP BETWEEN SOIL C AND P: NOVEL EVIDENCE

Using the same approach as described in Box 6.1, we derived C:P ratios to investigate the relationship between soil C and P at different scales. Our analysis of microsite to global scales suggested a weak relationship between total soil C and P (Fig. 6.3), which is in line with other findings that have found weaker relationships between soil C and P than C and N at the global scale (Hartman and Richardson, 2013). Within the dryland dataset, soil C:P ratio varied significantly between different dryland classifications and was highest in semiarid zones and lowest in arid zones. The low C:P ratios in the arid soils are in line with previous observations that nutrients concentrate relative to C in dry environments (Xu et al., 2013). In our regional dataset, soil C:P ratios varied greatly across vegetation type (Fig. 6.3B) and was lowest in croplands and improved grassland (presumably as a result of fertilizer P inputs) and highest in bogs, in line with previous findings (Hartman and Richardson, 2013).

Additionally, we examined the relationships between soil C, N and P further within our regional dataset using two modeling approaches (multimodel inference and structural equation modeling (SEM)) to identify the main predictors of soil C, direct and indirect impact of soil N and P in soil C, and how they are affected by soil properties and climate variables. First, we used multimodel inference analysis to explore whether addition of N and P alone and in combination improve prediction for soil C storage. Our analysis suggested the best models to predict soil C always included total N and P, bulk density, and soil pH (Fig. 6.4A). Notably, these models improved predictions of soil C significantly and more substantially than when N or P were used alone. We then employed SEM to identify direct and indirect effects of variables on soil C. SEM confirmed our findings regarding the direct relationship between total soil C and N, but the effect of soil P on soil C was indirect and influenced via its effect on soil N (Fig. 6.4B). For example, soil P will exert influence on the interaction between soil C and N in N_2-fixing systems, where P demand is high (Houlton et al., 2008). From a mechanistic point of view, bulk density has a directly strong effect on soil C, N, and P while pH effected both soil C and P, but not N. Climate conditions have a strong direct effect on soil P, and to a lesser extent, on soil N. This is important, as to date most climate experiments have focused on the impacts of climatic drivers on N cycling and as highlighted by our analysis and others (e.g., Reed et al., 2015), incorporating P into earth systems model is important in predicting the fate of soil C under future predicted change. Results found here aligned with previous work where climatic condition is reported to impact soil P concentration and hence decouple coupling with other elemental cycles (Delgado-Baquerizo et al., 2013).

The C:N ratio of SOM seems to be an important predictor of N impact on SOC dynamics. Recent experimental and metaanalysis studies support this assumption. For example, a metaanalysis suggested that addition of N increased SOC decomposition in N-limited ecosystems (forest and grasslands), but inhibits decomposition in N-rich ecosystems (arable) (Lu et al., 2011b). Further, there is growing evidence that the differential impact of increasing N availability on SOC storage across different studies is also driven by the chemical composition of the SOM—i.e., labile versus recalcitrant OM—and the nutritional status of ecosystems. N availability increases decomposition of the light carbon fraction (decadal turnover time, e.g., cellulose) while further stabilizing and inhibiting heavier mineral-associated carbon and highly lignified SOC, which often have multidecadal to century turnover times (Neff et al., 2002). If such differential impacts of N availability on the degradation of labile and recalcitrant C fractions is consistent across ecosystems or the globe, it can be argued that plants that produce highly lignified litter will promote C storage under N addition, but that the opposite should occur for plants which produce cellulose-rich litter. Experimental support for this argument includes the observation that, when N fertilization increased, SOC storage also increased in oak-dominated forests (with highly lignified litter), but decreased in sugar maple-dominated forests (with less lignified litter) (Waldrop et al., 2004).

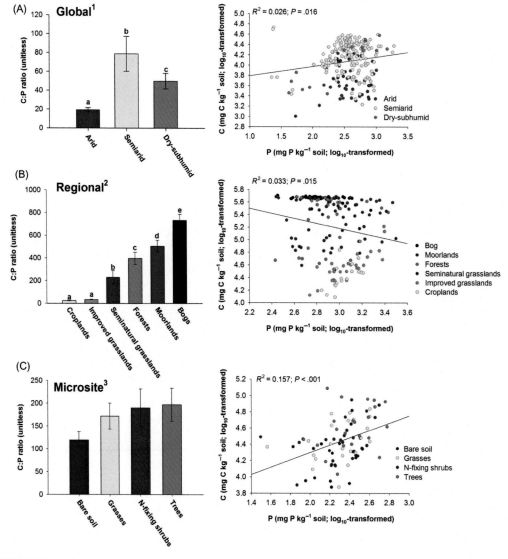

FIGURE 6.3

Mean soil ± SE C:P stoichiometry and the relationship between soil C and N for our (A) Global dataset encompassing different aridity zones (data from De Vries, F.T., Shade, A., 2013. Controls on soil microbial community stability under climate change. Front. Microbiol. 4, 265 (De Vries and Shade, 2013)) and belongs to the BIOCOM project (Maestre et al., 2012), (B) regional Scottish dataset encompassing different vegetation types (data from Delgado-Baquerizo, M., Reich, P.B., Khachane, A.N., Campbell, C.D., Thomas, N., Freitag, T.E., et al., 2017. It is elemental: soil nutrient stoichiometry drives bacterial diversity. Environ. Microbiol. 19, 1176−1188 (Delgado-Baquerizo et al., 2017)), and (C) microsite dataset (data from Delgado-Baquerizo, M., Maestre, F.T., Eldridge, D.J., Singh, B.K., 2016a. Microsite differentiation drives the abundance of soil ammonia oxidizing bacteria along aridity gradients. Front. Microbiol. 7, 505 (Delgado-Baquerizo et al., 2016a)). Bars with different letters are significantly different. *Solid lines* on the right hand graphs represent the fitted linear regression. Details of datasets and experiments are available from the references listed above.

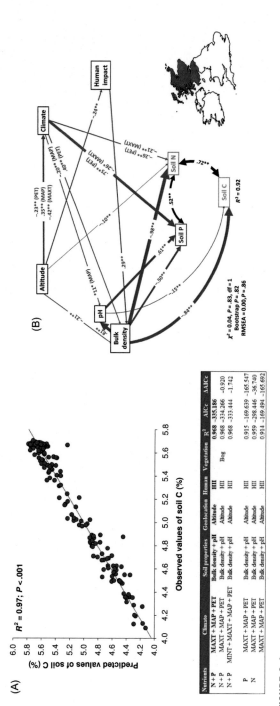

FIGURE 6.4

Results from (A) multimodel inference modeling and (B) structural equation modeling used to identify the main predictors of soil C and direct and indirect impact of soil N and P on soil and how they are affected by soil properties and climatic variables. In (B), numbers adjacent to arrows are standardized path coefficients, analogous to relative regression weights, and indicative of the effect size of the relationship. Positive numbers represent a positive relationship and negative numbers indicate a negative relationship between two parameters. A *double headed black arrow* between soil N and P and between soil C and N indicate that each parameter can affect the other and vice versa. The width of the arrow is proportional to the strength of the coefficients. The proportion of variance of SOC explained (R^2 is 0.92). * $P < .05$, ** $P < .01$. Details of datasets and experiments can be found in Delgado-Baquerizo et al. (2017).

Beyond direct effects on the decomposition of SOC, N additions can indirectly modulate soil carbon via impacts on aboveground net primary production, plant diversity, litter quality, and belowground C allocation. For example, when N is limiting, positive priming of SOM brought about by exudation of labile carbon from roots negatively impact soil C storage. Such indirect effects are discussed in more detail in the context of plant diversity (Chapter 2: Plant Communities as Modulators of Soil Carbon Storage) and nutrient additions (Section 6.4.1).

Clearly, while there is extensive understanding of the relationship between N availability and soil C, the relationship is complex and context dependant, whereby it is shaped by numerous physical, chemical and biological factors, and interactions. A system-level understanding on how these factors influence N availability under global change drivers (discussed later) will help inform earth systems models (ESMs) to better improve predictions of SOC stocks under future global change scenarios.

6.2.2 PHOSPHORUS

Like N, P is essential for growth and thus coupled to C (and N) cycling in both plant and microbial cells. Soil P is primarily derived from bedrock and the weathering of primary minerals. External inputs are largely negligible—with the exception of some areas that receive P inputs from dust deposition or fertilization—and because leaching losses are typically low, in natural terrestrial ecosystems the cycling of P is essentially closed, with new mineral inputs occurring over geological timescales. Even when total soil P levels are high, availability can be low owing to incomplete weathering of primary minerals and strong adsorption onto mineral surfaces of occlusion in secondary minerals. Thus, the mineralization of organic P is a vital component of ecosystem P cycling. Understanding the rate at which P moves between these different pools is important to be able to understand the interactions between soil C and P cycles. However, competition between biological demand for P and sorption onto mineral surfaces (Sollins et al., 1988) complicates our understanding of movement of P between pools. Conventional understanding dictates that the occluded P-fraction is biologically unavailable, however it is now emerging that occluded P becomes biologically available over short timescales (hours to decades) (Richter et al., 2006; Huang et al., 2014) which creates further uncertainties regarding our understanding of P-sorption and mineralization rates.

Soil P availability is strongly related to geology and soil age—old soils being depleted in P relative to younger soils (McGill and Cole, 1981; Walker and Syers, 1976). This is supported by data from long-term chronosequences whereby there is a shift toward P limitation as soils age (Peltzer et al., 2010; Selmants and Hart, 2008, 2010). Because of this, soil P is often the major limiting nutrient in the tropics, where soils are much older than those in the mid-latitudes—a consequence of the lack of glaciations in this region—and N is the major limiting nutrient in the mid-latitudes (Delgado-Baquerizo et al., 2016b; Reich and Oleksyn, 2004). Soil type is also a major factor in determining the P nutrient status of an ecosystem. Vegetation growing on Oxisols and Utisols in tropical regions is likely to be P limited because of the high P-binding of these soils. Similarly, carbonate-rich soils have high P-binding capacity and are often P limited, e.g., Yucatan Peninsula (Krasilnikov et al., 2013; Lugo and Murphy, 1986; Read and Lawrence, 2003). In acidic soils, P is bound to Al and Fe oxides and adsorption/desorption processes control plant supply (Sanyal and De Datta, 1991). As such, P limitation often dominates at low latitudes, while at mid latitudes N and P limitation is frequent, and at high latitudes N limitation dominates (Elser et al., 2007).

Terrestrial P limitation, and/or N-P colimitation, for plant production and microbial biomass is significant (Elser et al., 2007; Wang et al., 2010), and is predicted to increase (Vitousek et al., 2010). The importance of the imbalance between N and P in terrestrial ecosystems is increasingly recognized as an important driver of ecosystem C sequestration (Peñuelas et al., 2012). As organic P is mineralized biochemically out with cell membranes (compared to the biological mineralization of C and N), by microbial and plant-derived enzymes and organic acids, the rate of P mineralization is driven by P demand rather than by the need for energy (McGill and Cole, 1981). Thus, the coupling of P with C and N cycling is not as strong as that between C and N. Further, given the different primary origins of N and P in terrestrial ecosystems, P (mineral) may explain a unique portion in the variation of C, which is not accounted for by N (biological) and this has highlighted the need to explicitly include P in models to more accurately predict soil-C storage potential (Peñuelas et al., 2012). This is because—in the absence of fertilizer inputs—the ability of terrestrial ecosystems to respond to increased P demand under future climate scenarios will be constrained by the slow rate at which new P enters the soil via mineral weathering (Cleveland et al., 2013; Peñuelas et al., 2013), or by the ability of the microbial community to alter the rate of organic P mineralization. Thus the importance of P in regulating the global C cycle could manifest itself (Reed et al., 2015).

In general, soil P stoichiometry is more variable than N across the globe (Hartman and Richardson, 2013) and differs considerably between, and within, biomes (Fig. 6.2). As soil C accumulates with time, soil P becomes more diluted (Hartman and Richardson, 2013), thus soil C:P ratios are high in wetlands (846:1), tundra (545:1) and boreal forests (1030:1) compared to temperate (318:1) and tropical forests (169:1), grasslands (143:1), croplands (64:1), and desert soils (6.6:1) (Xu et al., 2013). Our analysis of data from microsite to global scale (Box 6.2) suggests that the relationship between soil C and P stocks, are comparatively stronger at the microsite scale compared to regional and global scales (Fig. 6.3) where this relationship is weaker. Hartman and Richardson (2013), in their global metaanalysis, also report a lower allometric slope of the relationship between soil C and P (0.29) than between soil C and N (0.88), which suggests a weak relationship between soil C and P cycling (Box 6.3).

The cycles of soil C and P are only coupled via microbial and root activity including the release of extracellular enzymes (e.g., phosphatase) that allow the mineralization of P from SOM. This key biological link between microbial and plant processes and P availability is however strongly threatened by climate change. For example, studies suggest that as aridity continues to increase in some regions (Huang et al., 2016), the role of biological activity in coupling the cycles of soil C and P will be strongly limited as a consequence of reductions in SOM and biological activity (Delgado-Baquerizo et al., 2013; Jiao et al., 2016; Yuan and Chen, 2015). The importance of water availability in soil C-nutrient relationships is also highlighted by Luo et al. (2016) and identifying a threshold at which biological processes exert stronger control on soil C:P over geochemical processes may help to better inform model predictions.

6.2.3 SULFUR

After N and P, sulfur (S) is the next most abundant cellular constituent. Soil S is generally coupled with C and N because of their stoichiometric requirements for biosynthesis (Burgin et al., 2011). However, the relationship between soil C and S at regional and global scales has been sparsely studied. In most soils, 80% or more total S is in organic form, and thus perhaps more tightly linked

---

BOX 6.3 NUTRIENT CONSTRAINTS IN TROPICAL LOWLAND AND MONTANE SYSTEMS

Tropical soils make a substantial contribution to the global C budget, accounting for a third of soil C stocks worldwide, despite representing only 15% of land surface area (Jobbágy and Jackson, 2000; Pan et al., 2013). Anthropogenic N deposition is increasing rapidly in the tropics as a consequence of biomass burning (Boy et al., 2008; Hietz et al., 2011), while aeolian sources of P will also increase the future availability of P in some tropical systems (Okin et al., 2004). However, the nature and extent to which N and P modulate soil C cycling in tropical ecosystems remains poorly understood, limiting our ability to predict how alteration of these nutrient cycles will impact future soil C storage in the tropics (Cusack et al., 2016; Townsend et al., 2011).

Theories of soil development, and the different sources with N and P are derived in terrestrial ecosystems, suggests that the nature of nutrient constraints to soil microorganisms (determining soil C turnover) may vary across gradients of tropical lowland to montane forests and grasslands. Old, strongly weathered and leached tropical lowland soils may be more deficient in rock-derived P (Reed et al., 2011b, Walker and Syers, 1976), while N is relatively high due to accumulation over time and rapid rates of internal N cycling (Templer et al., 2008). Younger, tropical montane soils, in contrast, may be more deficient in N because of slow rates of decomposition and biological N-fixation (due to cooler temperatures) (Tanner et al., 1998), while P remains abundant due to landslide events (Clark et al., 2016) and near-surface weathering (Porder et al., 2007).

Although there is some support for this shift from P to N limitation with increasing elevation (Nottingham et al., 2015), studies of the tropics remain very limited, evidence is inconclusive and considerable uncertainty remains as to whether increased nutrient availability will serve to either increase or decrease future rates of soil C turnover. In some cases, fertilization with N and P, in different montane and lowland forests respectively, has been shown to increase microbial biomass (Cusack et al., 2011; Liu et al., 2013; Turner and Wright, 2014), decomposition (Chen et al., 2015; Cleveland et al., 2002), and soil respiration (Cleveland and Townsend, 2006; Fisher et al., 2013; Homeier et al., 2012) which is consistent with basic stoichiometric decomposition (BSD) theory. However, in a lowland forest in Southern China, P addition reduced leaf-litter decomposition, proffered as a result of reduced microbial mining when P was externally supplied (Chen et al., 2013). Likewise, increased N availability in Puerto Rican montane forest soils decreased heterotrophic respiration and oxidative enzyme activities, resulting in increased soil C storage (Cusack et al., 2016).

In contrast to the aforementioned studies, other studies have reported no difference in leaf decomposition rates between control and fertilized plots (Davidson et al., 2004; Hobbie and Vitousek, 2000; McGroddy et al., 2004), perhaps suggesting that the quality of available C was a greater constraint compared to nutrient availability, or that nutrients other than N or P were limiting, in line with BSD theory. Most studies have primarily focused on N and P, so the role of S in tropical lowland and montane systems is largely unknown. However, in a lowland forest in French Guiana, fertilization with a combined nutrient treatment, including S (K, Ca, Mg, B, Cu, Fe, Mn, Mo, S, and Zn) inhibited leaf-litter decomposition (Barantal et al., 2012), potentially leading to increased SOC. Furthermore, Kaspari et al. (2008) demonstrated that decomposition in lowland Panama was limited by multiple nutrients. These two studies highlight the need to also consider the potential of multiple nutrients modulating soil C storage in the tropics.

---

to the C cycle than both N and P (Schlesinger, 1996). Geochemical processes such as atmospheric deposition and evaporation may be an important driver in the relationship between soil S and C in some environments. For example, in one study across a 3500 km aridity gradient in northern China, soil S was generally found to be strongly positively correlated to soil C. However, the relationship became negative when an aridity threshold (0.91) was reached, suggesting a decoupling of soil C and S cycling in hyper arid environments (Luo et al., 2016).

Of all elemental cycles, the S cycle has been most altered by anthropogenic activity (Likens et al., 1981), with widespread S emissions and acid precipitation across Europe and North America (Dentener et al., 2006; Schlesinger, 1996). Atmospheric inputs of S can be significant (>80 kg S ha^{-1} per year) close to industrial areas and $2-15$ kg ha^{-1} per year in more remote areas (Bunemann and Condron, 2007), but are typically equaled by leaching losses (Mitchell et al., 1992). S-deposition has an acidifying effect that retards decomposition either by directly affecting microbial activity (Baath et al., 1980) and enzyme production (Sinsabaugh, 2010), indirectly by affecting microbial C supply (Pennanen et al., 1998; Persson et al., 1989; Scheel et al., 2007), or causing imbalances in N and P cycling as a result of increased leaching of nutrients from soils and foliage (Johnson, 1979). However, since the mid-1990s, declining levels of S-deposition in Europe and North America (Driscoll et al., 2001; Kopáček and Veselý, 2005) have been observed, which has reduced leaching of N, P, and other elements in many areas affected by historic S-deposition. Subsequent "recovery" from S-deposition has also seen simultaneous increases in losses of dissolved organic carbon (DOC) from soils (e.g., Driscoll et al., 2003; Evans et al., 2005; Hejzlar et al., 2003; Monteith et al., 2007; Skjelkvåle et al., 2005). It has been proposed that increases in soil pH following declines in S-deposition are, in-part, responsible for the observed DOC losses (Evans et al., 2007; Oulehle et al., 2011). This may also have significant implications for N cycling. For example, Oulehle et al. (2011) reported increased loss of dissolved organic N, reduced net N-mineralization rates, and increased litter and soil C:N ratios following recovery from S-deposition in European boreal forest soils. The consequences of such changes remain unclear. Higher litter C:N ratio's may favor accumulation of SOC, while a drive toward N limitation may enhance fine root growth (Lamersdorf and Borken, 2004) and stimulate microbial priming of SOM (Phillips et al., 2011), which could serve to limit C sequestration in recovering boreal soils. Further uncertainty surrounds the impacts of changes in the soil S cycle and soil acidity on other geochemical cycles (e.g., Aluminum) (Oulehle et al., 2011; Scheel et al., 2007).

6.2.4 OTHER ELEMENTS

While N and P are typically considered to be the primary nutrients that will influence the ability of an ecosystem to sequester carbon, several studies have demonstrated that, in highly weathered soils, decomposition of litter inputs in forest soils can be limited by low calcium (Ca) or potassium (K) availability (Cuevas and Medina, 1988; Kaspari et al., 2008). Indeed, in a study of a Panamanian forest growing on highly weathered oxisols, litterfall and decomposition was controlled by at least four different nutrients (Kaspari et al., 2008). In some wet tropical soils, base cation leaching leads to Ca, Mg, and K limitation, but this can be overcome in coastal areas where deposition of base cations can alleviate such limitation and push toward P limitation (Chadwick et al., 1999). Knowledge on the role of other elements (e.g., iron) in modulating SOC storage is limited, although theoretically they can play a significant role and need to be further investigated.

6.3 THEORIES AND MECHANISMS TO EXPLAIN NUTRIENT IMPACTS ON SOC STORAGE

6.3.1 THEORETICAL FRAMEWORK

There are two leading, and competing, theories to explain the role of nutrients in SOC degradation (Chen et al., 2014) and hence SOC storage, namely: (1) Microbial nutrient mining (MNM) theory which predicts that when a nutrient is limiting, the microbial community "mines" SOM in order to meet their nutritional requirement, potentially leading to loss of soil C via increased microbial activity (e.g., microbial respiration) (Moorhead and Sinsabaugh, 2006); and (2) Basic stoichiometric decomposition (BSD) theory which proposes that decomposition of SOM is governed by the stoichiometry of substrates and microbiota (Craine et al., 2007), and that the maximum rate of SOM decomposition occurs when there is sufficient supply of nutrients (N and P) to match the availability of C in SOM (Cleveland et al., 2002). These theories are contrasting because increased nutrient availability under the BSD theory will allow more rapid loss of SOM due to an increase in the utilization of SOM, while the MNM theory proposes that increasing availability of nutrients can discourage mining and encourage the use of labile C (Fierer et al., 2007; Fontaine et al., 2003) (Fig. 6.5). It is believed that these theories can operate both individually as well as in combination, depending on the availability of C and other nutrients (Chen et al., 2014) (Fig. 6.5).

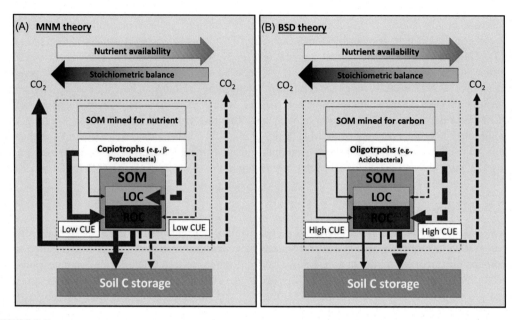

FIGURE 6.5

Theoretical framework for nutrient impacts on soil carbon storage (SOC). (A) According to microbial nutrient mining (MNM) theory, under high nutrient availability, copiotroph dominated communities with a low carbon use efficiency (CUE) mine soil organic matter (SOM), resulting in significant CO_2 loss. Addition of N (+P) reduce SOM mining by copiotrophs and promote SOC storage. (B) According to basic stoichiometric decomposition (BSD) theory, under sufficient nutrient availability, oligotrophs mine SOM for carbon to balance their homeostatic cellular stoichiometry, but have a high CUE and low metabolic rate so loss of carbon to the atmosphere is low. Addition of N (+P) further promotes SOM decomposition, but CO_2 loss remains comparatively low because of high CUE and low metabolic rate. LOC is labile organic carbon and ROC is recalcitrant organic carbon. *Blue arrows* indicate a positive effect and *red arrows* indicate a negative effect. *Dashed lines* represent the scenario under nutrient additions. It is possible that both mechanisms operate at the same site, but at different times. For example, addition of nutrients will promote copiotrophs initially and reduce decomposition of recalcitrant SOC. However, once nutrients and/or LOC are exhausted, there will be a shift in microbial processes which can promote ROC utilization. In another scenario, once nutrients are exhausted by sufficient LOC is present copiotrophs will mine ROC for nutrients (N and P). However, when LOC is exhausted, copiotrophs will be replaced by oligotrophs, which in the presence of high nutrients will mine ROC for growth and energy.

6.3.2 **MECHANISTIC KNOWLEDGE**

Growing evidence suggests that increasing N availability can increase SOC storage through impacts on soil microbial communities. For example, two metaanalyses (of 82 and 111 published papers respectively) assessing the impact of N addition across a range of biomes found, on average, 15% and 20% (respectively) declines in microbial biomass under N fertilization (Liu and Greaver, 2010; Treseder, 2008). Accompanied with reported declines in soil respiration under N addition (e.g., Liu and Greaver, 2010; Ramirez et al., 2010; Treseder, 2008), these observations imply a negative effect of N addition on decomposition, thus facilitating potential soil C-storage. Other studies provided either modest, zero or negative impacts of N addition on SOC storage (see Section 6.2.1). Several factors may drive these inconsistent effects of N on SOC storage. Some of these are considered below.

6.3.2.1 *Enzyme inhibition*

MNM theory supports the idea that degradation of recalcitrant SOM yields no net energy (Couteaux et al., 1995), and microbes and plant roots only mine SOM under nutrient limiting conditions. Under increased nutrient availability, the need for nutrient mining is limited, the production of SOM-mining enzymes is suppressed, and the use of labile C as a source of energy is encouraged (Craine et al., 2007), thus allowing accumulation of recalcitrant SOC. Indeed, high availability of N directly inhibits a number of enzymes which are critical to degrade SOM, particularly recalcitrant fractions of SOM (e.g., lignolytic enzymes; Berg and Matzner, 1997; Fog, 1988) and it has been demonstrated that fungi exposed to chronic N addition are less effective at litter decay (van Diepen et al., 2017). Under the microbial N mining hypothesis, decomposers may primarily degrade lignin to acquire N from cell membrane proteins that are shielded by lignin and other recalcitrant compounds (Craine et al., 2007; Rinkes et al., 2016). Thus, when N is available, the need for lignolytic enzyme production is removed. High availability of N may instead promote the activity of labile C degrading enzymes (e.g., cellulases; Sjöberg et al., 2004). Therefore, the impact of N addition on SOC storage may depend on the chemical composition of SOM. For example, if SOM is cellulose-rich, addition of N will enhance SOC loss, but the opposite effect would occur if SOM is highly lignified (Berg and Matzner, 1997; Fog, 1988; Knorr et al., 2005). Complex interactions between SOM chemistry, site biology, and environmental conditions may, however, occur, modifying the impact of N on soil C (see Sections 6.4 and 6.5).

The expression of lignin-degrading enzymes in some white rot fungi is known to be suppressed by N (Hammel, 1997). This is supported by some field studies, whereby the activity of lignolytic enzymes was significantly lower (5%−73%) in temperate forest soil under N addition (Carreiro et al., 2000; Gallo et al., 2004). Evidence suggests a consistent inhibitory effect of N on lignin-degrading enzymes across biomes and soil types (Ramirez et al., 2012). However, this observation is not universal. For example, Hobbie (2008) reported no impact of N addition on lignin decomposition rates in forest and grassland soils. It has been suggested that the initial lignin content of litter could determine the direction of the impact of N addition on decomposition, whereby decomposition is retarded in high lignin litter, but stimulated in low-lignin litter (Knorr et al., 2005). Additionally, as decomposition progresses, there is a shift from relatively labile litter in the early stages of decomposition to more recalcitrant lignin-rich forms in the latter stages (Berg and Matzner, 1997), thus stimulatory and/or inhibitory effects of N on decomposition are unlikely to be constant over time. Furthermore, N could also interact with the breakdown products of lignin degradation to form more recalcitrant compounds, thus facilitating C sequestration (Dijkstra et al., 2004; Fog, 1988). As such, while the enzyme inhibition theory offers a mechanism for how N might negatively affect some aspects of decomposition under certain conditions, additional mechanisms controlling the role of N in modulating C sequestration are evident.

6.3.2.2 Changes in microbial community structure

Microbial community composition as a modulator of soil C sequestration is discussed in detail elsewhere in this book (Chapter 3: Microbial Modulators and Mechanisms of Soil Carbon Storage), but it is worth mentioning here in the context of N (and P) impacts on the microbial community structure. In support of the MNM theory, N mediates soil C storage by causing shifts in microbial community structure. When N is limiting, oligotrophic microbes (i.e., K strategists) capable of mining N from SOM are favored, but when N is sufficient, copiotrophic microbes (r-strategists) thrive (Fierer et al., 2007; Fontaine et al., 2003) and more labile forms of C are utilized over recalcitrant SOM (Fig. 6.4A). However, SOM mining will only be suppressed if sufficient labile C is available to favor copiotrophic growth over oligotrophic growth (Chen et al., 2014). A shift from an oligotroph-dominated community to a copiotrophic community has been observed in response to N additions across a broad range of biomes (Leff et al., 2015; Ramirez et al., 2010, 2012), as well as in soils under different management practices that influence N availability, and potential soil C storage (Figuerola et al., 2012; Hartmann et al., 2012; İnceoğlu et al., 2011). Although experimental evidence is lacking with respect to shifts in microbial community structure between copiotrophic and oligotrophic growth under increasing P availability, when N is sufficient and P is limiting to microbial growth, a similar scenario is likely. For P, the majority of evidence suggests either neutral (e.g., Barantal et al., 2012; Chen et al., 2013; Cleveland et al., 2006; McGroddy et al., 2004), or positive (e.g., Fisk et al., 2015; Hobbie and Vitousek, 2000; Kaspari et al., 2008) effects of P addition on decomposition. This may be explained with the growth-rate hypothesis (Elser et al., 2003) whereby high levels of P availability could promote fast-growing organisms that require high levels of P for energy and thus accelerate the decomposition of SOM. Nevertheless, negative effects of P addition on decomposition have also been less frequently reported (Chen et al., 2013). Suppression of MNM was suggested as a mechanism for the observed reduction in decomposition, whereby greater P availability could also promote utilization of more labile C (Marklein and Houlton, 2012; Olander and Vitousek, 2000).

The strong relationship between fungal:bacterial (F:B) ratio and soil C:N ratio observed globally (Fierer et al., 2009), and the shift in dominance between fungi and bacteria, has been correlated with soil C. This relationship, in part, is driven by tighter stoichiometric constraints, and lower carbon use efficiencies (CUE—discussed later) of bacteria compared to fungi (i.e., bacteria require more N per unit biomass than fungi; Bardgett and McAlister, 1999; De Deyn et al., 2008; Kuijper et al., 2005), which leads to higher C inputs to soil (Austin et al., 2004). Fungal-dominated soils typically have higher soil C and C:N ratios (Bailey et al., 2002; Fierer et al., 2009; Guggenberger et al., 1999), and thus the F:B ratio has been suggested as an important factor for soil C storage (Jastrow et al., 2007; Strickland and Rousk, 2010). Thus, nutrient additions that cause a shift toward a bacterial-dominated decomposition pathway could impact SOC storage under the same principal as oligotrophic (fungi) and copiotrophic (bacterial) growth. However, lower soil C in fungal-dominated sites has also been reported (Busse et al., 2009; Mulder and Elser, 2009) although at least one study challenged the widely held hypothesis of a fungal-dominated pathway of decomposition in the presence of more complex/recalcitrant organic material (Rousk and Frey, 2015). It may be that shifts in the composition of the fungal community could also be important in determining the fate of soil C in response to nutrient additions (van Diepen et al., 2017; Wurzburger and Brookshire, 2017) which may, in part, account for the unexpected observations observed by Rousk and Frey (2015).

Fungi are also better adapted to scavenging nutrients from SOM than bacteria, particularly at depth (Allen, 2011). So greater abundance of fungi could equally translate, in some cases, to higher mining of SOM and greater losses of soil C under the mining hypothesis, when N is limiting. However, fungi may also play an important role in N mining of SOM (Rousk et al., 2016) whereby they selectively target N-rich compounds which overall could decreases microbial SOM-use (Rousk et al., 2016) and thus potentially increase soil C storage. Shifts in dominance within fungal communities in response to N may explain some of the inconsistencies reported earlier (van Diepen et al., 2017; Wurzburger and Brookshire, 2017). Application of N results in a shift from Basidiomycetes- to Ascomycetes-dominated fungal community (Nemergut et al., 2008). This is important because Basidiomycetes possess an arsenal of lignin-degrading enzymes and, therefore, the dominance of Ascomycetes in N-rich ecosystems can promote SOC storage.

It is also interesting to note here that there may be important interactions between nutrient availability and mycorrhizal type that influence SOC storage. For example, a metaanalysis demonstrated that under elevated CO_2, when N is limiting, increased plant biomass is only seen in plants associated with ectomycorrhizal (EM) and not with arbuscular mycorrhizal (AM) fungi (Terrer et al., 2016). This is likely to stem from the fact that EM fungi produce enzymes capable of obtaining N from SOM, whereas AM fungi scavenge available N via exploration (Smith and Read, 2008). The effect on soil carbon storage in EM-dominated communities will be a function of the relative response of increased litter inputs resulting from increased biomass versus the rate at which SOM is decomposed by the EM community to access N. Mycorrhizal type may also be an important factor in P-limited systems under eCO_2, where AM association may benefit the plant more than EM association because of the greater efficiency of AM fungi to acquiring P (Johnson et al., 2015). In this case, AM associations could facilitate SOC storage by promoting greater litter inputs. Together these examples suggest that community dynamics within fungal communities should be considered when predicting relationships between fungal dominance and soil C storage. This is also true for the bacterial community, as highlighted earlier and discussed in further detail in Chapter 3, Microbial Modulators and Mechanisms of Soil Carbon Storage.

6.3.2.3 Carbon and nutrient use efficiency

An important regulator of the fate of detrital C entering soil is the CUE of the microbial biomass (Bradford and Crowther, 2013; Manzoni et al., 2012; Six et al., 2006) which is influenced by (among other factors) substrate quality and nutrient availability (Keiblinger et al., 2010). At high substrate C:N ratio, CUE is low as microbial decomposition is nutrient limited and excess C is lost via overflow respiration. Generally, CUE increases with nutrient availability (Ågren et al., 2001) as substrate stoichiometry becomes more aligned to microbial stoichiometric demand. Thus, there is potential for higher rates of soil C storage as the amount of C lost via overflow respiration is lowered. In soils with decreasing C:N ratio (i.e., C-limited, N-sufficient) CUE has been shown to increase (Devêvre and Horwáth, 2000; Thiet et al., 2006; Ziegler and Billings, 2011). Conversely in N-limited soils (high C:N ratio), CUE is lower (Sinsabaugh et al., 2013). Under BSD theory, N additions may stimulate the production of N-rich extracellular enzymes, and accelerate SOM decomposition (Fig. 6.4B) as there are sufficient nutrients available to maintain stoichiometric balance. In support of this, several studies demonstrated increased decomposition with increasing N (e.g., Bragazza et al., 2006; Hobbie, 2005; Mack et al., 2004) and P (e.g., Cleveland et al., 2006; Cleveland and Townsend, 2006) availabilities. However, increased CUE (i.e., lower respiration rates) under N additions may compensate this loss of soil C. Clearly several mechanisms can be at interplay here, and predicting how nutrient-induced changes in CUE will affect C sequestration is not straightforward. For example, Manzoni et al. (2012), in their synthesis paper, draw on theory and empirical evidence to highlight how increased CUE (i.e., increased nutrient availability) could delay the mineralization of plant residues by reducing overflow respiration, leading to a build-up of organic N and reduced availability of inorganic N for plant uptake. On the one hand, this could negatively affect soil C sequestration by reducing plant growth, leading to reduced litter inputs, and increased litter C:N ratio which may drive mining of SOM. Alternatively, increased CUE could facilitate decomposition of SOM, as resources for enzyme production are increased (Allison et al., 2010), which could increase NPP and litter inputs as nutrients become available and positively influence soil C storage. This circular argument highlights the difficulty in predicting the impacts of altered N availability on soil C storage and highlights the uncertainties that still need to be addressed before parameters, such as CUE, can effectively be used in models to predict soil C storage.

Microbes can regulate their nutrient use efficiencies in response to imbalances in substrate stoichiometry (Mooshammer et al., 2014; Sterner and Elser, 2002). However, we only have limited knowledge of the regulation of microbial nitrogen use efficiency (NUE) and even less on phosphorus use efficiency (Zechmeister-Boltenstern et al., 2015). The majority of studies aimed at improving our mechanistic understanding of the N cycle and interactions with other biogeochemical cycles have focused on the process involved in mineralization of amino acids to ammonium in controlling inorganic N availability. However, it is now recognized that the initial depolymerization of proteins to oligopeptides and amino acids is the rate-limiting step of decomposition (Jan et al., 2009; Schimel and Bennett, 2004). As well as providing substrate for subsequent mineralization to ammonium, these depolymerization products (i.e., oligopeptides and amino acids) represent an energy and nutrient source that the microbial community can use directly (Jones et al., 2004, 2009). Thus, there is increasing recognition that improved understanding of factors controlling NUE—the partitioning of organic N into microbial biomass versus mineralization and release of inorganic N to the environment—is needed in order to better understand the regulation of ecosystem functions including soil C sequestration. As with CUE, microbial NUE within soil varies considerably (0.15−1; Mooshammer et al., 2014) and has been demonstrated to be flexible in response to resource stoichiometry. For example, higher microbial NUE in plant litter (0.89) and organic soil (0.83) compared to mineral soil (0.70) suggests that in N-limited environments (i.e., high C:N ratio) efficient retention of organic N within the microbial biomass (i.e., immobilization) reduces the amount of inorganic N for plant uptake (Mooshammer et al., 2014). One the one hand, this could serve to limit NPP which could feed back into reduced litter C inputs into soil and limit increases in soil C stocks, and on the other hand, could promote increased rhizodeposition and induce MNM of SOM mining with subsequent decreases in soil C stocks. In reality, both high NUE and MNM can operate in parallel when nutrients are limiting. Nutrient status of an environment is an important factor influencing whether soil C is lost or gained. For example, if both C and N are limiting, N addition will not result in significant C loss as a result of enzyme inhibition (Ramirez et al., 2012). However, if C is abundant and N is limiting, N addition may enhance C loss as MNM mechanisms may operate. Future research needs to focus on what circumstances lead to a net loss or gain in soil C under nutrient limited conditions.

6.4 MANAGEMENT EFFECTS ON NUTRIENT MODULATORS AND SOC

6.4.1 NUTRIENT ADDITIONS

6.4.1.1 Nitrogen fertilization

Global anthropogenic nitrogen deposition and fertilizer application equivalent to more than 200 Tg N each year has been reported (Galloway et al., 2008; Solomon, 2007), which far exceeds natural N inputs. This is further projected to increase by 2.5 times by 2100 (Lamarque et al., 2005), with the potential to strongly influence future soil C dynamics. Increased soil C stocks under N addition have been corroborated for some ecosystems in a metaanalysis (Lu et al., 2011b) which concluded that soil C increased on average by 3.5% in agricultural soils, but not in nonagricultural systems under N addition. Other attempts to identify broad scale patterns of N-induced changes have concluded that, overall, soil C levels are likely to increase between 3.5% and 8% under N additions (e.g., Liu and Greaver, 2010; Yue et al., 2016). However, there are several notable exceptions where soil C losses have been reported under elevated N (Khan et al., 2007; Mack et al., 2004; Neff et al., 2002). Such variation in responses between studies arises because soil C dynamics are affected by numerous factors including climate, soil type, ecosystem type, plant species composition, and nutritional status of the ecosystem (Laganiere et al., 2010; Lal, 2004a). Indeed, the initial nutrient status of the environment is likely to be a key factor here as it has been reported that, for agricultural soils at least, soil C sequestration will not occur unless there are sufficient nutrients (N, P and S) to facilitate microbial processing of C into stable SOC (Kirkby et al., 2014). Further, in a pioneering study, Townsend et al. (1996) argued that N may stimulate C sequestration in some ecosystems, but with a number of potentially harmful effects including increased N leaching, N_2O emission, acid rain, and soil acidification which will offset any potential benefits for SOC accumulation.

Identifying and understanding the mechanisms that lead to a positive response of SOC to N additions is necessary in order to better predict the future C-sequestration potential of different ecosystems. Key to this is identifying microbial responses to N addition, and subsequent consequences for SOC. A metaanalysis of ecosystem studies concluded that N additions reduced microbial biomass (-15%) and soil respiration to a similar degree (Treseder, 2008), suggesting reduced rates of litter decomposition and potential increased soil C-storage under N addition. This was supported, to some extent, by another study (Liu and Greaver, 2010) which found similar declines in soil microbial biomass (-20%) and soil respiration (-8%), leading to increased C storage ($+17\%$) in the organic layer. When potential confounding effects of multinutrient additions were removed, a mean 5.8% increase in total soil carbon owing to N addition was reported (Yue et al., 2016), highlighting that N-addition concentration and soil depth were important moderators of soil C responses to N addition (Lu et al., 2011b; Yue et al., 2016).

Nitrogen additions also alter the quantity and quality of plant input (i.e., carbon) entering the decomposition pathway, and thus can indirectly influence soil C sequestration. For example, higher plant productivity under higher N availability (e.g., Vitousek et al., 1997; Xia and Wan, 2008) could lead to greater litter inputs, but this may not translate to greater soil C as a consequence of more rapid turnover of higher quality litter. N additions are known to decrease the lignin:N ratio of litter (Laskowski and Berg, 2006), reducing the recalcitrance of litter and leading to faster C release during decomposition. Furthermore, N additions decrease belowground C allocation (Lu et al., 2011b), which is critical for soil carbon storage, as: (1) The belowground C pool represents as much as 50% of the C fixed annually (Liu and Greaver, 2010); and (2) roots typically decompose more slowly compared to aboveground vegetation (Berg and McClaugherty, 2014). Thus, soil C storage may be negatively affected by high N availability over the long-term. The stimulatory effect of N additions on fresh C supply to the soil from aboveground may serve to decrease SOC stocks through stimulation of microbial mineralization of mineral associated (i.e., "stable") carbon (Fontaine et al., 2007). Increased DOC in both organic ($+19.69\%$) and mineral ($+12.32\%$) soils, owing to N additions, may also act to facilitate SOM priming (Yue et al., 2016). However, this may be compensated to some extent by a reduction in rhizodeposition (e.g., Dijkstra et al., 2005; Högberg et al., 2010), which may serve to limit positive priming of SOM.

6.4.1.2 Phosphorus fertilization

Fertilizer P inputs have quadrupled relative to preindustrial levels (Falkowski et al., 2000; Vitousek et al., 1997), significantly altering soil stoichiometry. As observed for studies that have addressed N constraints on soil C-storage, those that have considered P reveal contrasting results with positive (Fisk et al., 2015), neutral (Li et al., 2010), and negative (Armitage and Fourqurean, 2016; Li et al., 2014) impacts reported or implied (Jing et al., 2017). Poeplau et al. (2016) reported that P additions lead to a depletion of soil C stocks across 10 long-term field trials (>45 years) in cropping systems covering a wide range of climatic and pedological conditions where N was limiting. This again highlights the importance of the initial nutrient status of an ecosystem in predicting the response of soil C stocks to P additions. Similarly, in a laboratory study, Bradford et al. (2008) demonstrated that P amendments increased decomposition, however this negative effect on SOC diminished slightly when N and P were added together. The study also demonstrated that different soil C fractions respond differently to nutrient addition, thus highlighting the need for empirical studies that aim to better understand interactions between C and other biogeochemical cycles, not only considering SOC as a single fraction (Ågren and Bosatta, 2002; Knorr et al., 2005).

Overall it is anticipated that soil C stocks will increase under N additions as a result of increased litter inputs, which will be preferentially utilized by decomposers over more stable soil C. However, the initial nutrient status of an ecosystem is a key factor to determining the magnitude of this response, and N additions may lead to other nutrient limitations which could limit the supply of fresh C inputs. Indeed, N additions have been shown to increase phosphatase activity across a wide variety of ecosystems (Marklein and Houlton, 2012), and thus potentially accelerate P cycling. However, if P is limiting, the ability of ecosystems to respond to increased P demand will be constrained by the slow rate at which new phosphorus enters the P-cycle via mineral weathering

(Cleveland et al., 2013; Peñuelas et al., 2013). This highlights the complex interactions between biogeochemical cycles that should be considered when seeking to improve the predictability of ESMs. Although P additions will also affect SOC stocks, at a global scale they may be less important than N additions, because atmospheric P inputs are low (Mahowald et al., 2008) and the use of P fertilizers in agriculture has stalled since 1989 while N inputs continue to increase (Peñuelas et al., 2012).

6.4.2 LAND-USE CHANGE—AFFORESTATION AND REFORESTATION

Soil nutrient dynamics vary with time following afforestation, typically declining over the short-term (<20 years), but increase over the long-term, via atmospheric N deposition and biological nitrogen fixation (BNF; Knops and Tilman, 2000; Morris et al., 2007; Yang et al., 2011). This can positively influence soil C stocks as NPP increases under the higher nutrient availabilities (e.g., Li et al., 2012). Although the planting of symbiotic N_2-fixing tree species have been demonstrated to increase N availability and promote soil C storage (Macedo et al., 2008), the high P demand of symbiotic N_2 fixation (Israel, 1987) could serve to potentially constrain soil C sequestration over the long-term in soils with low P status—as C-inputs are either limited by reduced NPP or MNM of SOM to facilitate increased P demand. Further, deeper rooting of trees in forested areas could mine soil N from deeper mineral layers than was possible under agricultural land use (Knops and Tilman, 2000) and thus threaten soil C stocks deeper in soil profiles, and merits further investigation.

In most ecosystems for P, unlike N, there are no/insignificant P inputs over short- or mid-term timescales. Thus, afforestation is likely to have negative effects on total soil P, as P mineralization initially increases and P is locked up in aboveground biomass. This is evidenced from long-term chronosequences that show reduced P availability relative to N availability with increasing age (Peltzer et al., 2010; Walker and Syers, 1976). The extent that this drives such afforested areas to P limitation and might invoke some of the mechanisms highlighted earlier is unknown, and subsequent consequences for soil C sequestration remain uncertain.

6.4.3 AGRICULTURAL PRACTICES

Agricultural soils have great potential to sequester C (e.g., 0.4−0.8 Pg C per year) by improving agricultural management practice (Lal, 2004b). Over the short-term, management practices such as reduced tillage and stubble retention that increase residue inputs typically facilitate SOC storage (Conant et al., 2001; Luo et al., 2010). However, over the long-term, in the absence of fertilizer inputs, depletion of available nutrients could limit the formation of stable SOM (Kirkby et al., 2011) and the accrual of soil C (Lu et al., 2011b). Indeed, it was demonstrated that nutrient additions to agricultural soils facilitated 8.7 t more C ha^{-1} compared to when they were not applied (Kirkby et al., 2016). The application of allochthonous plant-derived C—usually in the form of compost or biochar—has also been shown to enhance soil C stocks in numerous studies (e.g., Sohi et al., 2010). The high surface area of porous biochar can increase the nutrient retention capacity of soil and could prevent the mining of stable SOC as well as increase litter inputs with higher C:N that will favor copiotrophic growth and protect SOC stocks. However, some studies also report apparent "priming" (enhanced mineralization) of existing soil C in response to biochar addition

(Zimmerman et al., 2011). In both agroforestry and cropping systems, there is evidence that increased N availability under mixed plantings with N-fixing plants may promote C sequestration above- and belowground, and also promote humification (Gärdenäs et al., 2011; Nair et al., 2009). Similarly, increased P inputs can increase nitrogen fixation and lead to increased N inputs (Batterman et al., 2013; Reed et al., 2011a).

Although there is great potential to increase soil carbon sequestration in agricultural soils, achieving this at national and global scales may be challenging because soils, climate, and management regimes vary locally and interact with each other to determine the direction and magnitude of change in soil C. Further, a full lifecycle analysis of the addition of such allochthonous C to soils needs to be considered (Paustian et al., 2016) (Box 6.4) if they are to be considered as management regimes that promote SOC storage.

6.5 HOW NUTRIENT MODULATORS ARE AFFECTED BY ENVIRONMENTAL VARIABLES AND CONSEQUENCES FOR SOIL C

In the context of global change, there remains considerable uncertainty around the impacts of elevated CO_2 (eCO_2), rising temperatures and altered precipitation on the fate of soil C. Equally, the response of soil N, and other biogeochemical cycles, including P and S, to such projected changes are poorly constrained, and feedbacks between C and N cycling under environmental change are not fully understood.

6.5.1 ELEVATED CO$_2$ AND NUTRIENT AVAILABILITY

Varied responses of soil C to eCO_2 have been reported (see Chapter 7: Agricultural Management Practices and Soil Organic Carbon Storage) which is, in part, likely to be mediated by the nutrient status of the ecosystem. Indeed, the magnitude and direction of the response of plant biomass and SOC stocks to rising atmospheric CO_2 is expected to be strongly affected by N (Hungate et al., 2003) and P (Ellsworth et al., 2017) availability, because of their influence on ecosystem productivity. Under the progressive N limitation hypothesis (PNL), soil N is expected to decline under eCO_2 as it is locked up in long lived pools (i.e., aboveground biomass) as a result of increased NPP (Luo et al., 2004), which over the long-term is likely to limit SOC accumulation due to reduced C inputs from aboveground. However, most long-term free-air CO_2 studies do not exhibited signs of PNL (Feng et al., 2015), and a recent metaanalysis found that overall, gross N-mineralization rates were unaffected by eCO_2 (Rütting and Andresen, 2015). However, when the authors considered only those ecosystems that were N limited, increased rates of gross N-mineralization were found under eCO_2. Enhanced rhizodeposition of labile C under eCO_2 (De Graaff et al., 2007; Phillips et al., 2011; Van Groenigen et al., 2014), and increased root exploration in response to increased plant N demand, are possible mechanisms that enhance gross N-mineralization rates (Dijkstra et al., 2008; Rütting et al., 2010). However, in forest soils, increased plant C supply belowground under eCO_2 has been reported to increase both mineralization and N immobilization (Norby et al., 2010).

Increased mineralization may be restricted to N-limited ecosystems (Dijkstra et al., 2013), and PNL, or indeed progressive phosphorus limitation, and may only threaten SOC stocks in nutrient limited ecosystems (Hu et al., 2006). Increased root production under eCO_2 could promote competition for available N between microbes and plants which could accelerate mining of SOM in N-poor systems and limit aboveground NPP under eCO_2 (Reay et al., 2008), limiting litter inputs and negatively affecting soil C sequestration.

Most global change experiments have been conducted in high latitude ecosystems which are considered largely N limited and thus less attention has been given to the impacts of global drivers on P cycling. One study in a P-limited grassland demonstrated increased P mobilization under eCO_2 (Dijkstra et al., 2013) brought about by either increase desorption of P from charged soil particles due to higher rhizodeposition of organic acids, or by increased P mineralization of SOM (Lloyd et al., 2001). Similarly, in an open-top chamber study in a P-limited model subtropical forest, P availability increased under eCO_2 (Huang et al., 2014). As P mineralization is not directly coupled with SOM decomposition (McGill and Cole, 1981), as suggested by our analysis that demonstrated a lack of linkage between soil P concentration and soil C content (Fig. 6.5B), it is expected that, although soil N and P can affect each other (Fig. 6.5B), in P-limited systems, increased P mineralization will not necessarily lead to increased N mineralization because they are not linked to soil C to the same degree as each other.

Increased N supply from BNF under eCO_2 (Hartwig and Sadowsky, 2006; Hungate et al., 2003; Liang et al., 2016) provides another possible mechanism to explain reduced rates of N mineralization under eCO_2, although the extent to which this alleviates PNL remains questionable (Rütting, 2017). Furthermore, in ecosystems where BNF may be significant, increased BNF may be short lived under eCO_2 if other nutrients such as Mo and P are not available to sustain increased activity of N_2-fixers (Hungate et al., 1999). As the uncertainty in BNF has been identified as a factor contributing to variability in projected terrestrial biodiversity response to eCO_2 (Meyerholt et al., 2016), a better understanding of the impacts the interaction between terrestrial C and N cycling is warranted.

6.5.2 WARMING AND NUTRIENT AVAILABILITY

Under warming, soil N and P availabilities are likely to increase because of higher rates of microbial activity. Although increased gross rates of N-mineralization (e.g., Cookson et al., 2007; Daebeler et al., 2017; Larsen et al., 2011) and increased N availability (Bai et al., 2013 and references therein) under warming have widely been reported, others argue that a warming-induced increase in N availability can lead to suppressed rates of gross and net N mineralization (Bengtson et al., 2005; Craine et al., 2007; Ramirez et al., 2010) resulting from reduced microbial investment in N-mineralizing enzymes. Such contrasting results may be due to differences in the degree of nutrient limitation between studies—i.e., when N is not limiting, N mineralization is unaffected by warming, but increases under warming when N is limiting as demonstrated in a warming $\times$ N fertilizer study in arctic soils (Daebeler et al., 2017). When N is limiting, increased plant nutrient (N and P) demand and higher levels of microbial activity under warming facilitate the mining of SOM and could lead to losses in soil C as was demonstrated in a 7-year warming study in a deciduous forest (Melillo et al., 2011). Because nutrient mineralization is dependent on soil moisture, the magnitude of a warming effect on nutrient availability and consequences for soil C-storage will be ecosystem dependant. Brzostek et al. (2012) suggested that warming stimulated proteolytic enzyme activity in mesic sites (i.e., temperate and boreal forests and arctic tundra), but suppressed its potential activity in dry grasslands. Reports of no effect of warming on P demand/limitation in grassland soil (Menge and Field, 2007) have also been noted. If water become limiting under warming, SOM stocks reduced plant demand and microbial activity will serve to protect soil C stocks. Thus, more long-term studies are needed to identify the point at which particular ecosystems are likely to be nutrient and/or water-limited under warming so that we can better predict soil C stocks under future climate scenarios.

6.6 NUTRIENT MODULATORS AND MODELING

6.6.1 PARAMETERS USED FOR MODELING (BOTH PREDICTIVE AND MECHANISTIC)

ESMs are widely used by policy makers and other stakeholders to inform the public, predict future terrestrial carbon sink, and feed back responses to climate change (Thomas et al., 2015). Nitrogen availability is the primary constraint used to predict terrestrial C sequestration in ESMs and Thomas et al. (2015) reports that ESMs typically use stoichiometric relationships in which supply and uptake is adjusted to maintain cellular stoichiometry. However, as discussed earlier, soil microbial communities exert some control on stoichiometric flexibility, while elemental cycling can become uncoupled under certain conditions, such as C:P in arid environments (Box 6.2) as a consequence of the strong reduction in biological activity. The stoichiometric relationships currently used in ESMs may not, therefore, offer the flexibility observed under field conditions to robustly predict SOC. Improved understanding of how, and to what extent, microbial communities exert flexibility (i.e., changing community structure and/or altered nutrient use efficiencies) in their stoichiometry in relation to changing resource stoichiometry will facilitate the inclusion of C:N:P stoichiometric relationships in ESMs.

6.6.2 PARAMETERS COULD/SHOULD BE INCLUDED TO IMPROVE PREDICTIONS/ SOIL C STORAGE

Given the strong links between soil nutrients and C storage, inclusions of N and P pools, availabilities and stoichiometric relationships between C, N, and P in plants (litter, root and rhizodeposits), soil, and microbial pools are a key to improve the capacity of ESM models to predict soil C from local to global scales, although some challenges exist in parameterizing such components for incorporating into ESM (Box 6.4). For instance, inclusion of the N cycle can radically reduce projected land carbon uptake in response to increasing atmospheric CO_2 concentration, while also reducing the sensitivity of the terrestrial carbon cycle to changes in temperature and precipitation (Thornton et al., 2007). However, evidence for N enrichment impacts on soil C storage and fluxes remains contradictory, with some studies suggesting that soil C may decrease with N enrichment, others suggesting no change, and some indicating that soil C sinks may increase (Reay et al., 2008).

Incorporation of P and S interactions with net soil C storage in models is arguably even less robust, with large uncertainties around global soil microbial biomass P cycling (e.g., Xu et al., 2013), as well as conflicting evidence on the importance of biological versus biochemical mineralization for soil S availability (Kopittke et al., 2016). Further, where P is considered in biogeochemical models, the parametrization of P mineralization is based on a limited number of empirical studies (Yang et al., 2014). To help resolve some of this uncertainty in a theoretical framework before applying in ESMs, there is a necessity for more studies that use contrasting soils across elevation gradients and chronosequences. Further, experiments should explicitly test interactive effects of different management and environmental variables (e.g., N addition × warming; warming × eCO$_2$ etc.) under different nutrient limitations (C, N and/or P) and consider together both the response of N and P in different pools—particularly for P.

In order to help meet such challenges for managing and enhancing global soil C storage, an extension of existing soil measurement networks and sites is required (Paustian et al., 2016). Currently our understanding of how soil nutrients modulate soil C storage in poorly resolved ecosystems and regions is limited. For example, despite the potential for strong regulatory effects of N and P on C storage in tropical soils (see Box 6.2), considerable uncertainty remains. Other poorly resolved areas that need to be addressed include some areas of Africa, South America, and Asia (Batjes, 2008). Aligned to this, enhanced knowledge of fundamental soil C processes (e.g., N, P, and S interactions, temperature dependence of soil C fraction turnover) and further field-scale experimental trials of soil C amendments, such as biochar, are needed. Together, such actions can better constrain existing soil C models (e.g., Shangguan et al., 2014) and allow improved estimation and projection of soil C fluxes and storage for previously underrepresented regions, land uses, and emerging "Greenhouse Gas Removal" strategies (Spencer et al., 2011).

As with all estimations derived from complex systems, there is a constant trade-off between the often-high cost (financial and person-hours) of direct measurements and the uncertainties inherent in modeled estimations. For much of the developed world, we now have sufficient directly measured data to underpin useful model estimates of soil C stocks and changes under future land uses, climates, and management regimes. For these regions, the emphasis is on reducing uncertainties around specific soil C turnover processes—especially those arising from N and P interactions—and to more robustly assess novel mitigation strategies. For large areas of the developing world, model estimations of soil C storage and change remain much more uncertain due to a relative paucity of soils data. Increased investment in soil C and nutrient measurement and mapping in these regions may serve to bridge the current data gap, allowing more robust modeling of soil C storage and understanding of interactions with biogeochemical cycles, potentially unlocking much needed financial support for AFOLU climate change mitigation projects in the developing world. The FAO's Global Soil Partnership (GSP) is precisely the kind of initiative that could help deliver this. With the aim of improving soils governance and promoting sustainable management of soils globally (FAO, 2017), a key facet of the GSP is its wide engagement with stakeholders and its aim to enhance accessibility to soils data all around the world. If successful, the GSP has the potential to open up access to a wealth of vital soils data for regions where current soil C modeling efforts are restricted by data quality.

BOX 6.4 CHALLENGES AND FUTURE DIRECTIONS

1. Our statistical modeling (Fig. 6.5), along with previous studies, (Manzoni et al., 2012) suggest that inclusion of nutrient data (both total pool and fraction—i.e., inorganic and organic) can improve prediction of soil C and should be adapted for earth systems models (ESMs). Clearly biogeochemical (e.g., C, N, P, and S) cycles do not operate independently from each other and our ability to better inform models relies on the challenge of a better understanding of the interaction between these biogeochemical cycles, how they are impacted by environmental factors, and the potential consequences for soil organic carbon (SOC) stocks.

2. Most biogeochemical models use carbon use efficiency (CUE) as a constant (0.6–0.15; Manzoni et al., 2012). However, using a constant limits the accuracy of ectomycorrhizal to predict SOC dynamics (Bradford and Crowther, 2013; Lee and Schmidt, 2014), because as decomposition progresses and labile C declines, CUE declines (Ziegler et al., 2005) as a result of increasing need to divert resources to nutrient scavenging (enzyme production) and maintenance respiration.

3. Currently there is a lack of data that allows the establishment of empirical relationships between nutrient (N and P) availabilities (among other environmental data, e.g., pH) and CUE that could facilitate the inclusion of CUE into ESMs. Additionally, as different microbial functional groups harbor different CUE values, changes in community structure also plays a role in altering soil microbial biomass CUE, and incorporating these changes into biogeochemical models is challenging. Identifying the importance of these varying mechanisms is critical in order to include these parameters in ESMs.

4. Although it is clear that the relative contribution of oligotrophic and copiotrophic microbes in a given environment will potentially affect SOC stock, the magnitude and ecosystem dependence of change is not fully understood and a better mechanistic understanding of the interactions between resource and biomass stoichiometry, CUE, and litter chemistry is needed (Bailey et al., 2002; Grandy et al., 2009; Six et al., 2006) to improve both the theoretical framework and simulation models.

5. Including P-specific pools and fluxes into C models will likely help to improve the predictability of such models (Reed et al., 2015), particularly in P-limited systems such as highly weathered soils typical to many lowland tropical forests (Palm et al., 2007). Furthering our understanding of P mineralization, sorption, limitation, and stoichiometry will greatly improve parameterization P-cycling into C-cycling models (Reed et al., 2015). Further empirical data is required to improve the uncertainty associated with modeling the global distribution of different soil P pools (Yang et al., 2013) to improve the predictability of coupled biogeochemistry climate models.

6. Although not discussed in this chapter, full lifecycle analysis is crucial when attempting to quantify the net climate forcing impact of nutrient modulators on soil C with consideration of non-CO_2 greenhouse gas impacts, plant growth response, processing, and transport of the allochthonous C, as well as the opportunity cost of its redirection to soils all being necessary (Paustian et al., 2016). For instance, a growing body of evidence indicates that biochar addition may serve to reduce N_2O emissions, as well as enhance soil C stocks (e.g., Case et al., 2015). However, some studies also report apparent "priming" (enhanced mineralization) of existing soil C in response to biochar addition (Zimmerman et al., 2011). Further, offsets of the potential benefits of N addition of soil C accumulation (e.g., N leaching, N_2O emissions, acid rain, and soil acidification) (Townsend et al., 1996) need to be considered.

7. Evidence of the impact of other nutrients (e.g., S, Fe, Ca) on soil C are extremely limited and future research should explicitly consider and quantify their contribution.

6.7 CONCLUDING REMARKS

The C cycle is intrinsically linked with nutrient (e.g., N, P and S) cycling. However, the mechanisms that govern these linkages are not fully understood. Two contrasting theories, MNM and BSD, were proposed as theoretical frameworks to explain nutrient impacts on soil C dynamics. Overwhelmingly experimental evidence from the micro- to global-scale support MNM theory as the main mechanism determining the impact of nutrients on soil C storage. Evidence was provided to suggest that increasing N availability inhibits enzymes responsible for recalcitrant C degradation and thus promote long-term C storage in soils. Increasing N availability was also reported to change SOC chemically and make it more stable. However, evidence is also provided that the inhibitory effect of N on SOC decomposition is dependent on SOC type (i.e., lignin vs cellulose rich) and ecosystem type (e.g., boreal vs tropical). Identifying the direction and magnitude of nutrient impacts on soil C storage, how it varies between different ecosystems, or SOC availability are critical knowledge gaps. Other critical knowledge gaps that hinder our ability to quantify nutrient impact, and identify mechanisms that would help improve the predictions of ESMs, are identified in Box 6.4. To address these knowledge gaps, future efforts need to include the development of a new theoretical framework that includes multiple pools of nutrients, quantify the magnitude of—and provide evidence for—the dominant mechanisms (MNM vs BSD) which could explain nutrient impacts on soil C storage as well as parameterize key variables (e.g., multiple pools of nutrients and copiotroph/oligotrophy growth). Challenges need concerted research effort, both at the fundamental and applied science level, in order for them to be overcome.

ACKNOWLEDGMENTS

We thank all the members of the EPES-BIOCOM network (ERC grant agreement no. 242658) for the collection of field data. We thank Pankaj Trivedi for permission to include microaggregate data in our analysis (Fig. 6.1D).

REFERENCES

Ågren, G.I., Bosatta, E., 2002. Reconciling differences in predictions of temperature response of soil organic matter. Soil Biol. Biochem. 34, 129–132.

Ågren, G.I., Bosatta, E., Magill, A.H., 2001. Combining theory and experiment to understand effects of inorganic nitrogen on litter decomposition. Oecologia 128, 94–98.

Allen, M.F., 2011. Linking water and nutrients through the vadose zone: a fungal interface between the soil and plant systems. J. Arid Land 3, 155–163.

Allison, S.D., Wallenstein, M.D., Bradford, M.A., 2010. Soil-carbon response to warming dependent on microbial physiology. Nat. Geosci. 3, 336–340.

Armitage, A., Fourqurean, J.W., 2016. Carbon storage in seagrass soils: long-term nutrient history exceeds the effects of near-term nutrient enrichment. Biogeosciences 13, 313.

Austin, A.T., Yahdjian, L., Stark, J.M., Belnap, J., Porporato, A., Norton, U., et al., 2004. Water pulses and biogeochemical cycles in arid and semiarid ecosystems. Oecologia 141, 221–235.

Baath, E., Berg, B., Lohm, U., Lundgren, B., Lundkvist, H., Rosswall, T., et al., 1980. Effects of experimental acidification and liming on soil organisms and decomposition in a Scots pine forest. Pedobiologia 20, 85–100.

Bai, E., Li, S., Xu, W., Li, W., Dai, W., Jiang, P., 2013. A meta-analysis of experimental warming effects on terrestrial nitrogen pools and dynamics. New Phytol. 199, 441–451.

Bailey, V.L., Smith, J.L., Bolton, H., 2002. Fungal-to-bacterial ratios in soils investigated for enhanced C sequestration. Soil Biol. Biochem. 34, 997–1007.

Barantal, S., Schimann, H., Fromin, N., Hättenschwiler, S., 2012. Nutrient and carbon limitation on decomposition in an Amazonian moist forest. Ecosystems 15, 1039−1052.

Bardgett, R.D., McAlister, E., 1999. The measurement of soil fungal: bacterial biomass ratios as an indicator of ecosystem self-regulation in temperate meadow grasslands. Biol. Fertil. Soils 29, 282−290.

Batjes, N.H., 2008. Mapping soil carbon stocks of Central Africa using SOTER. Geoderma 146, 58−65.

Batterman, S.A., Wurzburger, N., Hedin, L.O., 2013. Nitrogen and phosphorus interact to control tropical symbiotic N_2 fixation: a test in *Inga punctata*. J. Ecol. 101, 1400−1408.

Bengtson, P., Falkengren-Grerup, U., Bengtsson, G., 2005. Relieving substrate limitation-soil moisture and temperature determine gross N transformation rates. Oikos 111, 81−90.

Berg, B., Matzner, E., 1997. Effect of N deposition on decomposition of plant litter and soil organic matter in forest systems. Environ. Rev. 5, 1−25.

Berg, B., McClaugherty, C., 2014. Decomposer Organisms, In: *Plant Litter.*, Springer, Berlin, Heidelberg.

Boy, J., Rollenbeck, R., Valarezo, C., Wilcke, W., 2008. Amazonian biomass burning-derived acid and nutrient deposition in the north Andean montane forest of Ecuador. Global Biogeochem. Cycles 22.

Bradford, M., Fierer, N., Reynolds, J., 2008. Soil carbon stocks in experimental mesocosms are dependent on the rate of labile carbon, nitrogen and phosphorus inputs to soils. Funct. Ecol. 22, 964−974.

Bradford, M.A., Crowther, T.W., 2013. Carbon use efficiency and storage in terrestrial ecosystems. New Phytol. 199, 7−9.

Bragazza, L., Freeman, C., Jones, T., Rydin, H., Limpens, J., Fenner, N., et al., 2006. Atmospheric nitrogen deposition promotes carbon loss from peat bogs. Proc. Natl. Acad. Sci. 103, 19386−19389.

Brzostek, E.R., Blair, J.M., Dukes, J.S., Frey, S.D., Hobbie, S.E., Melillo, J.M., et al., 2012. The effect of experimental warming and precipitation change on proteolytic enzyme activity: positive feedbacks to nitrogen availability are not universal. Global Change Biol. 18, 2617−2625.

Bunemann, E.K., Condron, L.M., 2007. Phosphorus and sulphur cycling in terrestrial ecosystems. In: Marschner, P., Rengel, Z. (Eds.), Nutrient Cycling in Terrestrial Ecosystems. Springer-Verlag, New York, NY.

Burgin, A.J., Yang, W.H., Hamilton, S.K., Silver, W.L., 2011. Beyond carbon and nitrogen: how the microbial energy economy couples elemental cycles in diverse ecosystems. Front. Ecol. Environ. 9, 44−52.

Busse, M.D., Sanchez, F.G., Ratcliff, A.W., Butnor, J.R., Carter, E.A., Powers, R.F., 2009. Soil carbon sequestration and changes in fungal and bacterial biomass following incorporation of forest residues. Soil Biol. Biochem. 41, 220−227.

Carreiro, M., Sinsabaugh, R., Repert, D., Parkhurst, D., 2000. Microbial enzyme shifts explain litter decay responses to simulated nitrogen deposition. Ecology 81, 2359−2365.

Case, S.D., Mcnamara, N.P., Reay, D.S., Stott, A.W., Grant, H.K., Whitaker, J., 2015. Biochar suppresses N_2O emissions while maintaining N availability in a sandy loam soil. Soil Biol. Biochem. 81, 178−185.

Chadwick, O.A., Derry, L.A., Vitousek, P.M., Huebert, B.J., Hedin, L.O., 1999. Changing sources of nutrients during four million years of ecosystem development. Nature 397, 491−497.

Chen, H., Dong, S., Liu, L., Ma, C., Zhang, T., Zhu, X., et al., 2013. Effects of experimental nitrogen and phosphorus addition on litter decomposition in an old-growth tropical forest. PLoS One 8, e84101.

Chen, R., Senbayram, M., Blagodatsky, S., Myachina, O., Dittert, K., Lin, X., et al., 2014. Soil C and N availability determine the priming effect: microbial N mining and stoichiometric decomposition theories. Global Change Biol. 20, 2356−2367.

Chen, Y., Sayer, E.J., Li, Z., Mo, Q., Li, Y., Ding, Y., et al., 2015. Nutrient limitation of woody debris decomposition in a tropical forest: contrasting effects of N and P addition. Funct. Ecol. 30, 295−304.

Clark, K., West, A., Hilton, R., Asner, G., Quesada, C., Silman, M., et al., 2016. Storm-triggered landslides in the Peruvian Andes and implications for topography, carbon cycles, and biodiversity. Earth Surf. Dyn. 4, 47.

Cleveland, C.C., Liptzin, D., 2007. C:N:P stoichiometry in soil: is there a "Redfield ratio" for the microbial biomass? Biogeochemistry 85, 235−252.

Cleveland, C.C., Townsend, A.R., 2006. Nutrient additions to a tropical rain forest drive substantial soil carbon dioxide losses to the atmosphere. Proc. Natl. Acad. Sci. 103, 10316–10321.

Cleveland, C.C., Townsend, A.R., Schmidt, S.K., 2002. Phosphorus limitation of microbial processes in moist tropical forests: evidence from short-term laboratory incubations and field studies. Ecosystems 5, 0680–0691.

Cleveland, C.C., Reed, S.C., Townsend, A.R., 2006. Nutrient regulation of organic matter decomposition in a tropical rain forest. Ecology 87, 492–503.

Cleveland, C.C., Houlton, B.Z., Smith, W.K., Marklein, A.R., Reed, S.C., Parton, W., et al., 2013. Patterns of new versus recycled primary production in the terrestrial biosphere. Proc. Natl. Acad. Sci. U.S.A. 110, 12733–12737.

Conant, R.T., Paustian, K., Elliott, E.T., 2001. Grassland management and conversion into grassland: effects on soil carbon. Ecol. Appl. 11, 343–355.

Cookson, W.R., Osman, M., Marschner, P., Abaye, D.A., Clark, I., Murphy, D.V., et al., 2007. Controls on soil nitrogen cycling and microbial community composition across land use and incubation temperature. Soil Biol. Biochem. 39, 744–756.

Couteaux, M.-M., Bottner, P., Berg, B., 1995. Litter decomposition, climate and liter quality. Trends Ecol. Evol. 10, 63–66.

Craine, J.M., Morrow, C., Fierer, N., 2007. Microbial nitrogen limitation increases decomposition. Ecology 88, 2105–2113.

Cuevas, E., Medina, E., 1988. Nutrient dynamics within Amazonian forests. Oecologia 76, 222–235.

Cusack, D.F., Silver, W.L., Torn, M.S., Burton, S.D., Firestone, M.K., 2011. Changes in microbial community characteristics and soil organic matter with nitrogen additions in two tropical forests. Ecology 92, 621–632.

Cusack, D.F., Karpman, J., Ashdown, D., Cao, Q., Ciochina, M., Halterman, S., et al., 2016. Global change effects on humid tropical forests: evidence for biogeochemical and biodiversity shifts at an ecosystem scale. Rev. Geophys. 54, 523–610.

Daebeler, A., Bodelier, P.L.E., Hefting, M.H., Rutting, T., Jia, Z., Laanbroek, H.J., 2017. Soil warming and fertilization altered rates of nitrogen transformation processes and selected for adapted ammonia-oxidising archaea in sub-arctic grassland soil. Soil Biol. Biochem. 107, 114–124.

Davidson, E.A., Reis de Carvalho, C.J., Vieira, I.C., Figueiredo, R.D.O., Moutinho, P., Yoko Ishida, F., et al., 2004. Nitrogen and phosphorus limitation of biomass growth in a tropical secondary forest. Ecol. Appl. 14, 150–163.

De Deyn, G.B., Cornelissen, J.H., Bardgett, R.D., 2008. Plant functional traits and soil carbon sequestration in contrasting biomes. Ecol. Lett. 11, 516–531.

De Graaff, M.A., Six, J., Van Kessel, C., 2007. Elevated CO_2 increases nitrogen rhizodeposition and microbial immobilization of root-derived nitrogen. New Phytol. 173, 778–786.

De Vries, F.T., Shade, A., 2013. Controls on soil microbial community stability under climate change. Front. Microbiol. 4, 265.

Delgado-Baquerizo, M., Maestre, F.T., Gallardo, A., Bowker, M.A., Wallenstein, M.D., Quero, J.L., et al., 2013. Decoupling of soil nutrient cycles as a function of aridity in global drylands. Nature 502, 672–676.

Delgado-Baquerizo, M., Maestre, F.T., Eldridge, D.J., Singh, B.K., 2016a. Microsite differentiation drives the abundance of soil ammonia oxidizing bacteria along aridity gradients. Front. Microbiol. 7, 505.

Delgado-Baquerizo, M., Reich, P.B., García-Palacios, P., Milla, R., 2016b. Biogeographic bases for a shift in crop C:N:P stoichiometries during domestication. Ecol. Lett. 19, 564–575.

Delgado-Baquerizo, M., Reich, P.B., Khachane, A.N., Campbell, C.D., Thomas, N., Freitag, T.E., et al., 2017. It is elemental: soil nutrient stoichiometry drives bacterial diversity. Environ. Microbiol. 19, 1176–1188.

Dentener, F., Drevet, J., Lamarque, J.F., Bey, I., Eickhout, B., Fiore, A.M., et al., 2006. Nitrogen and sulfur deposition on regional and global scales: a multimodel evaluation. Global Biogeochem. Cycles 20, GB4003. Available from: https://doi.org/10.1029/2005GB002672.

Devêvre, O.C., Horwáth, W.R., 2000. Decomposition of rice straw and microbial carbon use efficiency under different soil temperatures and moistures. Soil Biol. Biochem. 32, 1773−1785.

Dijkstra, F., Pendall, E., Mosier, A., King, J., Milchunas, D., Morgan, J., 2008. Long-term enhancement of N availability and plant growth under elevated CO_2 in a semi-arid grassland. Funct. Ecol. 22, 975−982.

Dijkstra, F., Carrillo, Y., Pendall, E., Morgan, J., 2013. Rhizosphere priming: a nutrient perspective. Front. Microbiol. 4, 216.

Dijkstra, F.A., Hobbie, S.E., Knops, J.M., Reich, P.B., 2004. Nitrogen deposition and plant species interact to influence soil carbon stabilization. Ecol. Lett. 7, 1192−1198.

Dijkstra, F.A., Hobbie, S.E., Reich, P.B., Knops, J.M., 2005. Divergent effects of elevated CO_2, N fertilization, and plant diversity on soil C and N dynamics in a grassland field experiment. Plant Soil 272, 41−52.

Driscoll, C.T., Lawrence, G.B., Bulger, A.J., Butler, T.J., Cronan, C.S., Eagar, C., et al., 2001. Acidic deposition in the Northeastern United States: sources and inputs, ecosystem effects, and management strategies. BioScience 51, 180−198.

Driscoll, C.T., Driscoll, K.M., Roy, K.M., Mitchell, M.J., 2003. Chemical response of lakes in the Adirondack Region of New York to declines in acidic deposition. Environ. Sci. Technol. 37, 2036−2042.

Ellsworth, D.S., Anderson, I.C., Crous, K.Y., Cooke, J., Drake, J.E., Gherlenda, A.N., et al., 2017. Elevated CO_2 does not increase eucalypt forest productivity on a low-phosphorus soil. Nat. Clim. Change 7, 279−282.

Elser, J.J., Acharya, K., Kyle, M., Cotner, J., Makino, W., Markow, T., et al., 2003. Growth rate−stoichiometry couplings in diverse biota. Ecol. Lett. 6, 936−943.

Elser, J.J., Bracken, M.E.S., Cleland, E.E., Gruner, D.S., Harpole, W.S., Hillebrand, H., et al., 2007. Global analysis of nitrogen and phosphorus limitation of primary producers in freshwater, marine and terrestrial ecosystems. Ecol. Lett. 10, 1135−1142.

Evans, C.D., Monteith, D.T., Cooper, D.M., 2005. Long-term increases in surface water dissolved organic carbon: observations, possible causes and environmental impacts. Environ. Pollut. 137, 55−71.

Evans, C.D., Freeman, C., Cork, L.G., Thomas, D.N., Reynolds, B., Billett, M.F., et al., 2007. Evidence against recent climate-induced destabilisation of soil carbon from 14C analysis of riverine dissolved organic matter. Geophys. Res. Lett. 34, L07407. Available from: https://doi.org/10.1029/2007GL029431.

Evans, S.E., Burke, I.C., 2013. Carbon and nitrogen decoupling under an 11-year drought in the shortgrass steppe. Ecosystems 16, 20−33.

Falkowski, P., Scholes, R., Boyle, E., Canadell, J., Canfield, D., Elser, J., et al., 2000. The global carbon cycle: a test of our knowledge of earth as a system. Science 290, 291−296.

FAO, 2017. Global soil partnership [Online]. Available: <http://www.fao.org/global-soil-partnership/en/> (accessed February 2017).

Feng, Z., Rütting, T., Pleijel, H., Wallin, G., Reich, P.B., Kammann, C.I., et al., 2015. Constraints to nitrogen acquisition of terrestrial plants under elevated CO2. Global Change Biol. 21, 3152−3168.

Fernández-Martínez, M., Vicca, S., Janssens, I.A., Sardans, J., Luyssaert, S., Campioli, M., et al., 2014. Nutrient availability as the key regulator of global forest carbon balance. Nat. Clim. Change 4, 471−476.

Fierer, N., Bradford, M.A., Jackson, R.B., 2007. Toward an ecological classification of soil bacteria. Ecology 88, 1354−1364.

Fierer, N., Strickland, M.S., Liptzin, D., Bradford, M.A., Cleveland, C.C., 2009. Global patterns in belowground communities. Ecol. Lett. 12, 1238−1249.

Figuerola, E.L., Guerrero, L.D., Rosa, S.M., Simonetti, L., Duval, M.E., Galantini, J.A., et al., 2012. Bacterial indicator of agricultural management for soil under no-till crop production. PLoS One 7, e51075.

Finzi, A.C., Austin, A.T., Cleland, E.E., Frey, S.D., Houlton, B.Z., Wallenstein, M.D., 2011. Responses and feedbacks of coupled biogeochemical cycles to climate change: examples from terrestrial ecosystems. Front. Ecol. Environ. 9, 61−67.

Fisher, J.B., Malhi, Y., Torres, I.C., Metcalfe, D.B., van de Weg, M.J., Meir, P., et al., 2013. Nutrient limitation in rainforests and cloud forests along a 3,000-m elevation gradient in the Peruvian Andes. Oecologia 172, 889–902.

Fisk, M., Santangelo, S., Minick, K., 2015. Carbon mineralization is promoted by phosphorus and reduced by nitrogen addition in the organic horizon of northern hardwood forests. Soil Biol. Biochem. 81, 212–218.

Fog, K., 1988. The effect of added nitrogen on the rate of decomposition of organic matter. Biol. Rev. 63, 433–462.

Fontaine, S., Mariotti, A., Abbadie, L., 2003. The priming effect of organic matter: a question of microbial competition? Soil Biol. Biochem. 35, 837–843.

Fontaine, S., Barot, S., Barré, P., Bdioui, N., Mary, B., Rumpel, C., 2007. Stability of organic carbon in deep soil layers controlled by fresh carbon supply. Nature 450, 277–280.

Franklin, O., Högberg, P., Ekblad, A., ÅGren, G.I., 2003. Pine forest floor carbon accumulation in response to N and PK additions: Bomb 14C Modelling and Respiration Studies. Ecosystems 6, 644–658.

Gallo, M., Amonette, R., Lauber, C., Sinsabaugh, R., Zak, D., 2004. Microbial community structure and oxidative enzyme activity in nitrogen-amended north temperate forest soils. Microb. Ecol. 48, 218–229.

Galloway, J.N., Townsend, A.R., Erisman, J.W., Bekunda, M., Cai, Z., Freney, J.R., et al., 2008. Transformation of the nitrogen cycle: recent trends, questions, and potential solutions. Science 320, 889–892.

Gärdenäs, A.I., ÅGren, G.I., Bird, J.A., Clarholm, M., Hallin, S., Ineson, P., et al., 2011. Knowledge gaps in soil carbon and nitrogen interactions—from molecular to global scale. Soil Biol. Biochem. 43, 702–717.

Goll, D.S., Brovkin, V., Parida, B.R., Reick, C.H., Kattge, J., Reich, P.B., et al., 2012. Nutrient limitation reduces land carbon uptake in simulations with a model of combined carbon, nitrogen and phosphorus cycling. Biogeosciences 9, 3547–3569.

Grandy, A.S., Strickland, M.S., Lauber, C.L., Bradford, M.A., Fierer, N., 2009. The influence of microbial communities, management, and soil texture on soil organic matter chemistry. Geoderma 150, 278–286.

Gruber, N., Galloway, J.N., 2008. An Earth-system perspective of the global nitrogen cycle. Nature 451, 293–296.

Guggenberger, G., Frey, S.D., Six, J., Paustian, K., Elliott, E.T., 1999. Bacterial and fungal cell-wall residues in conventional and no-tillage agroecosystems. Soil Sci. Soc. Am. J. 63, 1188–1198.

Hammel, K., 1997. Fungal degradation of lignin. Driven by Nature: Plant Litter Quality and Decomposition. CAB International, Wallingford, pp. 33–46.

Hartman, W.H., Richardson, C.J., 2013. Differential nutrient limitation of soil microbial biomass and metabolic quotients (qCO$_2$): is there a biological stoichiometry of soil microbes? PLoS One 8, e57127.

Hartmann, M., Howes, C.G., Vaninsberghe, D., Yu, H., Bachar, D., Christen, R., et al., 2012. Significant and persistent impact of timber harvesting on soil microbial communities in Northern coniferous forests. ISME J. 6, 2199–2218.

Hartwig, U.A., Sadowsky, M.J., 2006. Biological nitrogen fixation: A key process for the response of grassland ecosystems to elevated atmospheric [CO$_2$]. In: Nösberger, J., Long, S.P., Norby, R.J., Stitt, M., Hendrey, G.R., Blum, H. (Eds.), Managed Ecosystems and CO$_2$. Ecological Studies (Analysis and Synthesis), vol. 187. Springer, Berlin, Heidelberg.

Hejzlar, J., Dubrovský, M., Buchtele, J., Růžička, M., 2003. The apparent and potential effects of climate change on the inferred concentration of dissolved organic matter in a temperate stream (the Malše River, South Bohemia). Sci. Total Environ. 310, 143–152.

Hietz, P., Turner, B.L., Wanek, W., Richter, A., Nock, C.A., Wright, S.J., 2011. Long-term change in the nitrogen cycle of tropical forests. Science 334, 664–666.

Hobbie, S.E., 2005. Contrasting effects of substrate and fertilizer nitrogen on the early stages of litter decomposition. Ecosystems 8, 644–656.

Hobbie, S.E., 2008. Nitrogen effects on decomposition: a five-year experiment in eight temperate sites. Ecology 89, 2633–2644.

Hobbie, S.E., Vitousek, P.M., 2000. Nutrient limitation of decomposition in Hawaiian forests. Ecology 81, 1867–1877.

Högberg, M.N., Briones, M.J., Keel, S.G., Metcalfe, D.B., Campbell, C., Midwood, A.J., et al., 2010. Quantification of effects of season and nitrogen supply on tree below-ground carbon transfer to ectomycorrhizal fungi and other soil organisms in a boreal pine forest. New Phytol. 187, 485–493.

Homeier, J., Hertel, D., Camenzind, T., Cumbicus, N.L., Maraun, M., Martinson, G.O., et al., 2012. Tropical Andean forests are highly susceptible to nutrient inputs—rapid effects of experimental N and P addition to an Ecuadorian montane forest. PLoS One 7, e47128.

Houlton, B.Z., Wang, Y.-P., Vitousek, P.M., Field, C.B., 2008. A unifying framework for dinitrogen fixation in the terrestrial biosphere. Nature 454, 327–330.

Hu, S., Tu, C., Chen, X., Gruver, J.B., 2006. Progressive N limitation of plant response to elevated CO_2: a microbiological perspective. Plant Soil 289, 47–58.

Huang, J., Yu, H., Guan, X., Wang, G., Guo, R., 2016. Accelerated dryland expansion under climate change. Nat. Clim. Change 6, 166–171.

Huang, W., Zhou, G., Liu, J., Duan, H., Liu, X., Fang, X., et al., 2014. Shifts in soil phosphorus fractions under elevated CO_2 and N addition in model forest ecosystems in subtropical China. Plant Ecol. 215, 1373–1384.

Huang, Z., Clinton, P.W., Baisden, W.T., Davis, M.R., 2011. Long-term nitrogen additions increased surface soil carbon concentration in a forest plantation despite elevated decomposition. Soil Biol. Biochem. 43, 302–307.

Hungate, B.A., Dijkstra, P., Johnson, D., Hinkle, C.R., Drake, B., 1999. Elevated CO_2 increases nitrogen fixation and decreases soil nitrogen mineralization in Florida scrub oak. Global Change Biol. 5, 781–789.

Hungate, B.A., Dukes, J.S., Shaw, M.R., Luo, Y.Q., Field, C.B., 2003. Nitrogen and climate change. Science 302, 1512–1513.

Inceoğlu, Ö., Al-Soud, W.A., Salles, J.F., Semenov, A.V., van Elsas, J.D., 2011. Comparative analysis of bacterial communities in a potato field as determined by pyrosequencing. PLoS One 6, e23321.

Israel, D.W., 1987. Investigation of the role of phosphorus in symbiotic dinitrogen fixation. Plant Physiol. 84, 835–840.

Jan, M., Roberts, P., Tonheim, S., Jones, D., 2009. Protein breakdown represents a major bottleneck in nitrogen cycling in grassland soils. Soil Biol. Biochem. 41, 2272–2282.

Jandl, R., Kopeszki, H., Bruckner, A., Hager, H., 2003. Forest soil chemistry and mesofauna 20 years after an amelioration fertilization. Restor. Ecol. 11, 239–246.

Janssens, I.A., Dieleman, W., Luyssaert, S., Subke, J.-A., Reichstein, M., Ceulemans, R., et al., 2010. Reduction of forest soil respiration in response to nitrogen deposition. Nat. Geosci. 3, 315–322.

Jarvis, P., Linder, S., 2000. Botany: constraints to growth of boreal forests. Nature 405, 904–905.

Jastrow, J.D., Amonette, J.E., Bailey, V.L., 2007. Mechanisms controlling soil carbon turnover and their potential application for enhancing carbon sequestration. Clim. Change 80, 5–23.

Jiao, F., Shi, X.-R., Han, F.-P., Yuan, Z.-Y., 2016. Increasing aridity, temperature and soil pH induce soil C-N-P imbalance in grasslands. Sci. Rep. 6, 19601.

Jing, Z., Chen, R., Wei, S., Feng, Y., Zhang, J., Lin, X., 2017. Response and feedback of C mineralization to P availability driven by soil microorganisms. Soil Biol. Biochem. 105, 111–120.

Jobbágy, E.G., Jackson, R.B., 2000. The vertical distribution of soil organic carbon and its relation to climate and vegetation. Ecol. Appl. 10, 423–436.

Johnson, D.W., Curtis, P.S., 2001. Effects of forest management on soil C and N storage: meta analysis. Forest Ecol. Manage. 140, 227–238.

Johnson, N.C., Wilson, G.W.T., Wilson, J.A., Miller, R.M., Bowker, M.A., 2015. Mycorrhizal phenotypes and the Law of the Minimum. New Phytol. 205, 1473–1484.

Johnson, N.M., 1979. Acid rain: neutralization within the hubbard brook ecosystem and regional implications. Science 204, 497–499.

Jones, D., Kielland, K., Sinclair, F., Dahlgren, R., Newsham, K., Farrar, J., et al., 2009. Soil organic nitrogen mineralization across a global latitudinal gradient. Global Biogeochem. Cycles 23.

Jones, D.L., Farrar, J.F., Newsham, K.K., 2004. Rapid amino acid cycling in Arctic and Antarctic soils. Water Air Soil Pollut.: Focus 4, 169–175.

Kaspari, M., Garcia, M.N., Harms, K.E., Santana, M., Wright, S.J., Yavitt, J.B., 2008. Multiple nutrients limit litterfall and decomposition in a tropical forest. Ecol. Lett. 11, 35–43.

Keiblinger, K.M., Hall, E.K., Wanek, W., Szukics, U., Hämmerle, I., Ellersdorfer, G., et al., 2010. The effect of resource quantity and resource stoichiometry on microbial carbon-use-efficiency. FEMS Microbiol. Ecol. 73, 430–440.

Khan, S., Mulvaney, R., Ellsworth, T., Boast, C., 2007. The myth of nitrogen fertilization for soil carbon sequestration. J. Environ. Qual. 36, 1821–1832.

Kirkby, C., Kirkegaard, J., Richardson, A., Wade, L., Blanchard, C., Batten, G., 2011. Stable soil organic matter: a comparison of C:N:P:S ratios in Australian and other world soils. Geoderma 163, 197–208.

Kirkby, C.A., Richardson, A.E., Wade, L.J., Passioura, J.B., Batten, G.D., Blanchard, C., et al., 2014. Nutrient availability limits carbon sequestration in arable soils. Soil Biol. Biochem. 68, 402–409.

Kirkby, C.A., Richardson, A.E., Wade, L.J., Conyers, M., Kirkegaard, J.A., 2016. Inorganic nutrients increase humification efficiency and C-sequestration in an annually cropped soil. PLoS One 11, e0153698.

Knops, J.M., Tilman, D., 2000. Dynamics of soil nitrogen and carbon accumulation for 61 years after agricultural abandonment. Ecology 81, 88–98.

Knorr, M., Frey, S.D., Curtis, P.S., 2005. Nitrogen additions and litter decomposition: a meta-analysis. Ecology 86, 3252–3257.

Kopáček, J., Veselý, J., 2005. Sulfur and nitrogen emissions in the Czech Republic and Slovakia from 1850 till 2000. Atmos. Environ. 39, 2179–2188.

Kopittke, P.M., Dalal, R.C., Finn, D., Menzies, N.W., 2016. Global changes in soil stocks of carbon, nitrogen, phosphorus, and sulphur as influenced by long-term agricultural production. Global Change Biol. 23, 2509–2519.

Krasilnikov, P., del Carmen Gutiérrez-Castorena, M., Ahrens, R.J., Cruz-Gaistardo, C.O., Sedov, S., Solleiro-Rebolledo, E., 2013. The Soils of Mexico. Springer, Netherlands.

Kuijper, L.D., Berg, M.P., Morriën, E., Kooi, B.W., Verhoef, H.A., 2005. Global change effects on a mechanistic decomposer food web model. Global Change Biol. 11, 249–265.

Laganiere, J., Angers, D.A., Pare, D., 2010. Carbon accumulation in agricultural soils after afforestation: a meta-analysis. Global Change Biol. 16, 439–453.

Lal, R., 2001. World cropland soils as a source or sink for atmospheric carbon. Adv. Agron. 71, 145–191.

Lal, R., 2004a. Soil carbon sequestration impacts on global climate change and food security. Science 304, 1623–1627.

Lal, R., 2004b. Soil carbon sequestration to mitigate climate change. Geoderma 123, 1–22.

Lamarque, J.F., Kiehl, J., Brasseur, G., Butler, T., CAMERON-Smith, P., Collins, W., et al., 2005. Assessing future nitrogen deposition and carbon cycle feedback using a multimodel approach: analysis of nitrogen deposition. J. Geophys. Res.: Atmos. 110, D19303. Available from: https://doi.org/10.1029/2005JD005825.

Lamersdorf, N.P., Borken, W., 2004. Clean rain promotes fine root growth and soil respiration in a Norway spruce forest. Global Change Biol. 10, 1351–1362.

Larsen, K.S., Andresen, L.C., Beier, C., Jonasson, S., Albert, K.R., Ambus, P.E.R., et al., 2011. Reduced N cycling in response to elevated CO_2, warming, and drought in a Danish heathland: synthesizing results of the CLIMAITE project after two years of treatments, Global Change Biol., 17. pp. 1884–1899.

Laskowski, R., Berg, B., 2006. Litter Decomposition: Guide to Carbon and Nutrient Turnover. Elsevier, Amsterdam.

LeBauer, D.S., Treseder, K.K., 2008. Nitrogen limitation of net primary productivity in terrestrial ecosystems is globally distributed. Ecology 89, 371−379.

Lee, Z.M., Schmidt, T.M., 2014. Bacterial growth efficiency varies in soils under different land management practices. Soil Biol. Biochem. 69, 282−290.

Leff, J.W., Jones, S.E., Prober, S.M., Barberán, A., Borer, E.T., Firn, J.L., et al., 2015. Consistent responses of soil microbial communities to elevated nutrient inputs in grasslands across the globe. Proc. Natl. Acad. Sci. 112, 10967−10972.

Li, D., Niu, S., Luo, Y., 2012. Global patterns of the dynamics of soil carbon and nitrogen stocks following afforestation: a meta-analysis. New Phytol. 195, 172−181.

Li, J.H., Yang, Y.J., Li, B.W., Li, W.J., Wang, G., Knops, J.M., 2014. Effects of nitrogen and phosphorus fertilization on soil carbon fractions in alpine meadows on the Qinghai-Tibetan Plateau. PLoS One 9, e103266.

Li, L.-J., Zeng, D.-H., Yu, Z.-Y., Fan, Z.-P., Mao, R., 2010. Soil microbial properties under N and P additions in a semi-arid, sandy grassland. Biol. Fertil. Soils 46, 653−658.

Liang, J.Y., Qi, X., Souza, L., Luo, Y.Q., 2016. Processes regulating progressive nitrogen limitation under elevated carbon dioxide: a meta-analysis. Biogeosciences 13, 2689−2699.

Likens, G., Bormann, F.H., Johnson, N., 1981. Interactions between major biogeochemical cycles in terrestrial ecosystems. Some Perspectives of the Major Biogeochemical Cycles. Wiley, New York, NY.

Liu, L., Greaver, T.L., 2010. A global perspective on belowground carbon dynamics under nitrogen enrichment. Ecol. Lett. 13, 819−828.

Liu, L., Zhang, T., Gilliam, F.S., Gundersen, P., Zhang, W., Chen, H., et al., 2013. Interactive effects of nitrogen and phosphorus on soil microbial communities in a tropical forest. PLoS One 8, e61188.

Lloyd, J., Bird, M., Veenendaal, E., Kruijt, B., 2001. 1.8 - Should phosphorus availability be constraining moist tropical forest responses to increasing CO_2 concentrations? In Global Biogeochemical Cycles in the Climate System. Academic Press, San Diego, pp. 95−114.

Lu, M., Yang, Y., Luo, Y., Fang, C., Zhou, X., Chen, J., et al., 2011a. Responses of ecosystem nitrogen cycle to nitrogen addition: a meta-analysis. New Phytol. 189, 1040−1050.

Lu, M., Zhou, X., Luo, Y., Yang, Y., Fang, C., Chen, J., et al., 2011b. Minor stimulation of soil carbon storage by nitrogen addition: a meta-analysis. Agric. Ecosyst. Environ. 140, 234−244.

Lugo, A.E., Murphy, P.G., 1986. Nutrient dynamics of a Puerto Rican Subtropical Dry Forest. J. Trop. Ecol. 2, 55−72.

Luo, W., Dijkstra, F.A., Bai, E., Feng, J., Lü, X.-T., Wang, C., et al., 2016. A threshold reveals decoupled relationship of sulfur with carbon and nitrogen in soils across arid and semi-arid grasslands in northern China. Biogeochemistry 127, 141−153.

Luo, Y., Su, B., Currie, W.S., Dukes, J.S., Finzi, A., Hartwig, U., et al., 2004. Progressive nitrogen limitation of ecosystem responses to rising atmospheric carbon dioxide. Bioscience 54, 731−739.

Luo, Z., Wang, E., Sun, O.J., 2010. Soil carbon change and its responses to agricultural practices in Australian agro-ecosystems: a review and synthesis. Geoderma 155, 211−223.

Macedo, M., Resende, A., Garcia, P., Boddey, R., Jantalia, C., Urquiaga, S., et al., 2008. Changes in soil C and N stocks and nutrient dynamics 13 years after recovery of degraded land using leguminous nitrogen-fixing trees. Forest Ecol. Manage. 255, 1516−1524.

Mack, M.C., Schuur, E.A., Bret-Harte, M.S., Shaver, G.R., Chapin, F.S., 2004. Ecosystem carbon storage in arctic tundra reduced by long-term nutrient fertilization. Nature 431, 440−443.

Maestre, F.T., Quero, J.L., Gotelli, N.J., Escudero, A., Ochoa, V., Delgado-Baquerizo, M., et al., 2012. Plant species richness and ecosystem multifunctionality in global drylands. Science 335, 214−218.

Mahowald, N., Jickells, T.D., Baker, A.R., Artaxo, P., Benitez-Nelson, C.R., Bergametti, G., et al., 2008. Global distribution of atmospheric phosphorus sources, concentrations and deposition rates, and anthropogenic impacts. Global Biogeochem. Cycles 22, GB4026. Available from: https://doi.org/10.1029/2008GB003240.

Manzoni, S., Taylor, P., Richter, A., Porporato, A., ÅGren, G.I., 2012. Environmental and stoichiometric controls on microbial carbon-use efficiency in soils. New Phytol. 196, 79–91.

Marklein, A.R., Houlton, B.Z., 2012. Nitrogen inputs accelerate phosphorus cycling rates across a wide variety of terrestrial ecosystems. New Phytol. 193, 696–704.

McGill, W.B., Cole, C.V., 1981. Comparative aspects of cycling of organic C, N, S and P through soil organic matter. Geoderma 26, 267–286.

McGroddy, M.E., Silver, W.L., de Oliveira, R.C., 2004. The effect of phosphorus availability on decomposition dynamics in a seasonal lowland Amazonian forest. Ecosystems 7, 172–179.

Melillo, J.M., Butler, S., Johnson, J., Mohan, J., Steudler, P., Lux, H., et al., 2011. Soil warming, carbon–nitrogen interactions, and forest carbon budgets. Proc. Natl. Acad. Sci. 108, 9508–9512.

Menge, D.N., Field, C.B., 2007. Simulated global changes alter phosphorus demand in annual grassland. Global Change Biol. 13, 2582–2591.

Meyerholt, J., Zaehle, S., 2015. The role of stoichiometric flexibility in modelling forest ecosystem responses to nitrogen fertilization. New Phytol. 208, 1042–1055.

Meyerholt, J., Zaehle, S., Smith, M.J., 2016. Variability of projected terrestrial biosphere responses to elevated levels of atmospheric CO_2 due to uncertainty in biological nitrogen fixation. Biogeosciences 13, 1491–1518.

Mitchell, M.J., Burke, M.K., Shepard, J.P., 1992. Seasonal and spatial patterns of S, Ca, and N dynamics of a Northern Hardwood forest ecosystem. Biogeochemistry 17, 165–189.

Monteith, D.T., Stoddard, J.L., Evans, C.D., de Wit, H.A., Forsius, M., Hogasen, T., et al., 2007. Dissolved organic carbon trends resulting from changes in atmospheric deposition chemistry. Nature 450, 537–540.

Moorhead, D.L., Sinsabaugh, R.L., 2006. A theoretical model of litter decay and microbial interaction. Ecol. Monogr. 76, 151–174.

Mooshammer, M., Wanek, W., Hämmerle, I., Fuchslueger, L., Hofhansl, F., Knoltsch, A., et al., 2014. Adjustment of microbial nitrogen use efficiency to carbon:nitrogen imbalances regulates soil nitrogen cycling. Nat. Commun. 5, 3694.

Morris, S.J., Bohm, S., Haile-Mariam, S., Paul, E.A., 2007. Evaluation of carbon accrual in afforested agricultural soils. Global Change Biol. 13, 1145–1156.

Mulder, C., Elser, J.J., 2009. Soil acidity, ecological stoichiometry and allometric scaling in grassland food webs. Global Change Biol. 15, 2730–2738.

Nair, P.R., Nair, V.D., Kumar, B.M., Haile, S.G., 2009. Soil carbon sequestration in tropical agroforestry systems: a feasibility appraisal. Environ. Sci. Policy 12, 1099–1111.

Neff, J.C., Townsend, A.R., Gleixner, G., Lehman, S.J., Turnbull, J., Bowman, W.D., 2002. Variable effects of nitrogen additions on the stability and turnover of soil carbon. Nature 419, 915–917.

Nemergut, D.R., Townsend, A.R., Sattin, S.R., Freeman, K.R., Fierer, N., Neff, J.C., et al., 2008. The effects of chronic nitrogen fertilization on alpine tundra soil microbial communities: implications for carbon and nitrogen cycling. Environ. Microbiol. 10, 3093–3105.

Norby, R.J., Warren, J.M., Iversen, C.M., Medlyn, B.E., Mcmurtrie, R.E., 2010. CO_2 enhancement of forest productivity constrained by limited nitrogen availability. Proc. Natl. Acad. Sci. U.S.A. 107, 19368–19373.

Nottingham, A., Turner, B.L., Whitaker, J., Ostle, N.J., Mcnamara, N., Bardgett, R.D., et al., 2015. Soil microbial nutrient constraints along a tropical forest elevation gradient: a belowground test of a biogeochemical paradigm. Biogeosciences 12, 6071–6083.

Okin, G.S., Mahowald, N., Chadwick, O.A., Artaxo, P., 2004. Impact of desert dust on the biogeochemistry of phosphorus in terrestrial ecosystems. Global Biogeochem. Cycles 18, GB2005. Available from: https://doi.org/10.1029/2003GB002145.

Olander, L.P., Vitousek, P.M., 2000. Regulation of soil phosphatase and chitinase activity by N and P availability. Biogeochemistry 49, 175–191.

Oulehle, F., Evans, C.D., Hofmeister, J., Krejci, R., Tahovska, K., Persson, T., et al., 2011. Major changes in forest carbon and nitrogen cycling caused by declining sulphur deposition. Global Change Biol. 17, 3115–3129.

Palm, C., Sanchez, P., Ahamed, S., Awiti, A., 2007. Soils: a contemporary perspective. Annu. Rev. Environ. Resour. 32, 99−129.

Pan, Y., Birdsey, R.A., Phillips, O.L., Jackson, R.B., 2013. The structure, distribution, and biomass of the world's forests. Annu. Rev. Ecol. Evol. Syst. 44, 593−622.

Paul, A.E., 2007. Soil Microbiology and Biochemistry. Academic Press, Amsterdam.

Paustian, K., Lehmann, J., Ogle, S., Reay, D., Robertson, G.P., Smith, P., 2016. Climate-smart soils. Nature 532, 49−57.

Peltzer, D.A., Wardle, D.A., Allison, V.J., Baisden, W.T., Bardgett, R.D., Chadwick, O.A., et al., 2010. Understanding ecosystem retrogression. Ecol. Monogr. 80, 509−529.

Pennanen, T., Fritze, H., Vanhala, P., Kiikkila, O., Neuvonen, S., Baath, E., 1998. Structure of a microbial community in soil after prolonged addition of low levels of simulated acid rain. Appl. Environ. Microbiol. 64, 2173−3180.

Peñuelas, J., Sardans, J., Rivas-Ubach, A., Janssens, I.A., 2012. The human-induced imbalance between C, N and P in Earth's life system. Global Change Biol. 18, 3−6.

Peñuelas, J., Poulter, B., Sardans, J., Ciaia, P., van der Velde, M., Bopp, L., et al., 2013. Human-induced nitrogen-phosphorus imbalances alter natural and managed ecosystems across the globe. Nat. Commun. 4, 2934.

Persson, T., Lundkvist, H., Wirén, A., Hyvönen, R., Wessén, B., 1989. Effects of acidification and liming on carbon and nitrogen mineralization and soil organisms in mor humus. Water Air Soil Pollut. 45, 77−96.

Phillips, R.P., Finzi, A.C., Bernhardt, E.S., 2011. Enhanced root exudation induces microbial feedbacks to N cycling in a pine forest under long-term CO_2 fumigation. Ecol. Lett. 14, 187−194.

Poeplau, C., Bolinder, M.A., Kirchmann, H., Kätterer, T., 2016. Phosphorus fertilisation under nitrogen limitation can deplete soil carbon stocks: evidence from Swedish meta-replicated long-term field experiments. Biogeosciences 13, 1119−1127.

Porder, S., Vitousek, P.M., Chadwick, O.A., Chamberlain, C.P., Hilley, G.E., 2007. Uplift, erosion, and phosphorus limitation in terrestrial ecosystems. Ecosystems 10, 159−171.

Ramirez, K.S., Craine, J.M., Fierer, N., 2010. Nitrogen fertilization inhibits soil microbial respiration regardless of the form of nitrogen applied. Soil Biol. Biochem. 42, 2336−2338.

Ramirez, K.S., Craine, J.M., Fierer, N., 2012. Consistent effects of nitrogen amendments on soil microbial communities and processes across biomes. Global Change Biol. 18, 1918−1927.

Read, L., Lawrence, D., 2003. Litter nutrient dynamics during succession in dry tropical forests of the Yucatan: regional and seasonal effects. Ecosystems 6, 747−761.

Reay, D.S., Dentener, F., Smith, P., Grace, J., Feely, R.A., 2008. Global nitrogen deposition and carbon sinks. Nat. Geosci. 1, 430−437.

Reed, S.C., Cleveland, C.C., Townsend, A.R., 2011a. Functional ecology of free-living nitrogen fixation: a contemporary perspective. Annu. Rev. Ecol. Evol. Syst. 42, 489−512.

Reed, S.C., Townsend, A.R., Taylor, P.G., Cleveland, C.C., 2011b. Phosphorus cycling in tropical forests growing on highly weathered soils. Phosphorus in Action. Springer-Verlag, Berlin, Heidelberg.

Reed, S.C., Yang, X., Thornton, P.E., 2015. Incorporating phosphorus cycling into global modeling efforts: a worthwhile, tractable endeavor. New Phytol. 208, 324−329.

Reich, P.B., Oleksyn, J., 2004. Global patterns of plant leaf N and P in relation to temperature and latitude. Proc. Natl. Acad. Sci. U.S.A. 101, 11001−11006.

Richter, D.D., Allen, H.L., Li, J., Markewitz, J., 2006. Bioavailability of slowly cycling phosphorus: major restructuring of soil P fractions over four decades in an aggrading forest. Oecologia 150, 259−271.

Rinkes, Z.L., Bertrand, I., Amin, B.A.Z., Grandy, A.S., Wickings, K., Weintraub, M.N., 2016. Nitrogen alters microbial enzyme dynamics but not lignin chemistry during maize decomposition. Biogeochemistry 128, 171−186.

Robertson, G.P., Groffman, P., 2007. Soil Microbiology, Biochemistry, and Ecology. Springer, New York, NY.

Rousk, J., Frey, S.D., 2015. Revisiting the hypothesis that fungal-to-bacterial dominance characterizes turnover of soil organic matter and nutrients. Ecol. Monogr. 85, 457−472.

Rousk, K., Michelsen, A., Rousk, J., 2016. Microbial control of soil organic matter mineralization responses to labile carbon in subarctic climate change treatments. Global Change Biol. 22, 4150−4161.

Rütting, T., 2017. Nitrogen mineralization, not N_2 fixation, alleviates progressive nitrogen limitation − Comment on "Processes regulating progressive nitrogen limitation under elevated carbon dioxide: a meta-analysis" by Liang et al. (2016). Biogeosciences 14, 751−754.

Rütting, T., Andresen, L.C., 2015. Nitrogen cycle responses to elevated CO_2 depend on ecosystem nutrient status. Nutr. Cycling Agroecosyst. 101, 285−294.

Rütting, T., Clough, T.J., Müller, C., Lieffering, M., Newton, P.C., 2010. Ten years of elevated atmospheric carbon dioxide alters soil nitrogen transformations in a sheep-grazed pasture. Global Change Biol. 16, 2530−2542.

Sanyal, S.K., De Datta, S.K., 1991. Chemistry of phosphorus transformations in soil. In: Stewart, B.A. (Ed.), Advances in Soil Science. Springer New York, New York, NY.

Scheel, T., Dörfler, C., Kalbitz, K., 2007. Precipitation of dissolved organic matter by aluminum stabilizes carbon in acidic forest soils abbreviations: DOC, dissolved organic carbon; DOM, dissolved organic matter; OM, organic matter; UV, ultraviolet. Soil Sci. Soc. Am. J. 71, 64−74.

Schimel, J.P., Bennett, J., 2004. Nitrogen mineralization: challenges of a changing paradigm. Ecology 85, 591−602.

Schlesinger, W.H., 1996. Biogeochemistry, an Analysis of Global Change. Academic Press, San Diego, CA.

Selmants, P.C., Hart, S.C., 2008. Substrate age and tree islands influence carbon and nitrogen dynamics across a retrogressive semiarid chronosequence. Global Biogeochem. Cycles 22, GB1021. Available from: https://doi.org/10.1029/2007GB003062.

Selmants, P.C., Hart, S.C., 2010. Phosphorus and soil development: does the Walker and Syers model apply to semiarid ecosystems? Ecology 91, 474−484.

Shangguan, W., Dai, Y., Duan, Q., Liu, B., Yuan, H., 2014. A global soil data set for earth system modeling. J. Adv. Model. Earth Syst. 6, 249−263.

Shaver, G.R., Chapin, F.S., 1980. Response to fertilization by various plant growth forms in an Alaskan Tundra: nutrient accumulation and growth. Ecology 61, 662−675.

Sinsabaugh, R.L., 2010. Phenol oxidase, peroxidase and organic matter dynamics of soil. Soil Biol. Biochem. 42, 391−404.

Sinsabaugh, R.L., Manzoni, S., Moorhead, D.L., Richter, A., 2013. Carbon use efficiency of microbial communities: stoichiometry, methodology and modelling. Ecol. Lett. 16, 930−939.

Six, J., Frey, S.D., Thiet, R.K., Batten, K.M., 2006. Bacterial and fungal contributions to carbon sequestration in agroecosystems. Soil Sci. Soc. Am. J. 70, 555−569.

Sjöberg, G., Nilsson, S., Persson, T., Karlsson, P., 2004. Degradation of hemicellulose, cellulose and lignin in decomposing spruce needle litter in relation to N. Soil Biol. Biochem. 36, 1761−1768.

Skjelkvåle, B.L., Stoddard, J.L., Jeffries, D.S., TØRseth, K., HØGåsen, T., Bowman, J., et al., 2005. Regional scale evidence for improvements in surface water chemistry 1990−2001. Environ. Pollut. 137, 165−176.

Smith, S.A., Read, D., 2008. Mycorrhizal Symbiosis, third ed. Academic Press, London.

Sohi, S., Krull, E., Lopez-Capel, E., Bol, R., 2010. A review of biochar and its use and function in soil. Adv. Agron. 105, 47−82.

Sollins, P., Robertson, G., Uehara, G., 1988. Nutrient mobility in variable- and permanent-charge soils. Biogeochemistry 6, 181−199.

Solomon, S., 2007. Climate Change 2007-The Physical Science Basis: Working Group I Contribution to the Fourth Assessment Report of the IPCC. Cambridge University Press, Cambridge.

Spencer, S., Ogle, S.M., Breidt, F.J., Goebel, J.J., Paustian, K., 2011. Designing a national soil carbon monitoring network to support climate change policy: a case example for US agricultural lands. Greenhouse Gas Meas. Manage. 1, 167–178.

Sterner, R.W., Elser, J.J., 2002. Ecological Stoichiometry: The Biology of Elements from Molecules to the Biosphere. Princeton University Press, Princeton, NJ.

Strickland, M.S., Rousk, J., 2010. Considering fungal: bacterial dominance in soils–methods, controls, and ecosystem implications. Soil Biol. Biochem. 42, 1385–1395.

Tanner, E., Vitousek, P.M., Cuevas, E., 1998. Experimental investigation of nutrient limitation of forest growth on wet tropical mountains. Ecology 79, 10–22.

Templer, P.H., Silver, W.L., Pett-Ridge, J., DeAngelis, K.M, Firestone, M.K., 2008. Plant and microbial controls on nitrogen retention and loss in a humid tropical forest. Ecology 89, 3030–3040.

Terrer, C., Vicca, S., Hungate, B.A., Phillips, R.P., Prentice, I.C., 2016. Mycorrhizal association as a primary control of the CO_2 fertilization effect. Science 353, 72–74.

Thiet, R.K., Frey, S.D., Six, J., 2006. Do growth yield efficiencies differ between soil microbial communities differing in fungal:bacterial ratios? Reality check and methodological issues. Soil Biol. Biochem. 38, 837–844.

Thomas, R.Q., Brookshire, E.N.J., Gerber, S., 2015. Nitrogen limitation on land: how can it occur in Earth system models? Global Change Biol. 21, 1777–1793.

Thornton, P.E., Lamarque, J.F., Rosenbloom, N.A., Mahowald, N.M., 2007. Influence of carbon-nitrogen cycle coupling on land model response to CO_2 fertilization and climate variability. Global Biogeochem. Cycles 21, GB4018. Available from: https://doi.org/10.1029/2006GB002868.

Townsend, A.R., Braswell, B.H., Holland, E.A., Penner, J.E., 1996. Spatial and temporal patterns in terrestrial carbon storage due to deposition of fossil fuel nitrogen. Ecol. Appl. 6, 806–814.

Townsend, A.R., Cleveland, C.C., Houlton, B.Z., Alden, C.B., White, J.W., 2011. Multi-element regulation of the tropical forest carbon cycle. Front. Ecol. Environ. 9, 9–17.

Treseder, K.K., 2008. Nitrogen additions and microbial biomass: a meta-analysis of ecosystem studies. Ecol. Lett. 11, 1111–1120.

Trivedi, P., Delgao-Baquerizo, M., Jeffries, T.C., Trivedi, C., Anderson, I.C., Lai, K., et al., 2017. Soil aggregation and associated microbial communities modify the impact of agricultural management of carbon content. Environ. Microbiol. Available from: https://doi.org/10.1111/1462-2920.13779.

Turner, B.L., Wright, S.J., 2014. The response of microbial biomass and hydrolytic enzymes to a decade of nitrogen, phosphorus, and potassium addition in a lowland tropical rain forest. Biogeochemistry 117, 115–130.

Van Diepen, L.T.A., Frey, S.D., Landis, E.A., Morrison, E.W., Pringle, A., 2017. Fungi exposed to chronic nitrogen enrichment are less able to decay leaf litter. Ecology 98, 5–11.

Van Groenigen, K.J., Qi, X., Osenberg, C.W., Luo, Y., Hungate, B.A., 2014. Faster decomposition under increased atmospheric CO_2 limits soil carbon storage. Science 344, 508–509.

Vitousek, P.M., 2004. Nutrient Cycling and Limitation: Hawaii as a Model System. Princeton University Press, Princeton, NJ.

Vitousek, P.M., Howarth, R.W., 1991. Nitrogen limitation on land and in the sea: how can it occur? Biogeochemistry 13, 87–115.

Vitousek, P.M., Aber, J.D., Howarth, R.W., Likens, G.E., Matson, P.A., Schindler, D.W., et al., 1997. Human alteration of the global nitrogen cycle: sources and consequences. Ecol. Appl. 7, 737–750.

Vitousek, P.M., Porder, S., Houlton, B.Z., Chadwick, O.A., 2010. Terrestrial phosphorus limitation: mechanisms, implications, and nitrogen–phosphorus interactions. Ecol. Appl. 20, 5–15.

Waldrop, M.P., Zak, D.R., Sinsabaugh, R.L., Gallo, M., Lauber, C., 2004. Nitrogen deposition modifies soil carbon storage through changes in microbial enzymatic activity. Ecol. Appl. 14, 1172–1177.

Walker, T.W., Syers, J.K., 1976. The fate of phosphorus during pedogenesis. Geoderma 15, 1–19.

Wang, Y.P., Law, R.M., Pak, B., 2010. A global model of carbon, nitrogen and phosphorus cycles for the terrestrial biosphere. Biogeosciences 7, 2261–2282.

Woodward, F.I., Bardgett, R.D., Raven, J.A., Hetherington, A.M., 2009. Biological approaches to global environment change mitigation and remediation. Curr. Biol. 19, R615–R623.

Wurzburger, N., Brookshire, E.N.J., 2017. Experimental evidence that mycorrhizal nitrogen strategies affect soil carbon. Ecology 98, 1491–1497.

Xia, J., Wan, S., 2008. Global response patterns of terrestrial plant species to nitrogen addition. New Phytol. 179, 428–439.

Xu, X., Thornton, P.E., Post, W.M., 2013. A global analysis of soil microbial biomass carbon, nitrogen and phosphorus in terrestrial ecosystems. Global Ecol. Biogeogr. 22, 737–749.

Yang, X., Post, W.M., Thornton, P.E., Jain, A., 2013. The distribution of soil phosphorus for global biogeochemical modeling. Biogeosciences 10, 2525–2537.

Yang, X., Thornton, P.E., Ricciuto, D.M., Post, W.M., 2014. The role of phosphorus dynamics in tropical forests – a modeling study using CLM-CNP. Biogeosciences 11, 1667–1681.

Yang, Y., Luo, Y., Finzi, A.C., 2011. Carbon and nitrogen dynamics during forest stand development: a global synthesis. New Phytol. 190, 977–989.

Yuan, Z.Y., Chen, H.Y.H., 2015. Decoupling of nitrogen and phosphorus in terrestrial plants associated with global changes. Nat. Clim. Change 5, 465–469.

Yue, K., Peng, Y., Peng, C., Yang, W., Peng, X., Wu, F., 2016. Stimulation of terrestrial ecosystem carbon storage by nitrogen addition: a meta-analysis. Sci. Rep. 6, 19895.

Zaehle, S., Medlyn, B.E., De kauwe, M.G., Walker, A.P., Dietze, M.C., Hickler, T., et al., 2014. Evaluation of 11 terrestrial carbon–nitrogen cycle models against observations from two temperate Free-Air CO_2 enrichment studies, New Phytol., 202. pp. 803–822.

Zechmeister-Boltenstern, S., Keiblinger, K.M., Mooshammer, M., Peñuelas, J., Richter, A., Sardans, J., et al., 2015. The application of ecological stoichiometry to plant–microbial–soil organic matter transformations. Ecol. Monogr. 85, 133–155.

Ziegler, S.E., Billings, S.A., 2011. Soil nitrogen status as a regulator of carbon substrate flows through microbial communities with elevated CO_2. J. Geophys. Res.: Biogeosciences 116, G01011. Available from: https://doi.org/10.1029/2010JG001434.

Ziegler, S.E., White, P.M., Wolf, D.C., Thoma, G.J., 2005. Tracking the fate and recycling of C-13-labeled glucose in soil. Soil Sci. 170, 767–778.

Zimmerman, A.R., Gao, B., Ahn, M.-Y., 2011. Positive and negative carbon mineralization priming effects among a variety of biochar-amended soils. Soil Biol. Biochem. 43, 1169–1179.

FURTHER READING

Barrow, N.J., 1984. Modelling the effects of pH on phosphate sorption by soils. J. Soil Sci. 35, 283–297.

Lu, M., Zhou, X., Yang, Q., Li, H., Luo, Y., Fang, C., et al., 2013. Responses of ecosystem carbon cycle to experimental warming: a meta-analysis. Ecology 94, 726–738.

Singh, B.K., Bardgett, R.D., Smith, P., Reay, D.S., 2010. Microorganisms and climate change: terrestrial feedbacks and mitigation options. Nat. Rev. Microbiol. 8, 779–790.

AGRICULTURAL MANAGEMENT PRACTICES AND SOIL ORGANIC CARBON STORAGE

7

Bhupinder P. Singh[1], Raj Setia[2], Martin Wiesmeier[3,4] and Anitha Kunhikrishnan[1]

[1]*Elizabeth Macarthur Agricultural Institute, Menangle, NSW, Australia* [2]*Punjab Remote Sensing Centre, Ludhiana, Punjab, India* [3]*Technical University of Munich, Freising, Germany* [4]*Bavarian State Research Center for Agriculture, Freising, Germany*

7.1 INTRODUCTION

World soils constitute the third largest carbon (C) pool after oceanic (38,000 petagram; Pg) (1 Pg = 10^{15} g C = 1 Gt = 1 billion tonnes) and geologic (5000 Pg) pools (Lal, 2004a). The soil C (organic plus inorganic) pool, estimated at 2500 Pg to 1-m depth (Eswaran et al., 1993; Lal, 2010), is about 3.3 times the atmospheric pool (760 Pg) and 4.5 times the biotic pool (560 Pg). The total amount of soil organic C (SOC) stored worldwide is estimated to be 1500 Pg C in the top one meter of soils, 2344 Pg C in the top three meters of soil (Guo and Gifford, 2002; Jobbagy and Jackson, 2000), or 2400 Pg C in the top two meters (Batjes, 1996). The amount of SOC, which represents almost 50% of soil organic matter (SOM) (Pribyl, 2010), is the net balance between C inputs (as growing plants, and decaying plant, animal, and microbial matter), and C outputs (as SOC mineralization by microbes and C losses via erosion) (Davidson and Janssens, 2006; Regnier et al., 2013; Tian et al., 2009). Increasing SOC in agricultural systems has been considered as a possible solution to mitigate climate change, e.g., via removing atmospheric carbon dioxide (CO_2) into the long-lived C pool (i.e., SOC, also referred to as soil C sequestration). On the other hand, a small decrease in SOC stocks provides feedback to atmospheric CO_2 and methane (CH_4) concentrations. Therefore, quantifying the changes in SOC stocks is important for predicting current and future changes in CO_2-equivalent (eq.) emissions.

Agricultural land systems (i.e., pastureland and cropland systems) occupy ~40% of the Earth's land surface (Smith et al., 2008). Croplands represent one-third of the total agricultural lands and they are highly susceptible to degradation processes, such as erosion, compaction, waterlogging, acidification, salinization, and chemical pollution of soils (Chambers et al., 2016; Utuk and Daniel, 2015). Further, unfavorable land use, land cover, and land management changes over the past several decades have caused substantial depletion of SOC pools stored in soils across the globe, with an estimated loss of about 78 Pg C, of which 26 Pg C resulted from erosion and 25 Pg C from mineralization, following conversion of native vegetation to cropland (Lal, 2004b). A metaanalysis by Guo and Gifford (2002) found a loss of SOC of 42% and 59% following land-use changes from forest to crop, and from pasture to crop, respectively. Because of depletion of SOC stocks,

Soil Carbon Storage. DOI: https://doi.org/10.1016/B978-0-12-812766-7.00007-X

agricultural and degraded soil systems have the potential to sequester significant amounts of C by improved management, which could significantly offset fossil fuel greenhouse gas (GHG) emissions, while enhancing soil fertility and food production (Lal, 2004b; Wang et al., 2013).

Understanding the dynamics of SOC in relation to land use and management strategies (including the biotic interactions between above- and belowground ecosystems) is of paramount importance to identify pathways of C sequestration in soils for: (1) Increasing SOC at a level that is essential for maintaining soil fertility; and (2) Responding to global warming effects. In this chapter, land use and management-induced changes in SOC storage, and traditional and emerging land management practices in agricultural systems as modulators of SOC storage will be examined. Given the scope, it will not be possible to present an exhaustive review of all agricultural management studies carried out in different continents of the world. Rather, the main objective of this chapter is to identify management practices that can influence SOC cycling and storage in agricultural soils, particularly under cropland systems. This chapter also explores new initiatives taken for increasing C content in soil to mitigate climate change, improve soil health, and maintain agricultural productivity.

7.2 MANAGEMENT-INDUCED CHANGES IN SOC STORAGE: PROCESSES AND MECHANISMS OF SOC LOSS AND GAIN

7.2.1 KEY PRINCIPLES AND PRACTICES FOR ENHANCING SOC STORAGE

SOC is in a continuous state of turnover, where it is mineralized, released, and lost from (or replaced by) new organic C inputs in soil systems. Hence, the SOC status reflects the net balance of organic C inputs and losses. Some of the processes that enhance SOC losses are fallowing, cultivation, and stubble burning or removal. These farming practices can decrease SOC by reducing organic C inputs to the soil, increasing the decomposition of SOM, or both. For instance, when soil is tilled, the soil aggregates are disturbed and the aggregate-protected SOM is exposed, thus accelerating the loss of SOC via mineralization. Further, losses of SOC from erosion of surface soil can also occur, with a large impact on the amount of SOC stored in soil. Keeping the soil bare during fallow periods was previously considered as a common cropping practice for weed control. However, SOC declines rapidly during fallow periods because there is no C input from growing plants and the increased decomposition of SOM following cultivation and/or stubble burning, and especially under high soil moisture conditions. In Australian agriculture, annual erosion losses of 8 Mg SOC ha^{-1} from cropping and up to 80 Mg SOC ha^{-1} from soils under bare fallow are likely (Kane, 2015). On the other hand, agricultural management practices that increase organic C inputs into soil systems, such as through increasing plant productivity and diversity, or through the application of external sources of organic C, and/or decreasing C losses, can increase SOC storage. To maintain the equilibrium between C inputs and outputs, a number of sustainable agricultural management practices have been recommended to increase SOC to an optimum level for adaptation to local conditions, which is one of the major components of the 17 Sustainable Development Goals of the 2030 Agenda for Sustainable Development (FAO, 2017). Some of the key management practices include conservation tillage, stubble retention, efficient landscape management via increasing crop rotational diversity and perenniality (cover crops and green manures (GMs) in crop rotations),

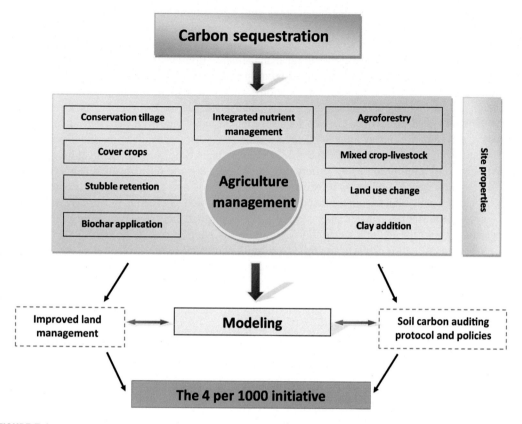

FIGURE 7.1

Overview of traditional and emerging management practices for carbon (C) sequestration in various agroecosystems. Management practices in arable lands such as no-till, cover crops, biochar, and clay addition have demonstrated potential to significantly increase soil organic carbon (SOC) storage (see details in the chapter). Integrating knowledge of the impact of improved management practices on SOC dynamics is critical to enhance model predictions and support important policy frameworks such as the "4 per 1000" initiative.

integrated application of organic amendments and inorganic nutrients, agroforestry, integrated "crop—livestock" mixed farming systems, biochar application to soils, and addition of clay to sandy soil (Fig. 7.1).

7.2.2 CARBON SEQUESTRATION IN AGRICULTURAL SOILS AS AFFECTED BY SITE CONDITIONS

The storage of SOC in agricultural soils generally depends on various site-specific conditions, including climate, topography, management, and soil characteristics. In this regard, two aspects are important—parameters that regulate the C input into agricultural soils, and factors controlling the stabilization of incorporated organic matter. Although comprehensive data on total C inputs by crops including roots and rhizodeposition are scarce, the level of crop C inputs is a function of net primary productivity (yield) and agricultural management (Bolinder et al., 2007; Wiesmeier et al., 2014a). Therefore, the C input is determined by crop type, climatic conditions (temperature, precipitation), soil characteristics (texture, aggregation), and management factors (fertilization, tillage, irrigation, return of harvest residues, etc.) (Hasibeder et al., 2015; Kong et al., 2005; Luo et al., 2010a; Sochorová et al., 2016). Carbon sequestration through improved agricultural management reaches a new equilibrium at high SOC levels after few years, depending on site-specific conditions. Han et al. (2016) estimated through a global metaanalysis study that the time required to reach a new SOC equilibrium under different organic and inorganic fertilizer treatments across agricultural systems was 30−69 years in warm temperate regions and 19−27 years in tropical regions with balanced application of nitrogen (N), phosphorus (P), and potassium (K) fertilizers. The fact that a new equilibrium is reached after several decades implies that there is an upper limit of SOC storage, relating to the hypothesis of soil C saturation (Chung et al., 2008; Goh, 2004; Six et al., 2002; Stewart et al., 2007, 2008). The time to reach a new equilibrium is also related to the limited potential of soils to stabilize SOC against microbial mineralization (Baldock and Skjemstad, 2000). Thus, besides the level of C input, the ability of agricultural soils to store C also depends on stabilization processes (see 7.2.3).

7.2.3 PHYSICAL, CHEMICAL, AND BIOLOGICAL SOC STABILIZATION PROCESSES

The stabilization of organic C in soils cannot be explained by a single mechanism. There are various mechanisms proposed for SOM protection against microbial oxidation (Six et al., 2002; von Lützow et al., 2006). The organic C in soil is mainly stabilized through the following mechanisms:

7.2.3.1 Physical protection

The C sequestration in soils through the process of aggregation is well understood (Six et al., 2002; Tisdall and Oades, 1982). Soil aggregates are clumps of soil particles that are adhered together by clay, fine roots, and microbial residues (Oades, 1984; Six et al., 2004). Tisdall and Oades (1982) proposed that roots and fungal hyphae, along with less degraded organic materials stabilize macroaggregates, and the oxidation of C in these aggregates is dependent on management practices. Conversely, highly decomposed organic components stabilize more C in microaggregates, facilitated by its high surface area and polyvalent cation bridging, and the oxidation of C in these aggregates is least dependent on management practices (Balesdent et al., 2000). It has been found that the turnover time of C in microaggregates (412 years) is much higher than C in macroaggregates (140 years) (Jastrow et al., 1996). This pattern seems to be related to the level of physical protection of organic matter across the aggregate-size classes, which in turn depends on the content and type of clay in soil (Hassink, 1994, 1997).

7.2.3.2 Chemical stabilization

The protection of organic matter increases with increasing silt and clay content (Chantigny et al., 1997; Guggenberger et al., 1999) due to the sorptive capacity of fine mineral surfaces (Nguyen and

Marschner, 2014). Hassink (1997) found a strong correlation between fine silt and clay fractions (<20 μm), and SOC stored in this fraction, though he did not find any correlation between texture and amount of C in sand-sized fractions across a wide range of uncultivated and grassland topsoils of temperate and tropical regions. Hassink's equation was used to estimate the C sequestration potential of agricultural soils in temperate regions (Angers et al., 2011; Carter et al., 2003; Sparrow et al., 2006; Wiesmeier et al., 2014b; Zhao et al., 2006). Similar relationships were found in other studies for soils under different land uses and environments (Feller and Beare, 1997; Feng et al., 2013; Liang et al., 2009; Six et al., 2002; Wiesmeier et al., 2015b). Moreover, clay mineralogy may also affect SOM stabilization (Hassink, 1997; Tisdall and Oades, 1982). Evidence was found that 2:1 clay minerals generally have a greater ability to protect OC than 1:1 clays (Barre et al., 2014; Feng et al., 2013; Six et al., 2002). Among 2:1 phyllosilicates, vermiculite, and smectite are probably more efficient for absorption of SOM due to higher specific surface areas compared to illite (Barre et al., 2014; Steffens, 2009; von Lützow et al., 2006). Nelson et al. (1997) found a greater C loss by mineralization and lower sorption of C in an illitic—kaolinitic soil than in a smectitic one. However, in the study of Hassink (1997), that included soils with various types of clay (also 1:1 clays), no effect of clay mineralogy on fine fraction SOC was observed. In a metaanalysis, Barre et al. (2014) concluded that the link between clay mineralogy and OC stabilization has not been clearly established yet.

In acid soils, reactive surfaces of amorphous iron (Fe) and aluminum (Al) oxides may have a stronger impact on SOM stabilization than clay minerals (Dümig et al., 2011; Kaiser et al., 2002; Kaiser and Zech, 2000; Kleber et al., 2005; Kögel-Knabner et al., 2008; Schöning et al., 2005; Spielvogel et al., 2008; Wiseman and Puttmann, 2005). Therefore, the C sequestration potential of more acidic, or also coarse-textured, agricultural soils could be related to the content of poorly crystalline mineral phases (Kleber et al., 2005; Wiesmeier et al., 2014b; Wiseman and Puttmann, 2005). In saline soils, the increasing concentration of exchangeable Ca increases C stabilization because bridging of organic ions with clay surfaces is increased by Ca^{2+} ions (Setia et al., 2013). These studies suggest that chemical stabilization of SOM is controlled by the quantity and characteristics of clay minerals, iron oxides, mono- and polyvalent cations, and the chemical composition of SOM.

7.2.3.3 Biochemical stabilization

Biochemical stabilization of SOM is related to an inherent chemical structure of residues or biomolecules produced during added residue and native SOM decomposition (Six et al., 2002). Biochemical stabilization is a function of: (1) Intra- and interstructural bond strengths; (2) the regular degree of occurrence of structural units; and (3) the degree of aromaticity (Krull et al., 2001). Due to their high aliphatic nature, the alkyl structures (e.g., nonhydrolyzable forms of C) are considered as chemically stable structures, such as lipids, waxes, insoluble polyesters, and microbial-synthesized macromolecules (Derenne and Largeau, 2001; Krull et al., 2001, 2003). Further, lignin is an aromatic compound and, thus, more resistant to decomposition than carbohydrates (Krull et al., 2003). Hence, alkyl and aromatic C compounds in soil are also considered as stable or passive pools with turnover times varying between 10 s and 100 s of years (Coleman et al., 1997).

7.3 TRADITIONAL AND EMERGING LAND MANAGEMENT PRACTICES IN AGRICULTURAL SYSTEMS AS MODULATORS OF SOC STORAGE

7.3.1 CONSERVATION TILLAGE (INCLUDING MINIMUM TILLAGE AND NO-TILL CROPPING SYSTEMS)

Conservation tillage is defined as any form of tillage that minimizes the number of tillage passes, where soil aggregate disruption is reduced, and a minimum of 30% of the soil surface covered with residues, with the aim to reduce soil erosion (CTIC, 2004). Hence, among different tillage methods, conservation tillage is one of the viable options to maintain or enhance SOM for improving soil fertility and ensuring sustainable food production (FAO, 2017). In general, SOM is more protected in no-till than conventional tillage due to higher aggregate stability, and a larger proportion of micropores and water stable aggregates, which decreases the decomposition of occluded SOM (Balesdent et al., 2000; Wander et al., 1998). Conventional tillage involves the mechanical manipulation of soil for obtaining conditions ideal for seed germination, seedling establishment, and crop growth. However, conventional tillage significantly affects certain soil characteristics such as soil structural stability, soil water conservation, soil temperature, infiltration, and evapotranspiration processes (Busari et al., 2015; Lal, 1993). In particular, from a SOC storage perspective, classic studies have shown that conventional tillage disrupts soil aggregates in surface layers and decreases the amount of total SOC, mainly in macroaggregates (Beare et al., 1994; Six et al., 1998) (Fig. 7.2).

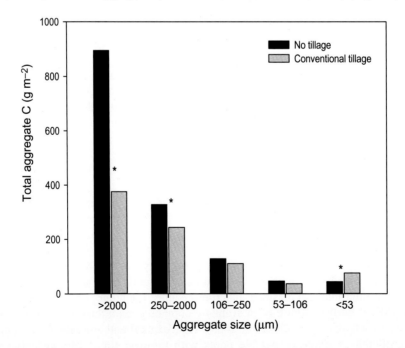

FIGURE 7.2

Impact of tillage on soil organic carbon storage in different soil aggregate sizes: Distribution of total aggregate C in no tillage and conventional tillage soils at 0−5 cm depth. The asterisk symbol (*) indicates significant differences at $P < .05$ level between tillage treatments within aggregate-size classes (Beare et al., 1994).

Among various tillage strategies, no-till provides one of the major credits in terms of increasing SOC storage (Lal et al., 2007). For example, West and Post (2002) found, from a global database, significantly higher SOC levels under no-till than conventional tillage, but similar SOC under conventional and reduced tillage. The average SOC sequestration rate (up to 30 cm depth) under no-till was 0.57 Mg ha^{-1} per year (West and Post, 2002). A metaanalysis study by Aguilera et al. (2013) showed an average increase of 0.44 Mg C ha^{-1} per year in the C sequestration rate to ~34 cm depth under no-till, and 0.32 Mg C ha^{-1} per year to ~27 cm depth under reduced tillage (cf. conventional management) in Mediterranean cropping systems. These results show that conservation tillage (particularly no-till) can be superior over conventional tillage for retaining SOC in surface layers and improving soil quality.

When considering different soil layers, a metaanalysis study by Luo et al. (2010b), which compared 69 sets of paired data for no-till and conventional tillage, showed a net gain in SOC stocks in the 0−10 cm layer under no-till (cf. conventional tillage). However, a net loss of SOC in no-till cf. conventional tillage was found in the 10−40 cm, while SOC stocks were similar between these contrasting tillage systems in deeper (40−60 cm) soil layers (Fig. 7.3). In another metaanalysis

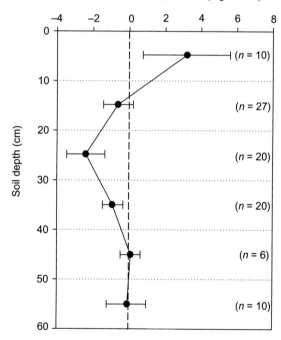

FIGURE 7.3

Impact of tillage on soil organic carbon storage at various soil depths: Mean difference of soil carbon (C) stock (Mg C ha^{-1}) under conventional tillage and no-tillage. Horizontal bars show the 95% confidence interval; numbers of observations are shown in the parenthesis.

Redrawn from Luo, Z.K., Wang, E.L., Sun, O.J., 2010b. Can no-tillage stimulate carbon sequestration in agricultural soils? A meta-analysis of paired experiments. Agric. Ecosyst. Environ. 139, 224–231.

study, Angers and Eriksen-Hamel (2008) also reported a greater accumulation of organic C under no-till than conventional tillage in the surface soil layers (0−10 cm), but conventional tillage (cf. no-till) had greater organic C content at the bottom of the plow layer ($\sim$25 cm). Further, in a 41-year long-term experiment in northern France, Dimassi et al. (2014) reported accumulation of SOC under conservation tillage versus conventional tillage in the surface layer (0−10 cm) that reached a plateau after 24 years. However, in the 10−28 cm soil layer, the SOC contents declined at 0.42%−0.44% per year, while the SOC contents were similar in the plowed (ca. 0−28 cm) or deeper (ca. 0−58 cm) soil layers under conservation versus conventional tillage after 41 years (Dimassi et al., 2014). These studies suggest that contrasting tillage practices may only be inducing a redistribution of C in the soil profile, e.g., the soil inversion by tillage would be translocating the surface SOC to lower depths (Olson and Al-Kaisi, 2015), and there may not be any apparent change in SOC stocks in the entire soil profile (Powlson et al., 2014). Hence, the entire soil profile should be considered to fully assess all the effects of tillage management practices on SOC gains or losses, rather than just considering the effect of tillage in topsoil layers only (Dimassi et al., 2014; Olson and Al-Kaisi, 2015; Powlson et al., 2008, 2011, 2014).

The extent of SOC retention and depletion not only depends on tillage practices, but also on soil texture, topography, and climate—which are also controlling factors of SOM formation. The effect of tillage on SOC sequestration is sometimes contradictory, particularly the short-term ($\leq$10 years) effect of conservation tillage because of variations in soil texture, climate, as well as biomass return and management (Olson and Al-Kaisi, 2015; VandenBygaart et al., 2003). After analyzing the relationships among SOC, texture, and climate for 500 rangeland and 300 cultivated soils in the US Central Plains, Burke et al. (1989) found that SOC losses due to cultivation increased with precipitation, and that relative SOC losses were lowest in clay-rich soils. The climate effect on SOC storage was high when a conventional tillage system was converted to no-till or reduced tillage in humid and subhumid climates (Francaviglia et al., 2017), whereas the effect of contrasting tillage intensity on SOC storage can be low in arid and semiarid regions (Fang et al., 2016; Francaviglia et al., 2017).

7.3.2 CROP STRAW RETENTION

Crop straw (stubble/residue) retention in fields is a well-known agricultural management practice which provides several positive effects including better soil structure, improved water retention, and less risk of erosion (Blanco-Canqui and Lal, 2009; Lal, 2005; Box 7.1). Straw return can also potentially improve the nutrient status of agricultural soils (Lal, 2008), which in turn enhances SOC due to increased crop rhizodeposition (Kuzyakov and Schneckenberger, 2004). Several studies have assessed the use of straw retention and/or incorporation as an option to enhance SOC storage, and have reported positive (Katterer and Andren, 1999; Lal, 2005; Mann et al., 2002; Powlson et al., 2008), negative (Ling-An et al., 2011; Wang et al., 2011), and negligible or no effects (Curtin and Fraser, 2003; Dimassi et al., 2014; Poeplau and Don, 2015). Nevertheless, a global metaanalysis using 176 field studies reported a $\sim$13% increase in SOC concentration in bulk soil following straw return (up to 50 cm depth) and the increases of SOC concentration were greater, in the range of $\sim$27 to $\sim$57%, for active (labile) C fractions (Liu et al., 2014).

The effects of straw/stubble incorporation or return on SOC storage and dynamics depend on many factors including rate of addition, climate, soil texture, and quality of the substrate. A strong

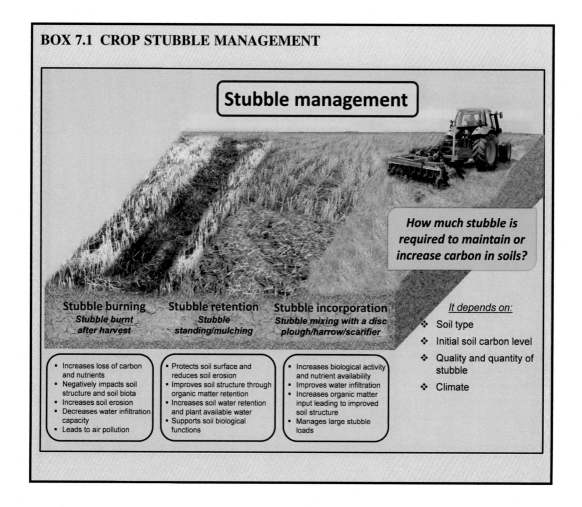

BOX 7.1 CROP STUBBLE MANAGEMENT

linear relationship between straw C input rate and SOC sequestration has been reported by many studies (Kong et al., 2005; Li et al., 2010; Liu et al., 2014). For example, Zhang et al. (2014) found the mean SOC storage of the 0–60 cm soil layers increased by 21.4%, 20.4%, and 8.2% with addition of 13.5, 9, and 4.5 Mg ha^{-1} of maize straw, respectively. Raffa et al. (2015) investigated varying degrees of crop residue removal across different climatic regions, soil types, and farming systems globally using a large database from 84 publications (660 observations) and observed 12% and 18% lower SOC contents when crop residues were removed (vs retained) in temperate and tropical climates, respectively. The study suggested that crop residue removal is not recommended, mainly in SOC-poor tropical and temperate soils, while partial residue removal can be considered in temperate climates in organic C-rich soils. Further, Peltre et al. (2016) demonstrated through a 100-year simulation study that loss or gain of SOC stocks at various straw incorporations in wheat cropping was related to initial SOC level, with potential for considerable SOC increases from full straw return in SOC-poor sandy loam soils (Fig. 7.4).

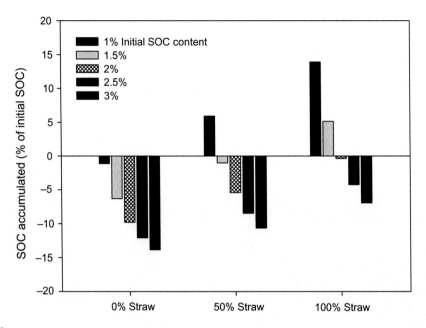

FIGURE 7.4

Impact of residue incorporation on soil organic carbon (SOC) storage: Simulated changes in SOC stocks (0−3 m) after 100 years of continuous wheat cropping with various levels of wheat straw incorporation and initial SOC concentrations at 0−25 cm depth.

Redrawn from Peltre, C., Nielsen, M., Christensen, B.T., Hansen, E.M., Thomsen, I.K., Bruun, S., 2016. Straw export in continuous winter wheat and the ability of oil radish catch crops and early sowing of wheat to offset soil C and N losses: a simulation study. Agric. Syst. 143, 195−202.

The SOC content is also influenced by the quality of crop straw (Chivenge et al., 2007; Lal, 2004a), which is partly determined by its C:N ratio (Blanco-Canqui and Lal, 2009). Crop straw with a low C:N ratio decomposes rapidly. For example, soybean residues decompose faster than maize residues due to a lower C:N ratio, thereby reducing the SOC content (Blanco-Canqui and Lal, 2009). When comparing two contrasting crop rotations, Khan et al. (2007) found that continuous maize cropping maintained a higher SOC content than soybean, which they attributed to relatively high C input from maize residues. Other factors, such as experimental duration (Lehtinen et al., 2014; Liu et al., 2014; Shen et al., 2007), N application rate (Jacinthe et al., 2002), and soil conditions (e.g., soil microorganisms, moisture, temperature, and aggregates) (Bandyopadhyay et al., 2010; Heikkinen et al., 2013; Lugato et al., 2006) are also important in explaining the large differences in SOC stocks and changes following straw return.

Burning crop straw is a common practice, especially in the tropics, to manage stubble loads, but it enhances nutrient mineralization as well as causing nutrient loss, air pollution, and reduced soil fertility. This practice can also be harmful to beneficial invertebrates and microorganisms (Hemwong et al., 2008; Jain et al., 2014; Wuest et al., 2005; Box 7.1). Furthermore, straw burning can cause a loss of SOC. For example, a loss of 1.75 Mg C ha^{-1} (0−10 cm layer) was reported over a period of 19 years in a field trial in south eastern Australia (Chan and Heenan, 2005). Similarly, Heenan et al. (2004) found a loss of 8.2 Mg C ha^{-1} (0−10 cm) after 21 years in a wheat/wheat rotation when stubble was burnt and soil tilled, versus a gain of 3.8 Mg C ha^{-1} (0−10 cm) in a clover/wheat rotation when stubble was retained and soil direct drilled. However, in a review of stubble retention systems in southern Australia, the higher SOC levels under stubble retention practices (in comparison to stubble burnt treatments) was not attributed to the sequestration of SOC, but rather to a slower rate of SOC loss (Scott et al., 2010).

When studying straw incorporation as a C sequestration option, it is also important to consider the fluxes of GHGs (Reiter, 2015). Liu et al. (2014) found an increase in CO_2 and nitrous oxide (N_2O) emissions from upland soils, but a decrease in CH_4 emission with the addition of straw. By considering all the fluxes, they concluded that straw incorporation still leads to a C sink in upland soils. Further, Jiang et al. (2017) found a decrease in CH_4 emissions following maize straw incorporation, but increased emission rates of N_2O. However, they found a significant linear relationship between annual straw C input rate and SOC sequestration rate over the 5-year cycle. These results imply that SOC accumulation counteracted the extra emissions of GHG.

Although it is widely assumed that straw/stubble retention increases SOC levels, the increase in SOC may be more pronounced during the first 10−20 years than in the longer term (Lehtinen et al., 2014; Liu et al., 2014; Luo et al., 2010a; Poeplau et al., 2015; Zhang et al., 2014; Zhao et al., 2013). Furthermore, straw return introduces large amounts of labile C to soils, which not only increases mineralization of active SOC pools (Liu et al., 2014), but also accelerates mineralization of stabilized SOC via positive priming effects (Guenet et al., 2012; Kirkby et al., 2014). On the other hand, Kirkby et al. (2011, 2013, 2014, 2016) highlighted the importance of applying supplementary nutrients to enhance SOC storage (with potential to minimize positive priming of SOC mineralization) during incorporation of C-rich crop residues into soil. In particular, Kirkby et al. (2014) reported increased net humification of wheat straw added to soil (net conversion of residue-C to fine fraction SOM-C) by two- to eightfold across a range of soils when inorganic-N, -P, and -S (sulfur) were added with the straw. Kirkby et al. (2016) observed that SOC stocks to 1.6 m depth increased by 5.5 Mg C ha^{-1} when supplementary nutrients in a balanced amount were applied along with incorporated wheat residues in soil, but decreased by 3.2 Mg C ha^{-1} without nutrient addition. These results draw attention to the need for a balanced supply of nutrients along with the input of crop residues to match the stoichiometry of resistant SOM to increase SOC sequestration.

7.3.3 ORGANIC AMENDMENTS AND FERTILIZERS (INTEGRATED NUTRIENT MANAGEMENT)

Intensive cultivation results in loss of soil via erosion, SOC depletion, and reduced productivity (Lal, 2001; Lal and Stewart, 1990). However, the use of farmyard manure (FYM), GM, composts, biosolids, and crop residues in agricultural systems—which represents a direct source of organic C and an important source of plant nutrients, is proving to be beneficial for improving soil quality and increasing SOC storage—especially on a long-term basis (Ghosh et al., 2012; Torri et al., 2014; Wei et al., 2016). Further, amendments of soil with recycled organic materials such as compost and biosolids, where organic C is in relatively stable forms, have proven more effective in increasing SOC storage than fresh plant residues and animal/plant manures (e.g., FYM and

GM), or in nonamended systems (Inbar et al., 1990; Zinati et al., 2001). Biosolids, e.g., consist of 40%−70% organic matter, with the C content ranging from 20% to 50%. Several studies and literature reviews have shown both short-and long-term increases in SOC from biosolids-derived organic matter (Jin et al., 2015; Tian et al., 2009; Torri et al., 2003, 2014). In various compost-treated (biowaste and green waste−sewage sludge) Luvisols, where 0.4 kg organic C m^{-2} was applied biannually for a period of 15 years, Paetsch et al. (2016) observed an increase in SOC stocks (0−28 cm depth) by ∼30% in bulk soil, and by ∼71%−155% in fine occluded particulate organic matter (<20 μm).

While a balanced supply of plant nutrients in soil enhances photosynthesis, leading to enhanced biomass production, C inputs, and SOC content (Kuzyakov and Gavrichkova, 2010), some studies have reported that unbalanced nutrients (van Groenigen et al., 2006) or excessive nutrient application (Qiu et al., 2010) may affect soil C and N storage. For example, soil acidification due to excessive N application (Guo et al., 2010) can lead to the loss of soil carbonates and further decrease soil total C storage (Qiu et al., 2010). On the other hand, integrated nutrient management, where the combined use of organic and inorganic amendments is aimed at optimizing nutrient inputs to meet crop demand, reducing nutrient losses and improving productivity, is becoming an important management concept to improve soil quality and long-term SOC accumulation, leading to a sustainable soil-plant system (Ghosh et al., 2012; Wang et al., 2015).

In a 22-year-old field experiment in a mollisol in China, Qiu et al. (2016) found that the combined application of exogenous C and fertilizer sources (straw + NPK, manure (M) + NPK, 1.5MNPK) decreased the inorganic C loss due to soil acidification, while favoring organic C and N storage in soil. Further, after long-term application of FYM alone or combined with chemical fertilizers (CFs) in soils, Su et al. (2006) showed that SOC sequestration was 3.6−9.7 Mg C ha^{-1} (average 5.9 Mg C ha^{-1}), with a sequestration rate of ∼0.3 Mg ha^{-1} per year. However, for soils that received no FYM, an average SOC loss of 5.8 Mg ha^{-1} was observed over a 23-year period. Using a global metaanalysis, Han et al. (2016) showed that topsoil organic C increased, on average, by 0.9 g kg^{-1} (10.0% relative increase cf. no fertilizer), 1.7 g kg^{-1} (15.4%), 2.0 g kg^{-1} (19.5%) and 3.5 g kg^{-1} (36.2%) under unbalanced application of chemical fertilizers (UCF), balanced application of CFs, CF plus straw (CFS), and CF plus manure (CFM), respectively. They estimated C sequestration occurred over 28−73 years under CFS application and 26−117 years under CFM application, but with high variability across climatic regions. Wei et al. (2016) studied 32 long-term (13−31 years) experiments in diverse cropping systems in China and found an increase in SOC content of 23% for organics plus fertilizers (15.2 g kg^{-1}), 18% for organics alone (14.5 g kg^{-1}), and 8% (13.3 g kg^{-1}) for fertilizers alone, when compared to control soils (12.3 g kg^{-1}). Further, the responses of SOC increase were higher in dry systems (wheat and/or maize cropping) compared to flooded systems (single, double, or triple rice cropping) (Wei et al., 2016). Furthermore, Chaudhary et al. (2017) observed increased SOC storage (0−15 cm) in the following order: FYM + NPK (5.3 Mg C ha^{-1}) > straw + NPK (5.1 Mg C ha^{-1}) > GM + NPK (4.1 Mg C ha^{-1}) > NPK (3.6 Mg C ha^{-1}) > control (no NPK, no amendments) (2.0 Mg C ha^{-1}) over 15 years in a rice−wheat cropping system (Fig. 7.5).

7.3.4 IMPROVED CROP ROTATION

Improvement of crop rotations by changing from a monoculture to more diverse cropping systems, integrating forages/ley and cover crops in rotations and increasing cropping intensity in drylands is

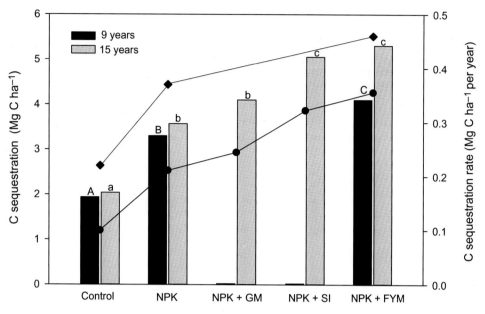

FIGURE 7.5

Impact of organic and inorganic amendments on soil organic carbon (SOC) storage: Long-term effect (9 years—Brar et al., 2013; 15 years—Chaudhary et al., 2017) of NPK fertilizers and organic amendments on C sequestration (Mg C ha^{-1}) and its rate (Mg C ha^{-1} per year), in surface soil (0–15 cm) in a rice–wheat cropping system. *FYM*, farmyard manure; *GM*, green manure; K, potassium; N, nitrogen; P, phosphorus; *SI*, straw incorporation. Values (small and capital letters) followed by different letters within each study are significantly different at $P \leq .05$ by Duncan's Multiple Range Test.

a promising strategy to increase SOC stocks in agricultural soils (Lal, 2004a). In general, crop rotations are more beneficial toward the build-up of SOM than monoculture systems (Jarecki and Lal, 2003). More recently, increasing crop rotational diversity has been shown to play a major role in increasing SOC storage and ecosystem functions, driven by enhanced root C input, soil microbial diversity, and soil aggregate stability (McDaniel et al., 2014; Tiemann et al., 2015). Crop rotations can be further improved by incorporation of perennial forages with extensive root systems to increase root C input and physical protection in soil aggregates, resulting in SOC sequestration rates up to 0.5 Mg C ha^{-1} per year (Dendoncker et al., 2004; Hutchinson et al., 2007; Paustian et al., 1997). The predicted rate of C sequestration from pasture in rotation (returning to crop after ca. 5 years) in dryland farming systems in Australia was 0.66 Mg C ha^{-1} per year at 0–10 cm and 0.78 Mg C ha^{-1} per year at 0–30 cm soil depth (Badgery et al., 2014).

An efficient build-up of SOC without a reduction of agricultural productivity can be achieved by cover crops—winter cover crops in temperate and tropical environments. Cover crops not only improve the soil structure and increase above- and belowground C input, but also reduce soil erosion and N leaching, control weeds, and increase water infiltration (Aertsens et al., 2013; Bronick and Lal, 2005; Cherr et al., 2006; Dabney et al., 2001; Fageria, 2007; Fageria et al., 2005;

Poeplau and Don, 2015). In addition, leguminous species can fix substantial amounts of atmospheric N. As a result, yield increases of main crops are often observed, which further increases the C input. A worldwide metaanalysis revealed a mean C sequestration rate of 0.3 Mg ha^{-1} per year by cover cropping without significant differences between legumes and nonlegumes and between temperate and tropical environments (Poeplau and Don, 2015). Other studies based on much smaller data sets reported higher mean C sequestration rates around 0.9 Mg ha^{-1} per year (Balkcom et al., 2013; Jarecki and Lal, 2003). In dry regions with precipitation <500 mm, cover crops could have an adverse effect on agricultural productivity as the available water for subsequent main crops may be reduced (Blanco-Canqui et al., 2011; Cherr et al., 2006; Unger and Vigil, 1998; Wortman et al., 2012). However, indications were found that species adapted to dry conditions such as hairy vetch are efficient for soil C sequestration without reducing yields under a continental climate (Wiesmeier et al., 2015a). In this regard, increased cropping frequency by a reduction of summer fallow also promotes SOC accumulation in semiarid agroecosystems (Lal, 2002; Paustian et al., 1997, 1998, 2000).

7.3.5 LAND-USE CHANGES

Land use largely affects SOC storage and land-use change is assumed to be the most dynamic factor affecting SOC changes. In the tropics, primary forests are converted to agricultural land on a large scale, whereas in temperate regions, both extensification (e.g., afforestation or abandonment of agricultural land) and intensification of land use (e.g., conversion of forest/grassland to cropland) can occur side by side (Schulp et al., 2008). A large number of studies and metaanalyses found evidence for a strong SOC decline of 30%−80% when forest/grassland were converted to cropland (Davidson and Ackerman, 1993; Guo and Gifford, 2002; Houghton and Goodale, 2004; Janssens et al., 2005; Murty et al., 2002; Poeplau et al., 2011; Sleutel et al., 2007; Wei et al., 2014). The loss of SOC in cultivated soils was attributed to erosion, lower C inputs, reduced SOM stabilization due to weakened aggregation, and subsequent mineralization promoted by increases in soil temperature and aeration (Balesdent et al., 2000; Guo and Gifford, 2002; Hamza and Anderson, 2005; Murty et al., 2002; Poeplau et al., 2011; Six et al., 1998). However, many studies failed to compare SOC stocks over the entire soil profile. Although the amount of SOC stored in subsoils exceeds 50% in most ecosystems (Batjes, 1996; Harrison et al., 2011; Jobbagy and Jackson, 2000; Rumpel and Kögel-Knabner, 2011), the majority of studies included only the first 30 cm of the soils (IPCC, 2003; Janssens et al., 2005; Jones et al., 2005; Smith et al., 2005; Stolbovoy et al., 2007). As large SOC differences among different land uses in topsoils cannot be assigned to subsoils, shallow soil sampling could lead to underestimation of cropland SOC stocks and thus to a biased evaluation of land use effects (Harrison et al., 2011). In a review by Murty et al. (2002), an average SOC loss of 30% after the conversion of forest into agricultural land was reported, but only by 19% when studies including soil depths deeper than 45 cm were considered. Similarly, Wiesmeier et al. (2015c) found evidence for only 20% lower total SOC stocks in cropland soils down to 1 m depth compared to forest soils in Germany, with a strong dependence on soil type. Furthermore, indications were found for increased proportions of stable mineral-associated organic C in temperate cropland soils, which may be attributed to an increase in soil-crop residue contact in the topsoil (Balesdent et al., 2000; Stemmer et al., 1999; Wiesmeier et al., 2014c; Wingeyer et al., 2012). Besides shallow soil sampling, the application of pedotransfer functions to derive missing soil parameters, particularly bulk density, probably caused underestimation of agricultural SOC stocks (Baker et al., 2007; Balesdent et al., 2000; Davidson and Ackerman, 1993; Wiesmeier et al., 2012).

Land-use extensification, particularly the conversion of cropland to grassland/forest, was proposed as a promising strategy for SOC sequestration (Del Galdo et al., 2003; Lal, 2004a; Post and Kwon, 2000; Schulp et al., 2008). For conversion of arable land to grassland, C sequestration rates of $0.3-1.9$ Mg ha^{-1} per year were reported (Dendoncker et al., 2004; Freibauer et al., 2004; Lugato et al., 2014; Post and Kwon, 2000; Smith, 2004; Soussana et al., 2004; Vleeshouwers and Verhagen, 2002). The absence of physical disturbance, improved aggregation, and increased belowground C inputs of grasses are the main factors influencing SOC accumulation in grassland soils. Conversion of cropland to forest resulted in SOC gains of $0.3-0.8$ Mg ha^{-1} per year with an increasing trend from temperate to subtropical/tropical regions (Freibauer et al., 2004; King et al., 2004; Paul et al., 2002; Post and Kwon, 2000; Smith, 2004). However, the build-up of SOC stocks after afforestation of cropland (particularly with coniferous species) is often associated with a shift from stable to labile SOC (Poeplau and Don, 2013; Wiesmeier et al., 2014c). For conversion of grassland to forest, no uniform effect was observed among different climatic zones (Paul et al., 2002; Post and Kwon, 2000). At least for temperate environments, a decreasing trend in SOC stocks was reported when grassland was afforested (Guo and Gifford, 2002; Poeplau et al., 2011).

7.3.6 AGROFORESTRY

Agroforestry has been widely practiced around the world for millennia to foster agricultural production (Jose, 2009; Young, 1989). In agroforestry systems, trees and shrubs together with cropped areas and/or grassland are integrated and form multifunctional agricultural production systems providing multiple benefits by optimized utilization of nutrients, water, and light. There is a huge variety of agroforestry systems that can be grouped under silvoarable systems (alley cropping, parklands), silvopastoral systems (e.g., Dehesas, Montados), protective systems (windbreaks, shelterbelts, riparian buffers), multistorey systems (e.g., home-gardens), rotational woodlots, and shifting cultivation (Kim et al., 2016; Lorenz and Lal, 2014; Nair et al., 2009; Plate 7.1). Besides providing agricultural crops, fodder, and firewood/timber, these systems sustain a number of environmental benefits and ecosystem services such as erosion control, improved soil fertility and water availability, increased habitat and species diversity, and improved aesthetics of agricultural landscapes (Jose, 2009; Stavi and Lal, 2013). However, the potential of agroforestry systems for soil C sequestration was only recognized in the past two decades (Albrecht and Kandji, 2003; Lorenz and Lal, 2014; Montagnini and Nair, 2004; Nair, 2012; Nair et al., 2009; Schoeneberger, 2009; Stavi and Lal, 2013). Besides C fixation in tree biomass, agroforestry systems efficiently accumulate C both in topsoils and subsoils in several ways. Quantitatively, the C inputs from trees, shrubs, and under storey vegetation in the form of litter fall, roots, and rhizodeposition make the most important contributions to increases in SOC stocks, mostly within woody components. Moreover, SOC stocks can be increased in adjacent agricultural land by agroforestry-induced yield increases associated with higher C inputs (Lorenz and Lal, 2014). Further studies are needed that investigate SOC storage in adjacent agricultural land using geostatistical methods (Stavi and Lal, 2013).

Although the increased quantity of C inputs may be the most important factor for C sequestration, there are further processes that positively affect the C balance of agroforestry systems, such as reduced soil erosion or even deposition of C-containing materials, lower decomposition by

PLATE 7.1

Different types of agroforestry systems: Shelterbelts with English walnut (*Juglans regia*), Republic of Moldova (above left, picture: R. Hübner); terracing and intercropping with saltcedar (*Tamarix ramosissima*) and alfalfa (*Medicago sativa*), Loess Plateau, China (above right, picture: J. Lu); Willow (*Salix* spp.) planted along water channels, Germany (below left, picture: R. Hübner); Alley cropping of English walnut (*J. regia*) and rapeseed (*Brassica napus*), France (below right, picture: R. Hübner).

recalcitrant litter, reduced soil disturbance, and improved physical protection of organic matter by aggregates (Lorenz and Lal, 2014). However, the C sequestration potential largely varies depending on the agroforestry system type, tree species, climate conditions, soil conditions, and management practices. In general, soil C sequestration rates are higher in tropical agroforestry systems than in temperate or arid/ semiarid environments (Hutchinson et al., 2007; Jose, 2009; Lorenz and Lal, 2014; Stavi and Lal, 2013), but comprehensive metaanalyses are still lacking. In the tropics and subtropics, soil C sequestration rates between 0.1 and 4.2 Mg ha^{-1} per year have been reported (Albrecht and Kandji, 2003; Lorenz and Lal, 2014; Montagnini and Nair, 2004; Oelbermann et al., 2004, 2006). In temperate agroforestry systems, soil C sequestration rates in the range of 0.1−6.4 Mg ha^{-1} per year have been observed (Bambrick et al., 2010; Cardinael et al., 2017; Chendev et al., 2015; Jarecki and Lal, 2003; Kim et al., 2016; Oelbermann et al., 2006; Palma et al., 2007; Udawatta and Jose, 2012). In a worldwide metaanalysis on soil C sequestration of different agroforestry systems, Kim et al. (2016) reported rates of $0.3–7.4$ Mg ha^{-1} per year for different systems. In order to give reliable estimates of the total global C sequestration potential, further systematic studies are needed that determine SOC accumulation for specific agroforestry systems and site conditions (climate, soil) and particularly should include subsoils (Cardinael et al., 2017; Lorenz and Lal, 2014; Nair et al., 2009).

7.3.7 **MIXED CROP–LIVESTOCK FARMING SYSTEMS**

Mixed crop–livestock farming systems integrate pasture, livestock, and crop production activities on the same farm, where the resources such as crop/pasture residues (feed base), animal manure (nutrient input), power, and cash are exchanged to gain benefits from the resulting crop–livestock interactions (Sumberg, 2003; Thornton and Herrero, 2015). Mixed farming systems are practiced on 2.5 billion hectares of land worldwide, of which 1.1 billion hectares are rainfed arable cropland, 0.2 billion hectares irrigated cropland, and 1.2 billion hectares grassland. Furthermore, these systems make large contributions to global livestock products, with $\sim 50\%$ and $\sim 90\%$ of global meat and milk production, respectively (Van Keulen and Schiere, 2004). Sulc and Tracy (2007) highlighted the following major benefits of integrating livestock and crop activities:

1. The use of crops produced on the farm to feed livestock, thereby reducing the cost of importing foodstuff for livestock production and the livestock act as a sink for agricultural production.
2. The organic waste of livestock including manures can serve as a primary source of nutrients for crop production.
3. Ruminant livestock encourage the establishment of perennial grass and legume forages as a primary foodstuff.

A significant enhancement of key soil quality parameters such as soil aggregation, SOC, and soil microbial biomass under a mixed crop–livestock system, cf. a pure cropping system has been reported (de Moraes et al., 2014). Conversely, mixed crop–livestock (cf. native forest) may decrease soil quality parameters, such as total SOC and particulate organic C, especially under high grazing intensity (Assmann et al., 2014). Further, light and moderate grazing in mixed crop–livestock systems showed an annual C increase of 0.96 Mg ha^{-1} per year (over 9 years), while the most intensive grazing showed no increase in SOC stocks, possibly due to lower C inputs from pasture roots, shoots, and manure (Assmann et al., 2014). The integration of a crop–livestock system in agricultural areas in Brazil, formerly cultivated under no-till crop succession, acted as a sink of C with accumulation rates ranging from 0.8 to 2.6 Mg ha^{-1} year, depending on the crops introduced, the edapho-climatic conditions, and the time periods (over $1–8$ years) of the crop–livestock integration (de Faccio Carvalho et al., 2010). Further assessments are still required on how crop–livestock integration in various agroecosystems influences SOC stocks and whether such impacts vary with climate, soil type, residue (plant plus animal) amount, and quality.

7.3.8 **BIOCHAR APPLICATION TO SOIL**

Biochar is a "man-made" pyrolyzed biomass that can be produced from different organic feedstock materials (such as woody residues, food and green wastes, biosolids, and animal manures) at temperatures between $400°C$ and $700°C$ under limited, or no, oxygen supply (Lehmann and Joseph, 2009).

This pyrolysis process stabilizes a significant proportion of organic C in the pyrolyzed biomass, which can have high stability (persistence) in soil, particularly compared to unburnt organic matter (Lehmann et al., 2015; Singh and Cowie, 2014). However, the inherent biochemical recalcitrance of biochars is primarily dependent on feedstock types and production conditions, with stabilities ranging from a few decades to thousands of years (Singh et al., 2012). In particular, biochars produced from woody material have greater mean residence times than those produced from manures (Singh et al., 2012) and this is likely to be related to the strong presence of inherent recalcitrant C compounds, such as lignin (Bird et al., 1999). Further, pyrolysis temperature could be altered to generate a highly stable biochar. For example, biochars produced at a high ($\sim 550°C$) temperature have greater persistence in soil than those produced at a low ($\sim 400°C$) temperature (Singh et al., 2012), mainly by increasing the proportion of nonaromatic and aromatic C, and the degree of condensation of aromatic C in biochar (McBeath and Smernik, 2009; McBeath et al., 2014).

Research has also analyzed biochar persistence in different soils, climatic conditions and agroecosystems (Fang et al., 2014a,b, 2017; Singh et al., 2015; Ventura et al., 2015; Weng et al., 2015). Biochar persistence was shown to be: (1) Higher in an iron-dominant clay-rich soil, even in a subtropical environment (Weng et al., 2015); and (2) lower in a clay-poor soil or a C-rich soil in temperate environments (Singh et al., 2015). Hence, organomineral associations induced by the interaction of biochar, soil type, and plant C inputs have the potential to enhance biochar C sequestration potential in certain clay-rich soils (Lehmann et al., 2015; Weng et al., 2015, 2017).

Biochar application can also enhance SOC storage through "negative priming," which can result from decreased mineralization rates of both existing (native) SOC and new plant-derived C, possibly facilitated by sorption of labile C on biochar and enhanced organomineral associations (Fang et al., 2015; Keith et al., 2011; Mandal et al., 2016; Singh and Cowie, 2014; Whitman et al., 2015). Further, a study in 2017 has shown that aged-biochar enhanced belowground recovery of root-derived C in a pasture system, as well as enhanced retention of root-derived C by accelerating organomineral interactions, which decreased SOC mineralization by 46 g CO_2-C m^{-2} per year in an iron-dominant, clay-rich soil (Weng et al., 2017). Conversely, biochar application can also increase mineralization of native SOC via "positive priming" (Luo et al., 2011; Singh and Cowie, 2014), by providing a small amount of labile C, and an environment, suitable for growth and survival of microorganisms (Lehmann et al., 2011; Mandal et al., 2016). However, this positive priming effect is generally small and short-lived, compared to the stabilization of biochar C and a long-term negative priming effect, particularly in clay-rich soils (Fang et al., 2015; Maestrini et al., 2014; Weng et al., 2017).

Biochar application in soil has gained recognition in the past few years as a useful strategy for both C sequestration-climate mitigation and soil improvement. The benefits from biochar application to soil are related to: (1) Reduction in biomass decay due to C stabilization in biochar, which can directly increase the stable C content of soil; (2) avoided emissions of N_2O and CH_4; and (3) enhanced productivity (Woolf et al., 2010). Further, biochar application can indirectly increase soil C sequestration potential through stabilization of native SOC (Fang et al., 2015) and root-derived C (Weng et al., 2015). In particular, Weng et al. (2017) have demonstrated a considerable C sequestration potential well in excess of recalcitrant C in the biochar, in a clay-rich soil. Because of these direct and indirect benefits, biochar application to soil can have a considerable negative emission potential, e.g., 0.7 Pg C eq. per year, with a maximum technical potential of 1.3 Pg C eq. per year (Smith, 2016; Woolf et al., 2010). As biochar application has been demonstrated to have a

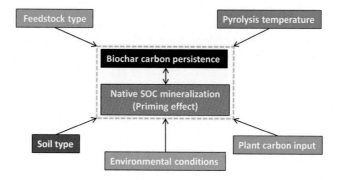

FIGURE 7.6

Schematic sketch showing the interactive influence of inherent biochar properties and external factors (soil, plant input, and environment) on biochar carbon persistence and "priming effects" of native soil organic carbon (SOC) mineralization with implications for carbon sequestration in biochar systems.

significant negative emission potential, there is a suggestion to include biochar within integrated assessment models to further assess its potential application in soil for C sequestration and climate mitigation (Smith, 2016). The interactive effects of inherent biochar properties and external factors (soil, plant, environment) on biochar C persistence and "priming effects"—as described earlier and summarized in Fig. 7.6—need further validation in various agroecosystems, soil types and environments (Fang et al., 2014b, 2017; Weng et al., 2017).

7.3.9 CLAY ADDITIONS TO SANDY SOIL

Improved management practices such as adoption of no-tiller stubble retention and improved rotations can sequester significant amounts of organic C compared to conventional management practices such as conventional tillage, stubble burning or monocropping in croplands (Minasny et al., 2017; Sanderman et al., 2010; Stockmann et al., 2013). However, it is difficult to increase the SOC content of sandy soils because of their low nutrient- and water-holding capacity which limits plant growth and C inputs (Baldock and Skjemstad, 2000; Hall et al., 2010). Since the organic C is not protected from microbial decomposers as observed in soils of heavier texture, sandy soils have low C sequestration capacity. While clay soils accumulate organic C relatively quickly, sandy soils may accumulate practically no C—even after 100 years of high organic C inputs (Christensen, 1996). Addition of clay to topsoils has the potential to dramatically increase organic C stored in soils. The technique is being implemented in Australia (particularly in South Australia and Western Australia), where subsoil clay is added to topsoils to overcome water repellency in sandy soils (Betti et al., 2015; Cann, 2000). The addition of clay changes the physical and chemical properties of the soil, enabling greater C sequestration. It also increases soil productivity, and reduces soil erosion, and nutrient leaching. Moreover, it has no negative environmental impact.

Though no-till has become increasingly popular and also has been shown to increase the amount of water and organic matter (nutrients) in the soil and decrease erosion, there are questions about the permanence of the resultant SOC sequestration, particularly in sandy soils (Blanco-Canqui and

Lal, 2009; VandenBygaart et al., 2003). No-till farming systems also involve the use of herbicides, which has an ongoing and not-inconsiderable cost. Soil enhancement through one-time addition of clay in sandy soils can provide a method to increase the ability of farming systems to retain and stabilize organic C inputs. The addition of clay can be achieved by several strategies: spreading, delving, and spading (Davenport et al., 2011; Schapel et al., 2017). These methods result in different distributions of the clay in the soil profile. Clay spreading is for deep sands where clay-rich subsoil is deeper than 70 cm from the surface. The method involves excavation of clay from the subsoil in a pit close to the deep sandy area, and spreading and incorporating the clay clods using a scraper or multispreader onto the soil surface. By contrast, clay delving is used where clay-rich subsoil is present within the top 70 cm depth (Desbiolles et al., 1997) and the clay digging implement is used to bring and mix clods of clay into the sandy surface. Spading, used since the late 2000s, is a modification method where clay-rich subsoil is within 30 cm of the surface. The spader can be used to both raise and incorporate the clay, thus creating an even mix of subsoil clay and topsoil sand.

Increasing clay content can increase the retention of organic C inputs, and consequently, SOC content that is physically protected from microbial decomposition (Hoyle et al., 2011). Clay modification increased SOC stocks (0−30 cm) by up to 14−22 Mg C ha^{-1} in various regions of South Australia (Schapel et al., 2017). Hall et al. (2010) reported an increase in SOC stocks of 2.2 Mg C ha^{-1} 8 years after addition of 300 Mg ha^{-1} of kaolinitic subsoil (32% clay) to the top 10 cm of a soil in Western Australia by delving. In studies undertaken on light-textured sands of North Thailand, where low quality (CEC: 28 cmol$_c$ kg^{-1}) smectitic clays were applied to soils at a rate of 50 Mg ha^{-1}, cumulative net primary productivity over 2 consecutive years increased from 0.22 Mg dry matter (DM) ha^{-1} in untreated soils to 23 Mg DM ha^{-1} with potential to increase SOC storage (Noble and Suzuki, 2005). Shi and Marschner (2013) showed that addition of a clay-rich subsoil (84% clay) to a loamy sand soil at 0%, 10%, and 30% (w/w) amended with shoot residues of native perennial grasses and barley individually or as mixtures, can increase SOC sequestration, while minimizing CO_2-C emission.

The organic C retention capacity of clay-incorporated sandy soils is also influenced by clay application rate and distribution (both depth of incorporation and size of clay clods), as well as time since modification. Hall et al. (2010) observed an increase in SOC concentration by 0.2% in the top 10 cm after 5 years of clay addition, although SOC at depths greater than 10 cm was not reported. While it is possible that higher SOC potentials can be achieved through clay addition, the actual increase in SOC and the time taken to achieve the equilibrium levels may vary. At this stage, there is limited data to confirm the SOC storage potential through clay addition/modification technology, although significant increases can be expected due to the role of clay in protecting organic C from microbial decomposers and by increasing plant growth.

7.4 SIMULATION OF SOC STORAGE POTENTIAL UNDER DIFFERENT MANAGEMENT SYSTEMS

The use of SOM models as research tools is important to enhance our understanding of SOC dynamics and to develop management strategies to optimize the C inputs and outputs for SOC sequestration. Models are mathematical representations of the real system and provide an opportunity to understand C and nutrient-cycling processes in soil-plant-atmosphere systems (Paustian et al., 1997; Skjemstad et al., 2004). The advantages of SOC modeling are:

1. Encapsulating our understanding of processes and relationships in soil-plant-atmosphere.
2. Improved understanding of SOC dynamics if process-based biogeochemical relationships among various factors of SOC cycling (such as interaction of farming practices, soil factors, and environment variables etc.) are used in constructing the model.
3. Provide time-series estimates without interpolation once the model is calibrated and validated for a region.
4. Flexible to improve the system by incorporating more measurements.
5. Lower costs compared to soil survey that is required for monitoring of SOC at regular intervals.
6. Useful to study the consequences of various climate and management scenarios (like changes in land use, management practices, and technology, etc.) on SOC storage.

To date, there are 250 models for predicting SOC dynamics and nutrient turnover for a wide range of environments and land use conditions (Manzoni and Porporato, 2009). Various simulation approaches, such as CENTURY (Parton et al., 1987), RothC (Jenkinson et al., 1990), and APSIM (Keating et al., 2003) are in use for predicting nutrient and SOC dynamics. These models differ in their underlying assumptions, and the processes responsible for SOM decomposition—detailed descriptions of their structures and functionality have been given elsewhere (Jenkinson et al., 1990; Keating et al., 2003; Parton et al., 1987). Among these three models, RothC is a simple process-based model because it requires few inputs to initialize while the model runs using a monthly time step. RothC does not have a functionality to simulate N dynamics, but it is used to simulate SOC only. However, SUNDIAL, a version of the Rothamsted Nitrogen Turnover model, can simulate both C and N in the soil-crop system (Bradbury et al., 1993; Smith and Bradbury, 1994). The CENTURY model is a general plant-soil ecosystem model that simulates plant production, C, N, P, and S cycling in the soil-plant system, as well as soil water and temperature. The model runs using a monthly "time step." APSIM (Agriculture Production System Simulator) is a framework of biophysical modules that simulate biological and physical processes in farming systems (Keating et al., 2003). The model simulates crop growth and soil processes on a daily time step in response to climate (temperature, rainfall, and radiation), soil water availability, and soil nutrient status (i.e., N and P).

Smith et al. (1997) evaluated nine SOC models against data from seven long-term experiments under arable, grassland, and woodland systems. Their results showed that six models (RothC, DNDC, CANDY, CENTURY, DAISY, and NCSOIL) performed well with no significant difference in accuracy—the residual mean square error of these models was lower than the other group of models (ITE, Verberne and SOMM). RothC (Bhattacharyya et al., 2011; Cerri et al., 2003; Wang et al., 2016), CENTURY (Cong et al., 2014; Kelly et al., 1997; Leite et al., 2004), and APSIM (Luo et al., 2011; Zhao et al., 2013) have mostly been successfully used for simulating the impact of different land use and management scenarios under different climatic zones on the assessment of SOC dynamics or stocks in different soils. Though RothC has been widely tested in different parts of the world, it does not include the loss of physically protected SOC. When land is converted from forest to agriculture, the losses of C during continuous cultivation of previously uncultivated land are not accurately simulated using RothC, however the incorporation of a factor accounting for loss of physically protected C may improve the predication accuracy (Gottschalk et al., 2010). Liu et al. (2009) tested the RothC model against data from a long-term cropping experiment with different tillage and management practices in Australia and found that RothC performed well in predicting SOC stocks in stubble burnt treatments, but failed to simulate the stubble retained treatments unless the stubble retention factor was reduced to $\leq 26\%$. RothC (RothC-26.3 commonly used in different parts of the world) cannot be used for subsoils, waterlogged and salt-affected soils. Jenkinson and Coleman (2008) developed a new model, RothPC-1, for modeling the turnover of organic C in the top meter of soil by including two extra parameters: (1) p, which moves organic C down the profile by an advective process; and (2) s, which slows decomposition with depth. Shirato et al. (2005) used RothC for predicting SOC stocks in paddy soils of Japan by modifying the decomposition rate constants of RothC by a factor of 0.2 in summer and 0.6 in winter. Setia et al. (2011) modified RothC for saline soils by incorporating a decomposition rate modifier and plant input modifier for salinity. Their results suggested that due to reduced plant inputs, SOC stocks in saline soils are predicted to be substantially lower than previously estimated. Cerri et al. (2007) simulated SOC changes in 11 land-use chronosequences of the Brazilian Amazon using the CENTURY model. They suggested that forest clearance and conversion to well-managed pastures would cause an initial decline in SOC stocks ($0-20$ cm depth), followed by a slow recovery to levels that could even exceed those under native forest in the majority of cases. Silver et al. (2000) found that the CENTURY model underestimated SOC stocks in coarse-textured soils and overestimated in clay-rich soils because the model default parameters might place too much emphasis on the effect of soil texture on passive C pool formation. However, Bricklemyer et al. (2007) found that CENTURY was insensitive to soil texture in predicting SOC stocks as a result of no-till management. Zhao et al. (2013) used APSIM to identify the key management practices affecting SOC dynamics in the wheat-cropping system in Australia and found that changes in SOC during the 122-year simulation were influenced by management practices such as fertilization (nonlinearly positive) and residue removal (linearly negative). Luo et al. (2011) used APSIM to predict long-term C dynamics under different management regimes in four semiarid sites across the wheat-belt of eastern Australia. They found that residue retention alone cannot reverse C loss in agricultural soils, but combining optimal N fertilizer application and whole residue retention can turn agricultural soils into C sinks.

These results show that the changes in SOC storage are clearly influenced not only by farming practices, but also by climate and soil, and their interactions, so that SOC modeling is useful to explore the management strategies that explain changes over time compared to those that rely on long-term field experiments. However, the results of modeling may not be reliable if: (1) Initial determination of SOC stocks is wrong; (2) the model is initialized without locally available data; (3) all the major components of the C cycle are not incorporated in the model; and (4) the uncertainties of outputs are not described.

7.5 THE 4 PER 1000 INITIATIVE

At the 21st conference of the parties to the United Nations Framework Convention on Climate Change (COP21) in Paris, the French Minister of Agriculture launched an ambitious program to increase the world's SOC stocks for climate change mitigation, the "4 per 1000 initiative." The basic idea of this initiative is that a yearly increase of global SOC stocks in the top 40 cm of soils by 0.4% would considerably counter-balance anthropogenic GHG emissions. In order to achieve the 4 per 1000 target, adoption of improved management practices on global agricultural land is needed, accompanied by restoration of degraded soils, and improved management of forest soils and wetlands. Over 170 countries and institutions have committed in a voluntary action plan to implement suitable practices to build-up SOC stocks. The 4 per 1000 initiative set a highly ambitious goal that requires close collaboration between the scientific community, famers/land managers, and policy makers on a global scale. However, for the first time, soil C sequestration has received attention as a strategy to mitigate climate change and contribute to food security on a global (political) level—"a major paradigm shift of historic significance" (Lal, 2016). Soil scientists around the world accepted this challenge and have already started to review the feasibility of the initiative (Chambers et al., 2016; Lal, 2016; Minasny et al., 2017). In a survey on SOC stock estimates and the feasibility of the 4 per 1000 target from 20 countries, a generally optimistic point of view was reported (Minasny et al., 2017). However, the different analyses varied greatly in detail, and most studies compared estimated target-derived C sequestration rates with observed rates of different management practices on a rather superficial level. For practical implementation of the initiative, more detailed feasibility studies are needed that include a precise registration of the status quo of SOC stocks at a defined reference year, an estimation of the organic C storage capacity of soils, and an analysis of existing land management practices together with an identification of potential areas for improving land management and building-up SOC at the field scale. This must be accompanied from the beginning by evaluating socioeconomic aspects, particularly costs, farmer/land manager attitudes, legal restrictions, conformity with existing subsidy systems, and provision of newly developed incentives (e.g., tradeable C credits). In this regard, there is a need for several specifications and revisions of the 4 per 1000 initiative. Chambers et al. (2016) claimed that the initiative should recognize existing soil health activities, maintain consistency with national GHG inventories (particularly in terms of a consistent soil depth), establish country-specific targets, prioritize specific land uses (particularly agricultural land), including restoration of degraded land, and commit to track progress by soil and GHG monitoring systems. Nevertheless, the initiative's framework should be flexible enough to consider different views of participating countries/regions on the method of implementation.

7.6 KEY TAKE-HOME MESSAGES AND RESEARCH PRIORITIES

A range of traditional and emerging agricultural management practices have been proving useful to increase SOC storage over time and slow down the rate of SOC loss. The key principles and practices rely on: (1) Decreasing C loss (output) by minimizing disturbance to soils from tillage, and eliminating fallowing, stubble burning, and overgrazing; and (2) increasing C inputs by retaining stubble and adding C-rich amendments, practicing integrated nutrient management, incorporating clay in sandy soils, changing cropland to mixed crop-pastures and agroforestry, and increasing crop diversity (Fig. 7.1). However, the effectiveness of the above practices is highly variable and the actual amount of SOC sequestration seems to be dependent on factors such as soil type, climate, topography, and the initial C level in soils. Nevertheless, considerable new research is required to identify a new generation of improved agricultural management practices and technologies that enhance SOC storage, and contribute to meeting the global "4 per 1000" initiative and offsetting GHG emissions. There is also a need to implement a reliable soil C auditing protocol for monitoring, reporting, and verifying C sequestration for incorporation into GHG inventories. Further, policy makers, farmers, and scientists need to promote a change of practices for C sequestration with additional benefits (e.g., climate change mitigation, soil health, food security). However, there could be "hidden costs" which need to be considered while implementing key practices and policies to foster soil C sequestration in agricultural systems.

Recommendations and research gaps:

- A need for more quantitative assessment of C sequestration potential of soils under improved land management practices for various soil types, climates, and agricultural systems to draw site-specific conclusions.
- Although organic C input is most active in surface soil layers, research is needed to identify management practices that enhance organic C input and storage below the soil surface ("subsoil") where SOC can be typically more stable than in surface soil. These practices may involve inversion of C-rich surface soil, or enhancement of plant C inputs in deep soil layers, e.g., via addressing subsoil constraints.
- Quality of C inputs is as important as quantity, and SOC stability varies according to the nature of added organic amendments, suggesting that selection of C amendment sources with high stability is important for long-term C sequestration in soil.
- While the amount of SOC storage and the duration of C sequestration is dependent on the C pools (labile vs recalcitrant) and their cycling and form of stabilization, more short- and long-term studies are required on this aspect under varying sites, crops, climates, and management practices.
- Recommendation of a range of land use and management practices for improved SOC storage and soil productivity is required, particularly in C-unsaturated soils to simultaneously mitigate the negative influence of soil GHG emissions.
- While improving parametrization of SOC models, due attention is needed that considers SOC saturation and the responses of priming effects and stable SOC fractions/pools to varying management and climate change scenarios.
- Inclusion of practices such as cover cropping and intercropping in monoculture croplands is needed, particularly species and varieties that enhance root C input and storage in soil, and focus on breeding crop plants with more extensive and deeper root systems.
- A due consideration is needed on appropriate management of soils that enhance belowground C inputs, in addition to the abundance and diversity of soil microorganisms, to greatly improve the C sequestration capacity of soils.
- Soil microbial contributions to climate change through the C cycle feedback and to SOC storage through enhancing soil aggregation are far from straight forward. Thus, questions such as: (1) How long microbial communities take to adapt to changes in soil microenvironments in response to agricultural management practices under changing climate? (2) Whether the biotic interactions between above- and belowground communities respond more to climate versus soil fertility; and (3) What implications these processes and microbial physiology have for SOC storage under different agroecosystems are still to be addressed.

- Stabilization mechanisms of microbial-derived residues through amendments such as biochar and clay inputs that facilitate soil organomineral interactions and aggregation require a thorough assessment. In particular, knowledge on the role of diverse microbial communities versus clay conditions (e.g., content and type) in relation to SOC storage is important to facilitate development of microbial-explicit C models.

ACKNOWLEDGMENT

BPS is grateful to The Grains Research & Development Corporation (GRDC) for funding a project (DAN00169) on soil organic matter functions in grain-based farming systems. MW is grateful to the Federal Ministry of Education and Research (BMBF) of Germany for funding the project "BonaRes-Centre for Soil Research, subproject C" (Grant 031A608C). We thank Dr. Yunying Fang of NSW Department of Primary Industries for assistance with drafting the biochar schematic diagram.

REFERENCES

Aertsens, J., De Nocker, L., Gobin, A., 2013. Valuing the carbon sequestration potential for European agriculture. Land Use Policy 31, 584–594.

Aguilera, E., Lassaletta, L., Gattinger, A., Gimeno, B.S., 2013. Managing soil carbon for climate change mitigation and adaptation in Mediterranean cropping systems: a meta-analysis. Agric. Ecosyst. Environ. 168, 25–36.

Albrecht, A., Kandji, S.T., 2003. Carbon sequestration in tropical agroforestry systems. Agric. Ecosyst. Environ. 99, 15–27.

Angers, D., Eriksen-Hamel, N., 2008. Full-inversion tillage and organic carbon distribution in soil profiles: a meta-analysis. Soil Sci. Soc. Am. J. 72, 1370–1374.

Angers, D.A., Arrouays, D., Saby, N.P.A., Walter, C., 2011. Estimating and mapping the carbon saturation deficit of French agricultural topsoils. Soil Use Manage. 27, 448–452.

Assmann, J.M., Anghinoni, I., Martins, A.P., de Andrade Costa, S.E.V.G., Cecagno, D., Carlos, F.S., et al., 2014. Soil carbon and nitrogen stocks and fractions in a long-term integrated crop–livestock system under no-tillage in southern Brazil. Agric. Ecosyst. Environ. 190, 52–59.

Badgery, W.B., Simmons, A.T., Murphy, B.W., Rawson, A., Andersson, K.O., Lonergan, V.E., 2014. The influence of land use and management on soil carbon levels for crop-pasture systems in Central New South Wales, Australia. Agric. Ecosyst. Environ. 196, 147–157.

Baker, J.M., Ochsner, T.E., Venterea, R.T., Griffis, T.J., 2007. Tillage and soil carbon sequestration – what do we really know? Agric. Ecosyst. Environ. 118, 1–5.

Baldock, J., Skjemstad, J., 2000. Role of the soil matrix and minerals in protecting natural organic materials against biological attack. Org. Geochem. 31, 697–710.

Balesdent, J., Chenu, C., Balabane, M., 2000. Relationship of soil organic matter dynamics to physical protection and tillage. Soil Tillage Res. 53, 215–230.

Balkcom, K.S., Arriaga, F.J., van Santen, E., 2013. Conservation systems to enhance soil carbon sequestration in the southeast U.S. coastal plain. Soil Sci. Soc. Am. J. 77, 1774–1783.

Bambrick, A.D., Whalen, J.K., Bradley, R.L., Cogliastro, A., Gordon, A.M., Olivier, A., et al., 2010. Spatial heterogeneity of soil organic carbon in tree-based intercropping systems in Quebec and Ontario, Canada. Agroforest. Syst. 79, 343–353.

Bandyopadhyay, P.K., Saha, S., Mani, P.K., Mandal, B., 2010. Effect of organic inputs on aggregate associated organic carbon concentration under long-term rice-wheat cropping system. Geoderma 154, 379–386.

Barre, P., Fernandez-Ugalde, O., Virto, I., Velde, B., Chenu, C., 2014. Impact of phyllosilicate mineralogy on organic carbon stabilization in soils: incomplete knowledge and exciting prospects. Geoderma 235, 382–395.

Batjes, N.H., 1996. Total carbon and nitrogen in the soils of the world. Eur. J. Soil Sci. 47, 151–163.

Beare, M., Hendrix, P., Coleman, D., 1994. Water-stable aggregates and organic matter fractions in conventional-and no-tillage soils. Soil Sci. Soc. Am. J. 58, 777–786.

Betti, G., Grant, C., Churchman, G., Murray, R., 2015. Increased profile wettability in texture-contrast soils from clay delving: case studies in South Australia. Soil Res. 53, 125–136.

Bhattacharyya, T., Pal, D., Deshmukh, A., Deshmukh, R., Ray, S., Chandran, P., et al., 2011. Evaluation of RothC model using four long term fertilizer experiments in black soils, India. Agric. Ecosyst. Environ. 144, 222–234.

Bird, M.I., Moyo, C., Veendaal, E.M., Lloyd, J., Frost, P., 1999. Stability of elemental carbon in a savanna soil. Global Biogeochem. Cycles 13, 923–932.

Blanco-Canqui, H., Lal, R., 2009. Crop residue removal impacts on soil productivity and environmental quality. Crit. Rev. Plant Sci. 28, 139–163.

Blanco-Canqui, H., Mikha, M.M., Presley, D.R., Claassen, M.M., 2011. Addition of cover crops enhances no-till potential for improving soil physical properties. Soil Sci. Soc. Am. J. 75, 1471–1482.

Bolinder, M.A., Janzen, H.H., Gregorich, E.G., Angers, D.A., VandenBygaart, A.J., 2007. An approach for estimating net primary productivity and annual carbon inputs to soil for common agricultural crops in Canada. Agric. Ecosyst. Environ. 118, 29–42.

Bradbury, N., Whitmore, A., Hart, P., Jenkinson, D., 1993. Modelling the fate of nitrogen in crop and soil in the years following application of 15N-labelled fertilizer to winter wheat. J. Agric. Sci. 121 (3), 363–379.

Brar, B.S., Singh, K., Dheri, G.S., Kumar, B., 2013. Soil carbon sequestration and soil carbon pools in rice–wheat cropping system: effect of long term use of inorganic fertilizers and organic manures. Soil Tillage Res. 128, 30–36.

Bricklemyer, R.S., Miller, P., Turk, P., Paustian, K., Keck, T., Nielsen, G., 2007. Sensitivity of the Century model to scale-related soil texture variability. Soil Sci. Soc. Am. J. 71, 784–792.

Bronick, C.J., Lal, R., 2005. Soil structure and management: a review. Geoderma 124, 3–22.

Burke, I.C., Yonker, C., Parton, W., Cole, C., Schimel, D., Flach, K., 1989. Texture, climate, and cultivation effects on soil organic matter content in US grassland soils. Soil Sci. Soc. Am. J. 53, 800–805.

Busari, M.A., Kukal, S.S., Kaur, A., Bhatt, R., Dulazi, A.A., 2015. Conservation tillage impacts on soil, crop and the environment. Int. Soil Water Conserv. Res. 3, 119–129.

Cann, M.A., 2000. Clay spreading on water repellent sands in the south east of South Australia–promoting sustainable agriculture. J. Hydrol. 231–232, 333–341.

Cardinael, R., Chevallier, T., Cambou, A., Beral, C., Barthes, B.G., Dupraz, C., et al., 2017. Increased soil organic carbon stocks under agroforestry: a survey of six different sites in France. Agric. Ecosyst. Environ. 236, 243–255.

Carter, M.R., Angers, D.A., Gregorich, E.G., Bolinder, M.A., 2003. Characterizing organic matter retention for surface soils in eastern Canada using density and particle size fractions. Can. J. Soil Sci. 83, 11–23.

Cerri, C., Coleman, K., Jenkinson, D., Bernoux, M., Victoria, R., Cerri, C., 2003. Modeling soil carbon from forest and pasture ecosystems of Amazon, Brazil. Soil Sci. Soc. Am. J. 67, 1879–1887.

Cerri, C.E., Easter, M., Paustian, K., Killian, K., Coleman, K., Bernoux, M., et al., 2007. Simulating SOC changes in 11 land use change chronosequences from the Brazilian Amazon with RothC and Century models. Agric. Ecosyst. Environ. 122, 46–57.

Chambers, A., Lal, R., Paustian, K., 2016. Soil carbon sequestration potential of US croplands and grasslands: implementing the 4 per Thousand Initiative. J. Soil Water Conserv. 71, 68A−74A.

Chan, K.Y., Heenan, D.P., 2005. The effects of stubble burning and tillage on soil carbon sequestration and crop productivity in southeastern Australia. Soil Use Manage. 21, 427−431.

Chantigny, M.H., Angers, D.A., Prévost, D., Vézina, L.-P., Chalifour, F.-P., 1997. Soil aggregation and fungal and bacterial biomass under annual and perennial cropping systems. Soil Sci. Soc. Am. J. 61, 262−267.

Chaudhary, S., Dheri, G.S., Brar, B.S., 2017. Long-term effects of NPK fertilizers and organic manures on carbon stabilization and management index under rice-wheat cropping system. Soil Tillage Res. 166, 59−66.

Chendev, Y.G., Sauer, T.J., Gennadiev, A.N., Novykh, L.L., Petin, A.N., Petina, V.I., et al., 2015. Accumulation of organic carbon in chernozems (Mollisols) under shelterbelts in Russia and the United States. Eurasian Soil Sci. 48, 43−53.

Cherr, C.M., Scholberg, J.M.S., McSorley, R., 2006. Green manure approaches to crop production: a synthesis. Agron. J. 98, 302−319.

Chivenge, P.P., Murwira, H.K., Giller, K.E., Mapfumo, P., Six, J., 2007. Long-term impact of reduced tillage and residue management on soil carbon stabilization: implications for conservation agriculture on contrasting soils. Soil Tillage Res. 94, 328−337.

Christensen, B.T., 1996. Matching measurable soil organic matter fractions with conceptual pools in simulation models of carbon turnover: revision of model structure. Evaluation of Soil Organic Matter Models. Springer, Berlin, Heidelberg, pp. 143−159.

Chung, H.G., Grove, J.H., Six, J., 2008. Indications for soil carbon saturation in a temperate agroecosystem. Soil Sci. Soc. Am. J. 72, 1132−1139.

Coleman, K., Jenkinson, D., Crocker, G., Grace, P., Klir, J., Körschens, M., et al., 1997. Simulating trends in soil organic carbon in long-term experiments using RothC-26.3. Geoderma 81, 29−44.

Cong, R., Wang, X., Xu, M., Ogle, S.M., Parton, W.J., 2014. Evaluation of the CENTURY model using long-term fertilization trials under corn-wheat cropping systems in the typical croplands of China. PLoS One 9, e95142.

CTIC, 2004. Conservation Tillage Information Center. National Crop Residue Management Survey. Conservation Technology Information Center; Springer, West Lafayette, IN, pp. 237−246.

Curtin, D., Fraser, P.M., 2003. Soil organic matter as influenced by straw management practices and inclusion of grass and clover seed crops in cereal rotations. Aust. J. Soil Res. 41, 95−106.

Dabney, S.M., Delgado, J.A., Reeves, D.W., 2001. Using winter cover crops to improve soil and water quality. Commun. Soil Sci. Plant Anal. 32, 1221−1250.

Davenport, D., Hughes, B., Davies, S., Hall, D., 2011. Spread, Delve, Spade, Invert: A Best Practice Guide to the Addition of Clay to Sandy Soils. Rural Solution SA, Agricultural Bureau of South Australia, Caring for our Country, Grains Research Development Corporation. Grains Research and Development Corporation (GRDC), No. 9781921779275. GRDC, Kingston, ACT.

Davidson, E.A., Ackerman, I.L., 1993. Changes in soil carbon inventories following cultivation of previously untilled soils. Biogeochemistry 20, 161−193.

Davidson, E.A., Janssens, I.A., 2006. Temperature sensitivity of soil carbon decomposition and feedbacks to climate change. Nature 440, 165−173.

Del Galdo, I., Six, J., Peressotti, A., Cotrufo, M.F., 2003. Assessing the impact of land-use change on soil C sequestration in agricultural soils by means of organic matter fractionation and stable C isotopes. Global Change Biol. 9, 1204−1213.

Dendoncker, N., Van Wesemael, B., Rounsevell, M.D.A., Roelandt, C., Lettens, S., 2004. Belgium's CO_2 mitigation potential under improved cropland management. Agric. Ecosyst. Environ. 103, 101−116.

Derenne, S., Largeau, C., 2001. A review of some important families of refractory macromolecules: composition, origin, and fate in soils and sediments. Soil Sci. 166, 833−847.

Desbiolles, J., Fielke, J., Chaplin, P., 1997. An application of tine configuration to obtain subsoil delving for the management of non-wetting sands. In: Proceedings Third International Conference on Soil Dynamics (ICSD III), pp. 201–210.

Dimassi, B., Mary, B., Wylleman, R., Labreuche, J., Couture, D., Piraux, F., et al., 2014. Long-term effect of contrasted tillage and crop management on soil carbon dynamics during 41 years. Agric. Ecosyst. Environ. 188, 134–146.

Dümig, A., Smittenberg, R., Kögel-Knabner, I., 2011. Concurrent evolution of organic and mineral components during initial soil development after retreat of the Damma glacier, Switzerland. Geoderma 163, 83–94.

Eswaran, H., Van Den Berg, E., Reich, P., 1993. Organic carbon in soils of the world. Soil Sci. Soc. Am. J. 57, 192–194.

de Faccio Carvalho, P.C., Anghinoni, I., de Moraes, A., de Souza, E.D., Sulc, R.M., Lang, C.R., et al., 2010. Managing grazing animals to achieve nutrient cycling and soil improvement in no-till integrated systems. Nutr. Cycling Agroecosyst. 88, 259–273.

Fageria, N.K., 2007. Green manuring in crop production. J. Plant Nutr. 30, 691–719.

Fageria, N.K., Baligar, V.C., Bailey, B.A., 2005. Role of cover crops in improving soil and row crop productivity. Commun. Soil Sci. Plant Anal. 36, 2733–2757.

Fang, Y., Singh, B., Singh, B.P., Krull, E., 2014a. Biochar carbon stability in four contrasting soils. Eur. J. Soil Sci. 65, 60–71.

Fang, Y., Singh, B.P., Singh, B., 2014b. Temperature sensitivity of biochar and native carbon mineralisation in biochar-amended soils. Agric. Ecosyst. Environ. 191, 158–167.

Fang, Y., Singh, B., Singh, B.P., 2015. Effect of temperature on biochar priming effects and its stability in soils. Soil Biol. Biochem. 80, 136–145.

Fang, Y., Singh, B.P., Badgery, W., He, X., 2016. In situ assessment of new carbon and nitrogen assimilation and allocation in contrastingly managed dryland wheat crop–soil systems. Agric. Ecosyst. Environ. 235, 80–90.

Fang, Y., Singh, B.P., Matta, P., Cowie, A.L., Van Zwieten, L., 2017. Temperature sensitivity and priming of organic matter with different stabilities in a Vertisol with aged biochar. Soil Biol. Biochem. 115, 346–356.

FAO, 2017. Soil Organic Carbon: The Hidden Potential. Food and Agriculture Organization of the United Nations, Rome.

Feller, C., Beare, M.H., 1997. Physical control of soil organic matter dynamics in the tropics. Geoderma 79, 69–116.

Feng, W.T., Plante, A.F., Six, J., 2013. Improving estimates of maximal organic carbon stabilization by fine soil particles. Biogeochemistry 112, 81–93.

Francaviglia, R., Di Bene, C., Farina, R., Salvati, L., 2017. Soil organic carbon sequestration and tillage systems in the Mediterranean Basin: a data mining approach. Nutr. Cycling Agroecosyst. 107, 125. Available from: https://doi.org/10.1007/s10705-016-9820-z.

Freibauer, A., Rounsevell, M.D.A., Smith, P., Verhagen, J., 2004. Carbon sequestration in the agricultural soils of Europe. Geoderma 122, 1–23.

Ghosh, S., Wilson, B., Ghoshal, S., Senapati, N., Mandal, B., 2012. Organic amendments influence soil quality and carbon sequestration in the Indo-Gangetic plains of India. Agric. Ecosyst. Environ. 156, 134–141.

Goh, K.M., 2004. Carbon sequestration and stabilization in soils: implications for soil productivity and climate change. Soil Sci. Plant Nutr. 50, 467–476.

Gottschalk, T.K., Dittrich, R., Diekötter, T., Sheridan, P., Wolters, V., Ekschmitt, K., 2010. Modelling land-use sustainability using farmland birds as indicators. Ecol. Indic. 10, 15–23.

van Groenigen, K.J., Six, J., Hungate, B.A., de Graaff, M.A., van Breemen, N., van Kessel, C., 2006. Element interactions limit soil carbon storage. Proc. Natl. Acad. Sci. U.S.A. 103, 6571–6574.

Guenet, B., Juarez, S., Bardoux, G., Abbadie, L., Chenu, C., 2012. Evidence that stable C is as vulnerable to priming effect as is more labile C in soil. Soil Biol. Biochem. 52, 43–48.

Guggenberger, G., Frey, S.D., Six, J., Paustian, K., Elliott, E.T., 1999. Bacterial and fungal cell-wall residues in conventional and no-tillage agroecosystems. Soil Sci. Soc. Am. J. 63, 1188–1198.

Guo, J.H., Liu, X.J., Zhang, Y., Shen, J.L., Han, W.X., Zhang, W.F., et al., 2010. Significant acidification in major Chinese croplands. Science 327, 1008–1010.

Guo, L.B., Gifford, R.M., 2002. Soil carbon stocks and land use change: a meta analysis. Global Change Biol. 8, 345–360.

Hall, D.J.M., Jones, H.R., Crabtree, W.L., Daniels, T.L., 2010. Claying and deep ripping can increase crop yields and profits on water repellent sands with marginal fertility in southern Western Australia. Aust. J. Soil Res. 48, 178–187.

Hamza, M.A., Anderson, W.K., 2005. Soil compaction in cropping systems – a review of the nature, causes and possible solutions. Soil Tillage Res. 82, 121–145.

Han, P., Zhang, W., Wang, G., Sun, W., Huang, Y., 2016. Changes in soil organic carbon in croplands subjected to fertilizer management: a global meta-analysis. Sci. Rep. 6, 27199.

Harrison, R.B., Footen, P.W., Strahm, B.D., 2011. Deep soil horizons: contribution and importance to soil carbon pools and in assessing whole-ecosystem response to management and global change. Forest Sci. 57, 67–76.

Hasibeder, R., Fuchslueger, L., Richter, A., Bahn, M., 2015. Summer drought alters carbon allocation to roots and root respiration in mountain grassland. New Phytol. 205, 1117–1127.

Hassink, J., 1994. Effect of soil texture on the size of the microbial biomass and on the amount of C and N mineralized per unit of microbial biomass in Dutch grassland soils. Soil Biol. Biochem. 26, 1573–1581.

Hassink, J., 1997. The capacity of soils to preserve organic C and N by their association with clay and silt particles. Plant Soil 191, 77–87.

Heenan, D.P., Chan, K.Y., Knight, P.G., 2004. Long-term impact of rotation, tillage and stubble management on the loss of soil organic carbon and nitrogen from a Chromic Luvisol. Soil Tillage Res. 76, 59–68.

Heikkinen, J., Ketoja, E., Nuutinen, V., Regina, K., 2013. Declining trend of carbon in Finnish cropland soils in 1974-2009. Global Change Biol. 19, 1456–1469.

Hemwong, S., Cadisch, G., Toomsan, B., Limpinuntana, V., Vityakon, P., Patanothai, A., 2008. Dynamics of residue decomposition and N_2 fixation of grain legumes upon sugarcane residue retention as an alternative to burning. Soil Tillage Res. 99, 84–97.

Houghton, R.A., Goodale, C., 2004. Effects of land-use change on the carbon balance of terrestrial ecosystems. In: DeFries, R., Asner, G., Houghton, R.A. (Eds.), Ecosystems and Land Use Change. American Geophysical Union, Washington, DC, pp. 85–98.

Hoyle, F.C., Baldock, J.A., Murphy, D.V., 2011. Soil organic carbon – role in rainfed farming systems. In: Tow, P., Cooper, I., Partridge, I., Birch, C. (Eds.), Rainfed Farming Systems. Springer Netherlands, Dordrecht, pp. 339–361.

Hutchinson, J.J., Campbell, C.A., Desjardins, R.L., 2007. Some perspectives on carbon sequestration in agriculture. Agric. Forest Meteorol. 142, 288–302.

Inbar, Y., Chen, Y., Hadar, Y., 1990. Humic substances formed during the composting of organic matter. Soil Sci. Soc. Am. J. 54, 1316–1323.

IPCC, 2003. Good Practice Guidance for Land Use, Land-Use Change and Forestry. Intergovernmental Panel on Climate Change (IPCC), Geneva.

Jacinthe, P.A., Lal, R., Kimble, J.M., 2002. Carbon budget and seasonal carbon dioxide emission from a central Ohio Luvisol as influenced by wheat residue amendment. Soil Tillage Res. 67, 147–157.

Jain, N., Bhatia, A., Pathak, H., 2014. Emission of air pollutants from crop residue burning in India. Aerosol Air Qual. Res. 14, 422–430.

Janssens, I.A., Freibauer, A., Schlamadinger, B., Ceulemans, R., Ciais, P., Dolman, A.J., et al., 2005. The carbon budget of terrestrial ecosystems at country-scale — a European case study. Biogeosciences 2, 15−26.

Jarecki, M.K., Lal, R., 2003. Crop management for soil carbon sequestration. Crit. Rev. Plant Sci. 22, 471−502.

Jastrow, J., Miller, R., Boutton, T., 1996. Carbon dynamics of aggregate-associated organic matter estimated by carbon-13 natural abundance. Soil Sci. Soc. Am. J. 60, 801−807.

Jenkinson, D., Andrew, S., Lynch, J., Goss, M., Tinker, P., 1990. The turnover of organic carbon and nitrogen in soil [and discussion]. Philos. Trans. R. Soc. Lond. B Biol. Sci. 329, 361−368.

Jenkinson, D.S., Coleman, K., 2008. The turnover of organic carbon in subsoils. Part 2. Modelling carbon turnover. Eur. J. Soil Sci. 59, 400−413.

Jiang, C., Yu, W., Ma, Q., Xu, Y., Zou, H., 2017. Alleviating global warming potential by soil carbon sequestration: a multi-level straw incorporation experiment from a maize cropping system in Northeast China. Soil Tillage Res. 170, 77−84.

Jin, V.L., Potter, K.N., Johnson, M.-V.V., Harmel, R.D., Arnold, J.G., 2015. Surface-applied biosolids enhance soil organic carbon and nitrogen stocks but have contrasting effects on soil physical quality. Appl. Environ. Soil Sci. 2015, 1−10.

Jobbagy, E.G., Jackson, R.B., 2000. The vertical distribution of soil organic carbon and its relation to climate and vegetation. Ecol. Appl. 10, 423−436.

Jones, R.J.A., Hiederer, R., Rusco, E., Montanarella, L., 2005. Estimating organic carbon in the soils of Europe for policy support. Eur. J. Soil Sci. 56, 655−671.

Jose, S., 2009. Agroforestry for ecosystem services and environmental benefits: an overview. Agroforest. Syst. 76, 1−10.

Kaiser, K., Zech, W., 2000. Dissolved organic matter sorption by mineral constituents of subsoil clay fractions. J. Plant Nutr. Soil Sci. 163, 531−535.

Kaiser, K., Eusterhues, K., Rumpel, C., Guggenberger, G., Kögel-Knabner, I., 2002. Stabilization of organic matter by soil minerals — investigations of density and particle-size fractions from two acid forest soils. J. Plant Nutr. Soil Sci. 165, 451−459.

Kane, D., 2015. Carbon Sequestration Potential on Agricultural Lands: A Review of Current Science and Available Practices. National Sustainable Agriculture Cooperation. Breakthrough Strategies and Solutions, LLC.

Katterer, T., Andren, O., 1999. Long-term agricultural field experiments in Northern Europe: analysis of the influence of management on soil carbon stocks using the ICBM model. Agric. Ecosyst. Environ. 72, 165−179.

Keating, B.A., Carberry, P.S., Hammer, G.L., Probert, M.E., Robertson, M.J., Holzworth, D., et al., 2003. An overview of APSIM, a model designed for farming systems simulation. Eur. J. Agron. 18, 267−288.

Keith, A., Singh, B., Singh, B.P., 2011. Interactive priming of biochar and labile organic matter mineralization in a smectite-rich soil. Environ. Sci. Technol. 45, 9611−9618.

Kelly, R., Parton, W., Crocker, G., Graced, P., Klir, J., Körschens, M., et al., 1997. Simulating trends in soil organic carbon in long-term experiments using the century model. Geoderma 81, 75−90.

Khan, S.A., Mulvaney, R.L., Ellsworth, T.R., Boast, C.W., 2007. The myth of nitrogen fertilization for soil carbon sequestration. J. Environ. Qual. 36, 1821−1832.

Kim, D.-G., Kirschbaum, M.U.F., Beedy, T.L., 2016. Carbon sequestration and net emissions of CH_4 and N_2O under agroforestry: synthesizing available data and suggestions for future studies. Agric. Ecosyst. Environ. 226, 65−78.

King, J.A., Bradley, R.I., Harrison, R., Carter, A.D., 2004. Carbon sequestration and saving potential associated with changes to the management of agricultural soils in England. Soil Use Manage. 20, 394−402.

Kirkby, C.A., Kirkegaard, J.A., Richardson, A.E., Wade, L.J., Blanchard, C., Batten, G., 2011. Stable soil organic matter: a comparison of C:N:P:S ratios in Australian and other world soils. Geoderma 163, 197−208.

Kirkby, C.A., Richardson, A.E., Wade, L.J., Batten, G.D., Blanchard, C., Kirkegaard, J.A., 2013. Carbon-nutrient stoichiometry to increase soil carbon sequestration. Soil Biol. Biochem. 60, 77−86.

Kirkby, C.A., Richardson, A.E., Wade, L.J., Passioura, J.B., Batten, G.D., Blanchard, C., et al., 2014. Nutrient availability limits carbon sequestration in arable soils. Soil Biol. Biochem. 68, 402−409.

Kirkby, C.A., Richardson, A.E., Wade, L.J., Conyers, M., Kirkegaard, J.A., 2016. Inorganic nutrients increase humification efficiency and C-sequestration in an annually cropped soil. PLoS One 11, e0153698.

Kleber, M., Mikutta, R., Torn, M.S., Jahn, R., 2005. Poorly crystalline mineral phases protect organic matter in acid subsoil horizons. Eur. J. Soil Sci. 56, 717−725.

Kögel-Knabner, I., Guggenberger, G., Kleber, M., Kandeler, E., Kalbitz, K., Scheu, S., et al., 2008. Organo-mineral associations in temperate soils: Integrating biology, mineralogy, and organic matter chemistry. J. Plant Nutr. Soil Sci. 171, 61−82.

Kong, A.Y.Y., Six, J., Bryant, D.C., Denison, R.F., van Kessel, C., 2005. The relationship between carbon input, aggregation, and soil organic carbon stabilization in sustainable cropping systems. Soil Sci. Soc. Am. J. 69, 1078−1085.

Krull, E., Baldock, J., Skjemstad, J., 2001. Soil texture effects on decomposition and soil carbon storage. In: Net Ecosystem Exchange CRC Workshop Proceedings. Citeseer, pp. 103−110.

Krull, E.S., Baldock, J.A., Skjemstad, J.O., 2003. Importance of mechanisms and processes of the stabilisation of soil organic matter for modelling carbon turnover. Funct. Plant Biol. 30, 207−222.

Kuzyakov, Y., Gavrichkova, O., 2010. REVIEW: Time lag between photosynthesis and carbon dioxide efflux from soil: a review of mechanisms and controls. Global Change Biol. 16, 3386−3406.

Kuzyakov, Y., Schneckenberger, K., 2004. Review of estimation of plant rhizodeposition and their contribution to soil organic matter formation. Arch. Agron. Soil Sci. 50, 115−132.

Lal, R., 1993. Tillage effects on soil degradation, soil resilience, soil quality, and sustainability. Soil Tillage Res. 27, 1−8.

Lal, R., 2001. Soil degradation by erosion. Land Degrad. Dev. 12, 519−539.

Lal, R., 2002. Soil carbon dynamics in cropland and rangeland. Environ. Pollut. 116, 353−362.

Lal, R., 2004a. Agricultural activities and the global carbon cycle. Nutr. Cycling Agroecosyst. 70, 103−116.

Lal, R., 2004b. Soil carbon sequestration to mitigate climate change. Geoderma 123, 1−22.

Lal, R., 2005. World crop residues production and implications of its use as a biofuel. Environ. Int. 31, 575−584.

Lal, R., 2008. Crop residues as soil amendments and feedstock for bioethanol production. Waste Manage. 28, 747−758.

Lal, R., 2010. Managing soils and ecosystems for mitigating anthropogenic carbon emissions and advancing global food security. Bioscience 60, 708−721.

Lal, R., 2016. Beyond COP21: potential and challenges of the "4 per Thousand" initiative. J. Soil Water Conserv. 71, 20A−25A.

Lal, R., Stewart, B.A., 1990. Soil Degradation. Springer-Verlag, New York, NY.

Lal, R., Reicosky, D., Hanson, J., 2007. Evolution of the plow over 10,000 years and the rationale for no-till farming. Soil Tillage Res. 93, 1−12.

Lehmann, J., Joseph, S., 2009. Biochar for environmental management: an introduction. In: Lehmann, J., Joseph, S. (Eds.), Biochar for Environmental Management: Science and Technology. Earthscan, London, pp. 1−9.

Lehmann, J., Rillig, M.C., Thies, J., Masiello, C.A., Hockaday, W.C., Crowley, D., 2011. Biochar effects on soil biota − a review. Soil Biol. Biochem. 43, 1812−1836.

Lehmann, J., Abiven, S., Kleber, M., Pan, G., Singh, B.P., Sohi, S., et al., 2015. Persistence of biochar in soil. In: Lehmann, J., Joseph, S. (Eds.), Biochar for Environmental Management: Science, Technology and Implementation, second ed. Routledge, USA, pp. 235−282.

Lehtinen, T., Schlatter, N., Baumgarten, A., Bechini, L., Kruger, J., Grignani, C., et al., 2014. Effect of crop residue incorporation on soil organic carbon and greenhouse gas emissions in European agricultural soils. Soil Use Manage. 30, 524−538.

Leite, L.F.C., de Sá Mendonça, E., de Almeida Machado, P.L.O., Fernandes Filho, E.I., Neves, J.C.L., 2004. Simulating trends in soil organic carbon of an Acrisol under no-tillage and disc-plow systems using the Century model. Geoderma 120, 283−295.

Li, Z.P., Liu, M., Wu, X.C., Han, F.X., Zhang, T.L., 2010. Effects of long-term chemical fertilization and organic amendments on dynamics of soil organic C and total N in paddy soil derived from barren land in subtropical China. Soil Tillage Res. 106, 268−274.

Liang, A.Z., Yang, X.M., Zhang, X.P., McLaughlin, N., Shen, Y., Li, W.F., 2009. Soil organic carbon changes in particle-size fractions following cultivation of Black soils in China. Soil Tillage Res. 105, 21−26.

Ling-An, N., Jin-Min, H., Zhang, B.-Z., Xin-Sheng, N., 2011. Influences of long-term fertilizer and tillage management on soil fertility of the North China Plain. Pedosphere 21, 813−820.

Liu, C., Lu, M., Cui, J., Li, B., Fang, C., 2014. Effects of straw carbon input on carbon dynamics in agricultural soils: a meta-analysis. Global Change Biol. 20, 1366−1381.

Liu, D.L., Chan, K.Y., Conyers, M.K., 2009. Simulation of soil organic carbon under different tillage and stubble management practices using the Rothamsted carbon model. Soil Tillage Res. 104, 65−73.

Lorenz, K., Lal, R., 2014. Soil organic carbon sequestration in agroforestry systems. A review. Agron. Sustain. Dev. 34, 443−454.

Lugato, E., Berti, A., Giardini, L., 2006. Soil organic carbon (SOC) dynamics with and without residue incorporation in relation to different nitrogen fertilisation rates. Geoderma 135, 315−321.

Lugato, E., Panagos, P., Bampa, F., Jones, A., Montanarella, L., 2014. A new baseline of organic carbon stock in European agricultural soils using a modelling approach. Global Change Biol. 20, 313−326.

Luo, Y., Durenkamp, M., De Nobili, M., Lin, Q., Brookes, P.C., 2011. Short term soil priming effects and the mineralisation of biochar following its incorporation to soils of different pH. Soil Biol. Biochem. 43, 2304−2314.

Luo, Z.K., Wang, E.L., Sun, O.J., 2010a. Soil carbon change and its responses to agricultural practices in Australian agro-ecosystems: a review and synthesis. Geoderma 155, 211−223.

Luo, Z.K., Wang, E.L., Sun, O.J., 2010b. Can no-tillage stimulate carbon sequestration in agricultural soils? A meta-analysis of paired experiments. Agric. Ecosyst. Environ. 139, 224−231.

von Lützow, M., Kögel-Knabner, I., Ekschmitt, K., Matzner, E., Guggenberger, G., Marschner, B., et al., 2006. Stabilization of organic matter in temperate soils: mechanisms and their relevance under different soil conditions − a review. Eur. J. Soil Sci. 57, 426−445.

Maestrini, B., Nannipieri, P., Abiven, S., 2014. A meta-analysis on pyrogenic organic matter induced priming effect. Global Change Biol. Bioenergy 7, 577−590.

Mandal, S., Sarkar, B., Bolan, N., Novak, J., Ok, Y.S., Van Zwieten, L., et al., 2016. Designing advanced biochar products for maximizing greenhouse gas mitigation potential. Crit. Rev. Environ. Sci. Technol. 46, 1367−1401.

Mann, L., Tolbert, V., Cushman, J., 2002. Potential environmental effects of corn (Zea mays L.) stover removal with emphasis on soil organic matter and erosion. Agric. Ecosyst. Environ. 89, 149−166.

Manzoni, S., Porporato, A., 2009. Soil carbon and nitrogen mineralization: theory and models across scales. Soil Biol. Biochem. 41, 1355−1379.

McBeath, A.V., Smernik, R.J., 2009. Variation in the degree of aromatic condensation of chars. Org. Geochem. 40, 1161−1168.

McBeath, A.V., Smernik, R.J., Krull, E., Lehmann, J., 2014. The influence of feedstock and production temperature on biochar carbon chemistry: a solid-state 13C NMR study. Biomass Bioenergy 60, 121–129.

McDaniel, M.D., Tiemann, L.K., Grandy, A.S., 2014. Does agricultural crop diversity enhance soil microbial biomass and organic matter dynamics? A meta-analysis. Ecol. Appl. 24, 560–570.

Minasny, B., Malone, B.P., McBratney, A.B., Angers, D.A., Arrouays, D., Chambers, A., et al., 2017. Soil carbon 4 per mille. Geoderma 292, 59–86.

Montagnini, F., Nair, P.K.R., 2004. Carbon sequestration: an underexploited environmental benefit of agroforestry systems. Agroforest. Syst. 61-2, 281–295.

de Moraes, A., de Faccio Carvalho, P.C., Anghinoni, I., Lustosa, S.B.C., de Andrade Costa, S.E.V.G., Kunrath, T.R., 2014. Integrated crop–livestock systems in the Brazilian subtropics. Eur. J. Agron. 57, 4–9.

Murty, D., Kirschbaum, M.U.F., McMurtrie, R.E., McGilvray, A., 2002. Does conversion of forest to agricultural land change soil carbon and nitrogen? A review of the literature. Global Change Biol. 8, 105–123.

Nair, P.K.R., 2012. Climate change mitigation: a low-hanging fruit of agroforestry. In: Nair, P.K.R., Garrity, D. (Eds.), Agroforestry – The Future of Global Land Use.. Springer, Dordrecht.

Nair, P.K.R., Kumar, B.M., Nair, V.D., 2009. Agroforestry as a strategy for carbon sequestration. J. Plant Nutr. Soil Sci. 172, 10–23.

Nelson, P.N., Barzegar, A.R., Oades, J.M., 1997. Sodicity and clay type: influence on decomposition of added organic matter. Soil Sci. Soc. Am. J. 61, 1052–1057.

Nguyen, T.-T., Marschner, P., 2014. Retention and loss of water extractable carbon in soils: effect of clay properties. Sci. Total Environ. 470, 400–406.

Noble, A.D., Suzuki, S., 2005. Improving the productivity of degraded cropping systems in Northeast Thailand: improving farmer practices with innovative approaches. In: PAWEES 2005 International Conference, Kyoto, Japan, September.

Oades, J.M., 1984. Soil organic matter and structural stability: mechanisms and implications for management. Biological Processes and Soil Fertility. Springer, Dordrecht, pp. 319–337.

Oelbermann, M., Voroney, R.P., Gordon, A.M., 2004. Carbon sequestration in tropical and temperate agroforestry systems: a review with examples from Costa Rica and southern Canada. Agric. Ecosyst. Environ. 104, 359–377.

Oelbermann, M., Voroney, R.P., Thevathasan, N.V., Gordon, A.M., Kass, D.C.L., Schlonvoigt, A.M., 2006. Soil carbon dynamics and residue stabilization in a Costa Rican and southern Canadian alley cropping system. Agroforest. Syst. 68, 27–36.

Olson, K., Al-Kaisi, M., 2015. The importance of soil sampling depth for accurate account of soil organic carbon sequestration, storage, retention and loss. Catena 125, 33–37.

Paetsch, L., Mueller, C.W., Rumpel, C., Houot, S., Kögel-Knabner, I., 2016. Urban waste composts enhance OC and N stocks after long-term amendment but do not alter organic matter composition. Agric. Ecosyst. Environ. 223, 211–222.

Palma, J.H.N., Graves, A.R., Bunce, R.G.H., Burgess, P.J., de Filippi, R., Keesman, K.J., et al., 2007. Modeling environmental benefits of silvoarable agroforestry in Europe. Agric. Ecosyst. Environ. 119, 320–334.

Parton, W., Schimel, D.S., Cole, C., Ojima, D., 1987. Analysis of factors controlling soil organic matter levels in Great Plains grasslands. Soil Sci. Soc. Am. J. 51, 1173–1179.

Paul, K.I., Polglase, P.J., Nyakuengama, J.G., Khanna, P.K., 2002. Change in soil carbon following afforestation. Forest Ecol. Manage. 168, 241–257.

Paustian, K., Andren, O., Janzen, H.H., Lal, R., Smith, P., Tian, G., et al., 1997. Agricultural soils as a sink to mitigate CO_2 emissions. Soil Use Manage. 13, 230–244.

Paustian, K., Cole, C.V., Sauerbeck, D., Sampson, N., 1998. CO_2 mitigation by agriculture: an overview. Climatic Change 40, 135−162.

Paustian, K., Six, J., Elliott, E.T., Hunt, H.W., 2000. Management options for reducing CO_2 emissions from agricultural soils. Biogeochemistry 48, 147−163.

Peltre, C., Nielsen, M., Christensen, B.T., Hansen, E.M., Thomsen, I.K., Bruun, S., 2016. Straw export in continuous winter wheat and the ability of oil radish catch crops and early sowing of wheat to offset soil C and N losses: a simulation study. Agric. Syst. 143, 195−202.

Poeplau, C., Don, A., 2013. Sensitivity of soil organic carbon stocks and fractions to different land-use changes across Europe. Geoderma 192, 189−201.

Poeplau, C., Don, A., 2015. Carbon sequestration in agricultural soils via cultivation of cover crops − a meta-analysis. Agric. Ecosyst. Environ. 200, 33−41.

Poeplau, C., Don, A., Vesterdal, L., Leifeld, J., Van Wesemael, B., Schumacher, J., et al., 2011. Temporal dynamics of soil organic carbon after land-use change in the temperate zone − carbon response functions as a model approach. Global Change Biol. 17, 2415−2427.

Poeplau, C., Katterer, T., Bolinder, M.A., Borjesson, G., Berti, A., Lugato, E., 2015. Low stabilization of aboveground crop residue carbon in sandy soils of Swedish long-term experiments. Geoderma 237, 246−255.

Post, W.M., Kwon, K.C., 2000. Soil carbon sequestration and land-use change: processes and potential. Global Change Biol. 6, 317−327.

Powlson, D.S., Riche, A.B., Coleman, K., Glendining, N., Whitmore, A.P., 2008. Carbon sequestration in European soils through straw incorporation: limitations and alternatives. Waste Manage. 28, 741−746.

Powlson, D.S., Stirling, C.M., Jat, M., Gerard, B.G., Palm, C.A., Sanchez, P.A., et al., 2014. Limited potential of no-till agriculture for climate change mitigation. Nat. Clim. Change 4, 678−683.

Powlson, D.S., Whitmore, A.P., Goulding, K.W.T., Glendining, N., Whitmore, A.P., 2011. Soil carbon sequestration to mitigate climate change: A critical re-examination to identify the true and the false. Eur. J. Soil Sci. 62, 42−55.

Pribyl, D.W., 2010. A critical review of the conventional SOC to SOM conversion factor. Geoderma 156, 75−83.

Qiu, S., Gao, H., Zhu, P., Hou, Y., Zhao, S., Rong, X., et al., 2016. Changes in soil carbon and nitrogen pools in a Mollisol after long-term fallow or application of chemical fertilizers, straw or manures. Soil Tillage Res. 163, 255−265.

Qiu, S.J., Ju, X.T., Ingwersen, J., Qin, Z.C., Li, L., Streck, T., et al., 2010. Changes in soil carbon and nitrogen pools after shifting from conventional cereal to greenhouse vegetable production. Soil Tillage Res. 107, 80−87.

Raffa, D.W., Bogdanski, A., Tittonell, P., 2015. How does crop residue removal affect soil organic carbon and yield? A hierarchical analysis of management and environmental factors. Biomass Bioenergy 81, 345−355.

Regnier, P., Friedlingstein, P., Ciais, P., Mackenzie, F.T., Gruber, N., Janssens, I.A., et al., 2013. Anthropogenic perturbation of the carbon fluxes from land to ocean. Nat. Geosci. 6, 597−607.

Reiter, L., 2015. Effect of Crop Residue Incorporation on Soil Organic Carbon Dynamics (Masters thesis). Swedish University of Agricultural Sciences, Uppsala.

Rumpel, C., Kögel-Knabner, I., 2011. Deep soil organic matter-a key but poorly understood component of terrestrial C cycle. Plant Soil 338, 143−158.

Sanderman, J., Farquharson, R., Baldock, J., 2010. Soil Carbon Sequestration Potential: A Review for Australian Agriculture. A Report Prepared for Department of Climate Change and Energy Efficiency, Australian Government. CSIRO Sustainable Agriculture Flagship, Canberra, ACT.

Schapel, A., Davenport, D., Marschner, P., 2017. Increases in organic carbon concentration and stock after clay addition to sands: validation of sampling methodology and effects of modification method. Soil Res. 55, 124.

Schoeneberger, M.M., 2009. Agroforestry: working trees for sequestering carbon on agricultural lands. Agroforest. Syst. 75, 27–37.

Schöning, I., Knicker, H., Kögel-Knabner, I., 2005. Intimate association between O/N-alkyl carbon and iron oxides in clay fractions of forest soils. Org. Geochem. 36, 1378–1390.

Schulp, C.J.E., Nabuurs, G.J., Verburg, P.H., 2008. Future carbon sequestration in Europe – effects of land use change. Agric. Ecosyst. Environ. 127, 251–264.

Scott, B., Eberbach, P., Evans, J., Wade, L., 2010. In: Clayton, E.H., Burns, H.M. (Eds.), EH Graham Centre Monograph No. 1: Stubble Retention in Cropping Systems in Southern Australia: Benefits and Challenges. Industry & Investment NSW, Orange. Available at: <http://www.csu.edu.au/research/grahamcentre>.

Setia, R., Smith, P., Marschner, P., Baldock, J., Chittleborough, D., Smith, J., 2011. Introducing a decomposition rate modifier in the Rothamsted carbon model to predict soil organic carbon stocks in saline soils. Environ. Sci. Technol. 45, 6396–6403.

Setia, R., Rengasamy, P., Marschner, P., 2013. Effect of exchangeable cation concentration on sorption and desorption of dissolved organic carbon in saline soils. Sci. Total Environ. 465, 226–232.

Shen, M.X., Yang, L.Z., Yao, Y.M., Wu, D.D., Wang, J.G., Guo, R.L., et al., 2007. Long-term effects of fertilizer managements on crop yields and organic carbon storage of a typical rice-wheat agroecosystem of China. Biol. Fertil. Soils 44, 187–200.

Shi, A., Marschner, P., 2013. Addition of a clay subsoil to a sandy top soil alters CO_2 release and the interactions in residue mixtures. Sci. Total Environ. 465, 248–254.

Shirato, Y., Paisancharoen, K., Sangtong, P., Nakviro, C., Yokozawa, M., Matsumoto, N., 2005. Testing the Rothamsted Carbon Model against data from long-term experiments on upland soils in Thailand. Eur. J. Soil Sci. 56, 179–188.

Silver, W.L., Neff, J., McGroddy, M., Veldkamp, E., Keller, M., Cosme, R., 2000. Effects of soil texture on below-ground carbon and nutrient storage in a lowland Amazonian forest ecosystem. Ecosystems 3, 193–209.

Singh, B.P., Cowie, A.L., 2014. Long-term influence of biochar on native organic carbon mineralisation in a low-carbon clayey soil. Sci. Rep. 4, 3687.

Singh, B.P., Cowie, A.L., Smernik, R.J., 2012. Biochar carbon stability in a clayey soil as a function of feedstock and pyrolysis temperature. Environ. Sci. Technol. 46, 11770–11778.

Singh, B.P., Fang, Y., Boersma, M., Collins, D., Van Zwieten, L., Macdonald, L.M., 2015. In situ persistence and migration of biochar carbon and its impact on native carbon emission in contrasting soils under managed temperate pastures. PLoS One 10, e0141560. Available from: https://doi.org/10.1371/journal.pone.0141560.

Six, J., Elliott, E.T., Paustian, K., Doran, J.W., 1998. Aggregation and soil organic matter accumulation in cultivated and native grassland soils. Soil Sci. Soc. Am. J. 62, 1367–1377.

Six, J., Conant, R., Paul, E., Paustian, K., 2002. Stabilization mechanisms of soil organic matter: implications for C-saturation of soils. Plant Soil 241, 155–176.

Six, J., Bossuyt, H., Degryze, S., Denef, K., 2004. A history of research on the link between (micro) aggregates, soil biota, and soil organic matter dynamics. Soil Tillage Res. 79, 7–31.

Skjemstad, J., Spouncer, L., Cowie, B., Swift, R., 2004. Calibration of the Rothamsted organic carbon turnover model (RothC ver. 26.3), using measurable soil organic carbon pools. Soil Res. 42, 79–88.

Sleutel, S., De Neve, S., Hofman, G., 2007. Assessing causes of recent organic carbon losses from cropland soils by means of regional-scaled input balances for the case of Flanders (Belgium). Nutr. Cycling Agroecosyst. 78, 265–278.

Smith, J., Smith, P., Wattenbach, M., Zaehle, S., Hiederer, R., Jones, R.J.A., et al., 2005. Projected changes in mineral soil carbon of European croplands and grasslands, 1990-2080. Global Change Biol. 11, 2141–2152.

Smith, J.U., Bradbury, N.J., 1994. Demonstration of sundial: simulation of nitrogen dynamics in arable land. In: Rounsevell, M.D.A., Loveland, P.J. (Eds.), Soil Responses to Climate Change. NATO ASI Series (Series I: Global Environmental Change), vol. 23. Springer, Berlin, Heidelberg.

Smith, P., 2004. Carbon sequestration in croplands: the potential in Europe and the global context. Eur. J. Agron. 20, 229−236.

Smith, P., 2016. Soil carbon sequestration and biochar as negative emission technologies. Global Change Biol. 22, 1315−1324.

Smith, P., Smith, J., Powlson, D., McGill, W., Arah, J., Chertov, O., et al., 1997. A comparison of the performance of nine soil organic matter models using datasets from seven long-term experiments. Geoderma 81, 153−225.

Smith, P., Martino, D., Cai, Z., Gwary, D., Janzen, H., Kumar, P., et al., 2008. Greenhouse gas mitigation in agriculture. Philos. Trans. R. Soc. B Biol. Sci. 363, 789−813.

Sochorová, L., Jansa, J., Verbruggen, E., Hejcman, M., Schellberg, J., Kiers, E.T., et al., 2016. Long-term agricultural management maximizing hay production can significantly reduce belowground C storage. Agric. Ecosyst. Environ. 220, 104−114.

Soussana, J.F., Loiseau, P., Vuichard, N., Ceschia, E., Balesdent, J., Chevallier, T., et al., 2004. Carbon cycling and sequestration opportunities in temperate grasslands. Soil Use Manage. 20, 219−230.

Sparrow, L.A., Belbin, K.C., Doyle, R.B., 2006. Organic carbon in the silt plus clay fraction of Tasmanian soils. Soil Use Manage. 22, 219−220.

Spielvogel, S., Prietzel, J., Kögel-Knabner, I., 2008. Soil organic matter stabilization in acidic forest soils is preferential and soil type-specific. Eur. J. Soil Sci. 59, 674−692.

Stavi, I., Lal, R., 2013. Agroforestry and biochar to offset climate change: a review. Agron. Sustain. Dev. 33, 81−96.

Steffens, M., 2009. Soils of a Semiarid Shortgrass Steppe in Inner Mongolia: Organic Matter Composition and Distribution as Affected by Sheep Grazing. Dissertation am Lehrstuhl für Bodenkunde, Technische Universität München, 133 pp.

Stemmer, M., Von Lutzow, M., Kandeler, E., Pichlmayer, F., Gerzabek, M.H., 1999. The effect of maize straw placement on mineralization of C and N in soil particle size fractions. Eur. J. Soil Sci. 50, 73−85.

Stewart, C.E., Paustian, K., Conant, R.T., Plante, A.F., Six, J., 2007. Soil carbon saturation: concept, evidence and evaluation. Biogeochemistry 86, 19−31.

Stewart, C.E., Paustian, K., Conant, R.T., Plante, A.F., Six, J., 2008. Soil carbon saturation: evaluation and corroboration by long-term incubations. Soil Biol. Biochem. 40, 1741−1750.

Stockmann, U., Adams, M.A., Crawford, J.W., Field, D.J., Henakaarchchi, N., Jenkins, M., et al., 2013. The knowns, known unknowns and unknowns of sequestration of soil organic carbon. Agric. Ecosyst. Environ. 164, 80−99.

Stolbovoy, V., Montanarella, L., Filippi, N., Jones, A., Gallego, J., Grassi, G., 2007. Soil Sampling Protocol to Certify the Changes of Organic Carbon Stock in Mineral Soil of the European Union. Office for Official Publications of the European Communities, Luxembourg.

Su, Y.-Z., Wang, F., Suo, D.-R., Zhang, Z.-H., Du, M.-W., 2006. Long-term effect of fertilizer and manure application on soil-carbon sequestration and soil fertility under the wheat−wheat−maize cropping system in northwest China. Nutr. Cycling Agroecosyst. 75, 285−295.

Sulc, R.M., Tracy, B.F., 2007. Integrated crop−livestock systems in the US Corn Belt. Agron. J. 99, 335−345.

Sumberg, J., 2003. Toward a dis-aggregated view of crop−livestock integration in Western Africa. Land Use Policy 20, 253−264.

Thornton, P.K., Herrero, M., 2015. Adapting to climate change in the mixed crop and livestock farming systems in sub-Saharan Africa. Nat. Clim. Change 5, 830−836.

Tian, G., Granato, T.C., Cox, A.E., Pietz, R.I., Carlson, C.R., Abedin, Z., 2009. Soil carbon sequestration resulting from long-term application of biosolids for land reclamation. J. Environ. Qual. 38, 61–74.

Tiemann, L.K., Grandy, A.S., Atkinson, E.E., Marin-Spiotta, E., McDaniel, M.D., 2015. Crop rotational diversity enhances belowground communities and functions in an agroecosystem. Ecol. Lett. 18, 761–771.

Tisdall, J.M., Oades, J.M., 1982. Organic matter and water-stable aggregates in soils. J. Soil Sci. 33, 141–163.

Torri, S., Alvarez, R., Lavado, R., 2003. Mineralization of carbon from sewage sludge in three soils of the Argentine pampas. Commun. Soil Sci. Plant Anal. 34, 2035–2043.

Torri, S.I., Corrêa, R.S., Renella, G., 2014. Soil carbon sequestration resulting from biosolids application. Appl. Environ. Soil Sci. 2014, 1–9.

Udawatta, R.P., Jose, S., 2012. Agroforestry strategies to sequester carbon in temperate North America. Agroforest. Syst. 86, 225–242.

Unger, P.W., Vigil, M.F., 1998. Cover crop effects on soil water relationships. J. Soil Water Conserv. 53, 200–207.

Utuk, I.O., Daniel, E.E., 2015. Land degradation: a threat to food security: a global assessment. J. Environ. Earth Sci. 5, 13–21.

Van Keulen, H., Schiere, J., 2004. Crop-livestock systems: old wine in new bottles. In: 'New directions for a diverse planet', Proceedings of the 4th International Crop Science Congress, 26 Sep–1 Oct 2004, Brisbane, Australia.

VandenBygaart, A., Gregorich, E., Angers, D., 2003. Influence of agricultural management on soil organic carbon: a compendium and assessment of Canadian studies. Can. J. Soil Sci. 83, 363–380.

Ventura, M., Alberti, G., Viger, M., Jenkins, J.R., Girardin, C., Baronti, S., et al., 2015. Biochar mineralization and priming effect on SOM decomposition in two European short rotation coppices. Global Change Biol. Bioenergy 7, 1150–1160.

Vleeshouwers, L.M., Verhagen, A., 2002. Carbon emission and sequestration by agricultural land use: a model study for Europe. Global Change Biol. 8, 519–530.

Wander, M., Bidart, M., Aref, S., 1998. Tillage impacts on depth distribution of total and particulate organic matter in three Illinois soils. Soil Sci. Soc. Am. J. 62, 1704–1711.

Wang, G., Luo, Z., Han, P., Chen, H., Xu, J., 2016. Critical carbon input to maintain current soil organic carbon stocks in global wheat systems. Sci. Rep. 6. Available from: https://doi.org/10.1038/srep19327.

Wang, H., Liu, S., Wang, J., Shi, Z., Lu, L., Zeng, J., et al., 2013. Effects of tree species mixture on soil organic carbon stocks and greenhouse gas fluxes in subtropical plantations in China. Forest Ecol. Manage. 300, 4–13.

Wang, J.B., Chen, Z.H., Chen, L.J., et al., 2011. Surface soil phosphorus and phosphatase activities affected by tillage and crop residue input amounts. Plant Soil Environ. 57, 251–257.

Wang, W., Lai, D., Wang, C., Pan, T., Zeng, C., 2015. Effects of rice straw incorporation on active soil organic carbon pools in a subtropical paddy field. Soil Tillage Res. 152, 8–16.

Wei, W., Yan, Y., Cao, J., Christie, P., Zhang, F., Fan, M., 2016. Effects of combined application of organic amendments and fertilizers on crop yield and soil organic matter: an integrated analysis of long-term experiments. Agric. Ecosyst. Environ. 225, 86–92.

Wei, X., Shao, M., Gale, W., Li, L., 2014. Global pattern of soil carbon losses due to the conversion of forests to agricultural land. Sci. Rep. 4. Available from: https://doi.org/10.1038/srep04062.

Weng, Z., Van Zwieten, L., Singh, B.P., Kimber, S., Morris, S., Cowie, A., et al., 2015. Plant-biochar interactions drive the negative priming of soil organic carbon in an annual ryegrass field system. Soil Biol. Biochem. 90, 111–121.

Weng, Z., Van Zwieten, L., Singh, B.P., Tavakkoli, E., Joseph, S., Macdonald, L.M., et al., 2017. Biochar built soil carbon over a decade by stabilizing rhizodeposits. Nat. Clim. Change 7, 371–376. Available from: https://doi.org/10.1038/nclimate3276.

West, T.O., Post, W.M., 2002. Soil organic carbon sequestration rates by tillage and crop rotation. Soil Sci. Soc. Am. J. 66, 1930–1946.

Whitman, T., Singh, B.P., Zimmerman, A., 2015. Priming effects in biochar-amended soils: implications of biochar-soil organic matter interactions for carbon storage. In: Lehmann, J., Joseph, S. (Eds.), Biochar for Environmental Management: Science, Technology and Implementation (, second ed. Routledge, New York, NY, pp. 455–488.

Wiesmeier, M., Spörlein, P., Geuss, U., Hangen, E., Haug, S., Reischl, A., et al., 2012. Soil organic carbon stocks in southeast Germany (Bavaria) as affected by land use, soil type and sampling depth. Global Change Biol. 18, 2233–2245.

Wiesmeier, M., Hübner, R., Dechow, R., Maier, H., Spörlein, P., Geuß, U., et al., 2014a. Estimation of past and recent carbon input by crops into agricultural soils of southeast Germany. Eur. J. Agron. 61, 10–23.

Wiesmeier, M., Hübner, R., Spörlein, P., Geuß, U., Hangen, E., Reischl, A., et al., 2014b. Carbon sequestration potential of soils in southeast Germany derived from stable soil organic carbon saturation. Global Change Biol. 20, 653–665.

Wiesmeier, M., Schad, P., von Lützow, M., Poeplau, C., Spörlein, P., Geuß, U., et al., 2014c. Quantification of functional soil organic carbon pools for major soil units and land uses in southeast Germany (Bavaria). Agric. Ecosyst. Environ. 185, 208–220.

Wiesmeier, M., Lungu, M., Hubner, R., Cerbari, V., 2015a. Remediation of degraded arable steppe soils in Moldova using vetch as green manure. Solid Earth 6, 609–620.

Wiesmeier, M., Munro, S., Barthold, F., Steffens, M., Schad, P., Kögel-Knabner, I., 2015b. Carbon storage capacity of semi-arid grassland soils and sequestration potentials in northern China. Global Change Biol. 21, 3836–3845.

Wiesmeier, M., von Lützow, M., Spörlein, P., Geuss, U., Hangen, E., Reischl, A., et al., 2015c. Land use effects on organic carbon storage in soils of Bavaria: the importance of soil types. Soil Tillage Res. 146, 296–302.

Wingeyer, A.B., Walters, D.T., Drijber, R.A., Olk, D.C., Arkebauer, T.J., Verma, S.B., et al., 2012. Fall conservation deep tillage stabilizes maize residues into soil organic matter. Soil Sci. Soc. Am. J. 76, 2154–2163.

Wiseman, C.L.S., Puttmann, W., 2005. Soil organic carbon and its sorptive preservation in central Germany. Eur. J. Soil Sci. 56, 65–76.

Woolf, D., Amonette, J.E., Street-Perrott, F.A., Lehmann, J., Joseph, S., 2010. Sustainable biochar to mitigate global climate change. Nat. Commun. 1, 56.

Wortman, S.E., Francis, C.A., Bernards, M.L., Drijber, R.A., Lindquist, J.L., 2012. Optimizing cover crop benefits with diverse mixtures and an alternative termination method. Agron. J. 104, 1425–1435.

Wuest, S.B., Caesar-TonThat, T.C., Wright, S.F., Williams, J.D., 2005. Organic matter addition, N, and residue burning effects on infiltration, biological, and physical properties of an intensively tilled silt-loam soil. Soil Tillage Res. 84, 154–167.

Young, A., 1989. Agroforestry for Soil Conservation. CAB International, Wallingford.

Zhang, P., Wei, T., Jia, Z., Han, Q., Ren, X., Li, Y., 2014. Effects of straw incorporation on soil organic matter and soil water-stable aggregates content in semiarid regions of northwest china. PLoS One 9, e92839.

Zhao, G., Bryan, B.A., King, D., Luo, Z., Wang, E., Song, X., et al., 2013. Impact of agricultural management practices on soil organic carbon: simulation of Australian wheat systems. Global Change Biol. 19, 1585–1597.

Zhao, L.P., Sun, Y.J., Zhang, X.P., Yang, X.M., Drury, C.F., 2006. Soil organic carbon in clay and silt sized particles in Chinese mollisols: relationship to the predicted capacity. Geoderma 132, 315–323.

Zinati, G.M., Li, Y., Bryan, H.H., 2001. Accumulation and fractionation of copper, iron, manganese, and zinc in calcareous soils amended with composts. J. Environ. Sci. Health, Part B 36, 229–243.

IMPACT OF GLOBAL CHANGES ON SOIL C STORAGE—POSSIBLE MECHANISMS AND MODELING APPROACHES

Iain P. Hartley[1] and Brajesh K. Singh[2]

[1]*University of Exeter, Exeter, United Kingdom* [2]*Western Sydney University, Penrith, NSW, Australia*

8.1 INTRODUCTION

Currently, terrestrial ecosystems are reducing rates of climate change by absorbing up to one third of all the carbon dioxide (CO_2) that human activity releases into the atmosphere (Ciais et al., 2013). Accurate predictions of how rates of uptake will change in the future are essential for predicting future rates of warming. United Nations Framework Convention on Climate Change (UNFCCC) agreements, aimed at limiting climate change to less than 2°C by 2100, rely heavily on terrestrial ecosystems continuing to absorb substantial amounts of CO_2. In this context, soil carbon (C) storage is critical. Soils, including permafrost, store up to 3600 billion metric tons of C, which is more C than all the world's plant biomass (600 billion tons) and the atmosphere (850 billion tons) put together (Ciais et al., 2013). Thus, relatively small proportional changes have the potential to influence atmospheric CO_2 concentrations substantially.

In recent years, there have been two major paradigm shifts in our understanding of how C storage in soils is controlled: (1) It is now considered that physicochemical stabilization mechanisms that regulate the accessibility of soil organic matter (SOM) to microbes, rather than the chemical recalcitrance of the SOM itself, control long-term SOM persistence (Dungait et al., 2012; Schmidt et al., 2011); and (2) microbial physiological responses, especially associated with changes in C-use efficiency (CUE; the efficiency with which organic matter assimilated by microbes is converted into biomass vs respired), have the potential to strongly influence the effects of global change on soil C storage (Cotrufo et al., 2013). Critically, models which have incorporated these processes have produced very different predictions of how global change will affect SOM storage compared with conventional multipool SOM models (Tang and Riley, 2015; Allison et al., 2010). This chapter aims to examine the extent to which interactions between soil microbial communities and physicochemical stabilization mechanisms determine how C storage in soils is affected by climate change, rising atmospheric CO_2 concentrations, and nitrogen (N) deposition.

Soil Carbon Storage. DOI: https://doi.org/10.1016/B978-0-12-812766-7.00008-1

8.2 WHAT ARE THE KEY GLOBAL CHANGE MODULATORS OF SOIL C STORAGE?

8.2.1 CLIMATE CHANGE

Rising global temperatures and changes in precipitation patterns will have major direct and indirect effects on soil microbes and soil processes (Davidson and Janssens, 2006). Soil temperature and moisture are key variables directly influencing rates of microbial activity. In addition, the indirect effects of climate change, mediated through changes in plant productivity and community composition, will likely affect the quality and quantity of organic matter inputs and therefore rates of C and nutrient cycling in soils (Hartley et al., 2012). To evaluate how C storage in soils will change it is necessary to consider the effects of climate change on both rates of C input and C output.

8.2.1.1 Rates of warming

In 2015 and 2016, global average temperatures were nearly 1°C greater than the 20th century average (NOAA, 2017), and more than 1°C warmer than preindustrial revolution temperatures. While there are international agreements to limit warming to less than 2°C by 2100, commitments made by governments to date seem unlikely to meet this target and thus warming may reach 3°C, or more, by the end of the century. The rate of warming is not equal across the planet, with greater temperature rises at high latitude and altitudes due to polar amplification associated with the snow and ice albedo feedback (Hartmann et al., 2013). These cold regions store globally significant quantities of organic matter and therefore understanding how cold-adapted microbial communities will respond to warming is particularly important.

8.2.1.2 Permafrost thaw

In the context of greater warming at high latitudes, the microbial breakdown of previously-frozen organic matter as permafrost thaws may be particularly important (Schuur et al., 2015). Permafrost is ground which has been frozen for at least 2 years, but most of the organic matter stored in permafrost soils has been frozen for thousands of years, effectively locked away out of the contemporary C cycle. Frozen layers of these soils store at least as much C as the atmosphere (Hugelius et al., 2014), so if decomposition rates increase substantially after thaw, then there is the potential for huge amounts of C to be released either as CO_2 or CH_4—with important implications for future rates of climate change. There are a growing number of studies investigating potential rates of C release through laboratory experiments and field observations (Schadel et al., 2016, 2014; Treat et al., 2015), and their findings are considered in detail in Section 8.3.2.1.3.

8.2.1.3 Changes in precipitation and soil moisture

Projections of how precipitation patterns will be affected by climate change are less certain than predictions of rates of warming. There is some disagreement between models in terms of which regions will get wetter or drier, and which areas will experience more extreme rainfall events, droughts and floods (Collins et al., 2013). Changes in precipitation may be extremely important, especially in both very wet (Freeman et al., 2001) and very dry ecosystems

(Liu et al., 2009), where soil moisture can be more important than temperature in controlling rates of microbial activity. Furthermore, while the effects of rising temperatures on microbial activity have received considerable attention, the impacts of changes in soil moisture are arguably less well understood. In particular, the role that fluctuations in soil water availability (drying and rewetting cycles) play in controlling microbial community composition and activity remains only partially understood, with controversy remaining regarding the relative importance of organic matter protection during droughts versus high rates of activity during rewetting (Borken and Matzner, 2009; Placella et al., 2012; Xiang et al., 2008). With both wetlands (Charman et al., 2013) and drylands (Ahlstrom et al., 2015; Poulter et al., 2014) potentially playing key roles in the global C cycle, improved understanding of soil moisture controls over microbial activity represent a key research priority.

8.2.1.4 Effects on plant communities

Impacts of climate change on soil microbial processes need to be put into the context of changes in the geographical distribution of particular species and communities, and changes in productivity. Any increase in decomposition rates, while releasing C, should also increase soil nutrient availability and it has been demonstrated that resultant increases in plant biomass production can partly offset soil C losses (Melillo et al., 2002, 2011; Crowther et al., 2016). On the other hand, colonization of high latitude and altitude areas by more productive plant communities can have major impacts on soil microbial communities. Counterintuitively, in these circumstances, greater plant productivity and organic matter inputs into soils can actually result in the net loss of SOM through the stimulation of decomposition, a process referred to as positive priming (Hartley et al., 2012; Parker et al., 2015). It is becoming increasingly clear that whole ecosystem approaches are required to understand the impacts of climate change on ecosystem and soil C storage, and studies that focus exclusively on above- or belowground potentially miss key responses.

8.2.2 ELEVATED ATMOSPHERIC CO_2

Atmospheric CO_2 concentrations have increased from $\sim$280 ppm prior to the industrial revolution to $\sim$400 ppm in 2016, and may increase further to between 700 and 1000 ppm by 2100 (Hartmann et al., 2013). However, in soils, CO_2 concentrations are regularly greater than 10,000 ppm (Andrews and Schlesinger, 2001), and therefore, it is unlikely that elevated atmospheric CO_2 concentrations (eCO_2) will directly impact on soil microbes. Rather, the impacts of eCO_2 will be mediated through the effects on plant productivity and the quantity and quality of organic matter inputs to soils.

8.2.2.1 Below-ground C allocation and priming effects

In the short-term, rates of photosynthesis increase substantially under eCO_2 at least in C3 plants. Greater substrate availability to the key Calvin Cycle enzyme RUBISCO, together with reductions in rates of photorespiration, promote greater rates of C fixation (Leakey et al., 2009). All things being equal, greater rates of photosynthesis should equal greater rates of plant growth. However, the availability of nutrients may limit the ability of plants to produce more biomass (Zaehle et al., 2015). To sustain a strong growth response, plants either have to produce biomass more efficiently (greater biomass production per unit nutrient) or acquire more nutrients from soils

(Zaehle et al., 2014). The latter may be achieved by allocating some of the additional C fixed under eCO_2 to roots and mycorrhizal fungi and thus potentially increasing access to limiting nutrients (Phillips et al., 2012; Terrer et al., 2016; Nie et al., 2013a). Increased belowground C allocation may have profound effects on soil microbial communities and C and nutrient cycling, and can result in the stimulation of SOM decomposition (positive priming; van Groenigen et al., 2014).

8.2.3 N DEPOSITION

N deposition is the input of reactive N species (ammonium and oxides of N) from the atmosphere to terrestrial ecosystems (Fleischer et al., 2013). Rates of N deposition have increased dramatically in most developed countries since the start of the industrial revolution, and are expected to continue to increase in the developing world through the 21st century (Galloway et al., 2004). The rates of input are now so high that, globally, they exceed preindustrial rates of natural N input into terrestrial ecosystems through biological N fixation and lightning (Fowler et al., 2013). Changes in N availability may have a wide range of effects on soil C storage, from directly influencing rates of decomposition, to changing ecosystem productivity and plant C allocation to roots and the rhizosphere.

8.3 WHAT ARE DOMINANT MECHANISMS THAT EXPLAIN GLOBAL CHANGE MODULATOR IMPACTS ON SOIL C STORAGE?

Until the past couple of decades, C storage in soils was viewed as being controlled predominately by the combination of: (1) The quantity and quality of organic matter inputs into the soils; and (2) climatic conditions including soil temperature and moisture (Schmidt et al., 2011). This is the case for soil horizons which are exclusively made up of organic material (e.g., thick organic horizons or peats), and the role of raised water tables in promoting anaerobic conditions that reduce decomposition rates in soils and sediments is well understood (Freeman et al., 2004). However, in oxygenated mineral soils, the view of how SOM storage is controlled has changed strongly. There is now the recognition that, under aerobic conditions, almost all the organic matter stored in soils can be degraded by soil microbes and that long-term storage is more related to physicochemical protection mechanisms that regulate how accessible SOM is to microbes (Schmidt et al., 2011; Dungait et al., 2012; von Luetzow et al., 2006). This is the context in which microbial responses to global change must be placed. To predict changes in soil C storage, it is essential to improve understanding of how soil microbes will respond to changes in temperature, soil moisture, the quality and quantity of organic matter inputs, and changes in the accessibility of SOM (Fig. 8.1). In addition, the extent to which microbes control the accessibility of SOM is a key research priority.

8.3.1 ORGANIC MATTER FORMATION AND STABILIZATION (INPUTS)

8.3.1.1 Quality and quantity of inputs

As mentioned earlier, the quality and quantity of C inputs has historically been considered as a key determinant of soil C storage. In peatlands, this viewpoint remains valid with studies having

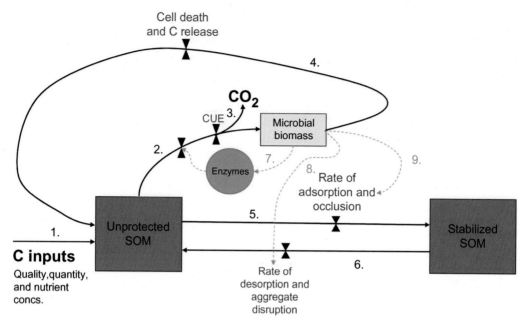

FIGURE 8.1

Schematic diagram showing the key processes considered to control C storage in soils. Organic matter inputs initially enter the unprotected SOM pool (1). Microbial physiology can then influence rates of unprotected SOM decomposition, and the release of CO_2 and nutrients. Rather than simply being another pool of organic matter, microbial biomass produces the enzymes that breakdown SOM (2, 7), and CUE controls the efficiency with which biomass and enzymes are then produced from the decomposition of unprotected SOM, as opposed to the C being released as CO_2 (3). When organic matter is released from microbial cells, or the cells die, they become unprotected SOM again (4), with these microbial products generally contributing more to SOM formation than directly plant-derived material. The capacity of soils to stabilize SOM is critical in determining how much C is stored in the long term, with the relative rates of adsorption onto soil particles and occlusion in aggregates (5), versus desorption and aggregate disruption (6) being key is this regard. The role of microbial communities in controlling the rates of these two processes (8, 9) is therefore extremely important to understand. The diagram concentrates on C, but it should be noted that the relative availabilities of C and nutrients are critical in determining CUE, the types of enzymes that the microbes produce and potentially the extent to which microbes can destabilize physicochemically-protected SOM.

identified plant productivity as a stronger predictor of accumulation rates than climatic factors that may control decomposition rates (Charman et al., 2013), at least when water tables do not drop substantially below the surface (Fenner and Freeman, 2011). Furthermore, rates of accumulation tend to be greater when the inputs are of a lower-quality, e.g., in *Sphagnum* moss-dominated rather than vascular-plant dominated peatlands (Larmola et al., 2013). The potential for N deposition to promote vascular plant dominance, at the expense of *Sphagnum*, could therefore have strong impacts on rates of C accumulation in these ecosystems.

In mineral soils, studies are beginning to question the long-held view that low-quality inputs promote greater C storage (Cotrufo et al., 2013, 2015). Although low-quality material may initially breakdown more slowly, microbial CUE (the efficiency with which organic matter assimilated by microbes is converted into biomass) may be reduced for the breakdown of low-quality material. This means that a greater proportion of the material will be released back to the atmosphere as CO_2 rather than converted into microbial biomass (Cotrufo et al., 2013). Furthermore, substrate-addition experiments have demonstrated that microbial processing of even very simple compounds can result in a wide range of chemically-varied products, and that the formation of new SOM is positively related to microbial CUE (Kallenbach et al., 2016). In this context, the stoichiometry (elemental ratios, e.g., C:N:P) of the organic matter inputs may be particularly important. The stoichiometry of microbial biomass tends to be less variable than the stoichiometry of the organic matter inputs, with the C:N and C:P ratios in microbial biomass tending to be considerably narrower than for SOM. Thus excess C may simply be respired, and negative relationships have been observed between the C:N ratio of readily decomposable SOM and microbial CUE (Sinsabaugh et al., 2016). In addition, latitudinal gradients in labile SOM C:N ratios suggest that, contrary to the results of short-term experiments, CUE may be positively related to mean annual temperature. However, latitudinal gradients in SOM C:N are the result of long-term soil and ecosystem development, with tropical soils having a gone through a much longer period of pedogenesis. These gradients may not, therefore, be informative for predicting responses to climate warming.

In terms of other effects of global change, changes in N availability through greater rates of N deposition may increase microbial CUE and therefore the formation of new SOM (Midgley and Phillips, 2016), while more C-rich inputs under eCO_2 could have the opposite effect (van Groenigen et al., 2015). In addition, changes in microbial community composition induced by the direct effects of any global change driver or change in vegetation community, may affect CUE. For example, the greater C:N ratios in fungi compared with bacteria, may reduce discrepancies between substrate and microbial biomass C:N ratios and promote greater CUE (Sinsabaugh et al., 2016). Therefore, an improved understanding of the how global change affects fungal to bacterial ratios could prove important for predicting changes in soil C storage.

To date, there is only preliminary evidence to support these hypotheses directly, but given that the efficiency of microbial processing of inputs is likely important for controlling rates of new SOM formation, it is critical to understand better this issue in the context of global change. Finally, hypotheses related to how the decomposability and stoichiometry of organic matter inputs affect microbial CUE and rates of new SOM formation, must also be placed into the context of priming theory and our developing understanding of how inputs of new organic matter can affect the decomposition of existing organic matter (Chen et al., 2014; Hartley et al., 2010; Kuzyakov, 2010). This is considered in Section 8.3.2.3.

8.3.1.2 Edaphic factors and stabilization (soil physical and chemical properties)

Long-term C storage in oxygenated, mineral soils is no longer considered to be driven by the chemical recalcitrance of SOM (von Luetzow et al., 2006). Rather, physicochemical protection mechanisms appear to be the key to long-term persistence (Schmidt et al., 2011; Dungait et al., 2012). The adsorption of organic matter on to charged clay particles and formation of organomineral complexes prevents organic matter binding with the active sites of SOM-degrading enzymes. In addition, the binding of soil particles together to form aggregates (producing occluded SOM), which can be promoted by the presence of organomineral complexes, results in the physical separation of microbes and organic matter. Therefore, adsorption and aggregate formation can control rates of SOM stabilization.

In terms of adsorption reactions, the products of microbial decomposition have been shown to be more likely to become stabilized in soils than the original plant compounds (Cotrufo et al., 2015; Schmidt et al., 2011). Therefore, the rate of inputs, and the efficiency with which microbes processes these inputs (see 8.3.1.1), should be key determinants of how rapidly new SOM is formed and stabilized (Castellano et al., 2015). On the other hand, the further microbial degradation of these microbial products versus the adsorption onto clay particles can be seen as competing processes and it is unclear to what extent which different microbial communities can directly control (reduce) rates of adsorption. It has, however, been argued that rates of stabilization by adsorption may be largely abiotically controlled by soil physical and chemical properties (Schimel and Schaeffer, 2012), and that the extent to which soil particles are C saturated may be key to determining rates of adsorption (Castellano et al., 2015; Campbell and Paustian, 2015). It seems probable that soils will have a maximum potential to stabilize SOM, dependent on textural or mineral properties. In soils in which this maximum potential has not been reached, microbial products may be adsorbed rapidly, with the rate of stabilization declining as the soil particles become saturated. The extent to which soils are currently saturated or under-saturated with SOM, and the extent to which environmental factors can influence levels of C saturation may be extremely important in determining in which soils C storage increases and in which soils C storage decreases in the future (Wiesmeier et al., 2014).

In contrast, there is evidence that the formation of aggregates and the developed of occluded SOM pools may be directly affected by microbial community composition. For example, arbuscular mycorrhizal fungi, and a key compound that they produce (glomalin), have been proposed to help bind soil particles together, and thus play a critical role in controlling rates of aggregation, and in increasing the average size of aggregates (Wilson et al., 2009). Overall, though, the role of different microbes in controlling rates of SOM stabilization remains poorly understood.

In terms of the destabilization of SOM, there have been some suggestions that microbes may play a direct role in the desorption of SOM from organomineral complexes and the disruption of aggregates. For example, the release of organic acids by roots or microbes is known to play a key role in liberating mineral-bound cations and phosphorus (P) (Dakora and Phillips, 2002), and oxalic acid has been shown to liberate mineral-bound organic matter (Keiluweit et al., 2015). In addition, priming studies (covered in detail in Section 8.3.2.3) have shown that the addition of labile SOM can stimulate microbial activity, promote the disruption of aggregates, and liberate occluded SOM (Tian et al., 2016b). However, overall, we still have relatively limited understanding of the extent to which roots, mycorrhizal fungi and different communities of microbes in soils control rates of aggregate disruption and the desorption of mineral-bound SOM.

Given that we now know that adsorption/desorption and occlusion reactions are critical in controlling long-term stabilization of SOM, the fact that we do not fully understand the extent to which microbial communities control the rates of these key processes represents a major knowledge gap. This contributes to the fact that, currently, it is difficult to predict how global change will affect rates of SOM formation and net stabilization (stabilization minus destabilization). For example, studies disagree regarding whether adsorption or desorption reactions should be more temperature sensitive (Thornley and Cannell, 2001; Conant et al., 2011). In addition, soil moisture contents, and fluctuations in moisture contents, could also influence rates of stabilization and destabilization, with drying potentially promoting stabilization, but rewetting potentially destabilizing aggregates and promoting desorption (Kaiser et al., 2015; Schimel et al., 2011). The net effects of wetting and drying cycles are very poorly understood. In terms of elevated CO_2, increased allocation of C to arbuscular mycorrhizal fungi could increase rates of aggregation and therefore SOM stabilization (Wilson et al., 2009). However, increased root and mycorrhizal activity have often been associated with losses of soil C (Cheng et al., 2012; van Groenigen et al., 2015), so it remains difficult to determine how important the potential increase in aggregation is. The impacts of N deposition remain similarly uncertain. Soil acidification caused by N addition, may increase mineral surface reactivity and thus the potential for adsorption onto clay particles (Riggs et al., 2015; Thornley and Cannell, 2001), yet there is limited evidence for this process increasing C storage. Again, because N deposition may have the opposite effect to eCO$_2$ in reducing C allocation to roots and mycorrhizal fungi, potential rates of aggregate formation may decline, although this has not always been supported by observations (Riggs et al., 2015). Our overall understanding of how the physicochemical stabilization of SOM will be affected by global change, and the role of different microbial communities in mediating responses, remains limited and should be considered a major research priority.

8.3.2 ENVIRONMENTAL CONTROL OVER DECOMPOSITION RATES (OUTPUTS)

In contrast to our limited understanding of how rates of SOM formation and stabilization are controlled, the effects of global change on decomposition rates have received considerable attention and knowledge has advanced substantially in recent decades. There has been extensive study of how factors such as temperature, moisture and N availability affect decomposition rates. Furthermore, plant-control of decomposition rates through the stimulation of microbial activity (priming effects) has also received growing attention, and this issue is no longer seen as being of only academic interest in controlled experiments, but, rather, is now considered to be a process which can control ecosystem C storage (Hartley et al., 2012; Kuzyakov, 2010).

8.3.2.1 Temperature
8.3.2.1.1 Different pools of soil organic matter
In the short-term, rising temperatures are known to result in strong increases in microbial activity and decomposition rates (Kirschbaum, 1995, 2006; Davidson and Janssens, 2006). Both leaf litter decomposition and the decomposition of SOM have been shown to increase with temperature, with the greatest proportional increases being observed at low temperatures. Considerable controversy has emerged regarding whether the decomposition of all the organic matter stored in soils is affected by temperature in the same way. Early studies suggested that it may only be a relatively small proportion of physicochemically unprotected SOM which decomposes more rapidly at higher temperatures, and that the decomposition of stabilized SOM is temperature insensitive (Giardina and Ryan, 2000; Liski et al., 1999). However, more recent studies have demonstrated that the breakdown of more slowly decomposing SOM may be more temperature sensitive than the breakdown of more readily decomposable SOM (Hartley and Ineson, 2008; Conant et al., 2008a,b;

Hopkins et al., 2012). These studies demonstrate that SOM which turns over on a decadal timescale is more sensitive to temperature than SOM which turns over more rapidly, consistent with kinetic theory. However, there remains considerable uncertainty regarding the temperature sensitivity of the most slowly decomposing SOM, which is important as an analysis indicated that, globally, the average age of SOM in the top 1 m of soils is greater than 3000 years old (He et al., 2016), and Earth systems models cannot reflect this slow turnover. As explained earlier, long-term storage in oxygenated soils is related to physicochemical stabilization of SOM (Schmidt et al., 2011; Dungait et al., 2012). It is thus extremely important to determine the extent to which rising temperatures could destabilize this SOM and make it more available to decomposers. We arguably still know little about what proportion of SOM stored in the world's soils may be vulnerable to the impacts of warming during the 21st century (Crowther et al., 2016).

8.3.2.1.2 Microbial community responses to temperature

In long-term warming experiments, the initial stimulation of activity caused by the higher soil temperature tends not to be sustained with respiration rates often declining back toward prewarming levels (Rustad et al., 2001). Two major hypotheses have been proposed to explain this observation. Firstly, it has been suggested that the loss of the most readily decomposable SOM may have resulted in overall rates of decomposition declining (Kirschbaum, 2004). Alternatively, it has been proposed that changes within the microbial community may have resulted in a down-regulation of activity (Bradford et al., 2008). The microbial respiration response appears analogous to the response of plant respiration to changes in temperature. In the short-term, plant respiration is highly temperature sensitive, but in the longer term, through a process referred to as thermal acclimation, plants can compensate for the effects of a temperature change by up-regulating respiration following cooling and down-regulating respiration in response to warming (Atkin and Tjoelker, 2003). There are, though, many fundamental differences between microbes and higher plants that need to be taken into account when considering how they respond to changes in temperature (see Box 8.1).

It is important to distinguish between the two competing hypotheses as with the loss of readily-decomposable SOM, the reduction in the effect of warming with time is simply a symptom of the ongoing loss of SOM. In contrast, a down-regulation of microbial activity would result in decomposition rates declining prior to the SOM being lost. There is now increasing evidence that the loss of readily decomposable SOM is involved in this observation. Reductions in total and labile SOM stocks have been observed in warming studies (Bradford et al., 2008; Hartley et al., 2007; Crowther et al., 2016). However, this does not mean that there is no potential for microbial community responses to reduce the effects of rising temperatures on decomposition rates in the longer term. Given the potential importance of this issue it has received considerable attention in the past 10 years.

There are multiple reasons why microbial activity may become gradually down-regulated at higher temperatures. For example, there is a strong trade-off between the affinity of enzymes for their substrates and their stability at high temperatures (Bradford et al., 2010). The shift to more stable enzymes with lower substrate affinities could explain the observed down-regulation in activity. Furthermore, in addition to direct physiological trade-offs, there is growing evidence that CUE may decline at higher temperature, with less biomass production per unit SOM decomposition (Frey et al., 2013). If this was to translate to a gradual reduction in microbial biomass production at

BOX 8.1 KEY DIFFERENCES BETWEEN HIGHER PLANTS AND MICROBES THAT MAY AFFECT RESPONSES TO GLOBAL WARMING

Overwhelming evidence suggests that thermal acclimation compensates for the effects of changes in temperature on plant respiration (Atkin and Tjoelker, 2003). Reports on the response of microbial respiration to warming vary in their conclusions (Bradford et al., 2008; Hartley et al., 2007; Knorr et al., 2005b), however, a significant number of studies suggest microbes can show enhancing responses, increasing the effects of temperature change on rates of respiration (Karhu et al., 2014; Nazaries et al., 2015; Vicca et al., 2009; Nie et al., 2013b). These enhancing effects do not seem to be explainable solely by changes in microbial biomass (Fig. 8.2), and must be related to changes in community composition or function (Auffret et al., 2016; Karhu et al., 2014). The differential responses between plant and microbial communities (compensatory vs enhancing) may be explained by the physiological, ecological, and evolutionary divergence between these taxa. For example, plants mainly utilize atmospheric carbon (C) for photosynthesis and the balance between C fixation and plant respiration is finely tuned. Although the rate of C accumulation by photosynthesis can be several fold higher than C loss via autotrophic respiration (Van Oijen et al., 2010), plants are entirely dependent on fixed C for their growth, tissue repair and acquisition of essential nutrients (e.g., N, P) through the release of exudates into the root zone to promote the growth of beneficial microbes (Singh et al., 2004). On the other hand, microbes mainly utilize soil organic matter (SOM) and do not have the physiological requirement to balance fixation with utilization. As a community, their growth is constrained by the availability of SOM, not by their physiological capability to fix C.

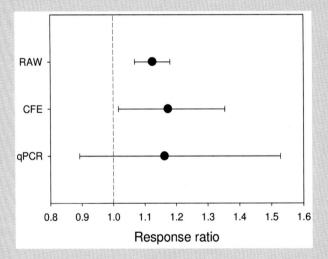

FIGURE 8.2

Summary of the results from a study investigating whether compensatory or enhancing microbial community responses are more common in soils sampled from across a major temperature gradient (Karhu et al., 2014). Response ratios less than 1 indicated a compensatory response, while response ratios greater than 1 indicated an enhancing response. Mean values and 95% confidence intervals are presented. The overall response ratio was significantly greater than 1 (uncorrected for changes in biomass, RAW), indicating that enhancing responses dominated. This conclusion remained after responses were corrected for changes in microbial biomass as measured by chloroform fumigation (CFE), and quantitative polymerase chain reaction (qPCR).

(Continued)

BOX 8.1 (CONTINUED)

Perennial plants live for decades to millennia and it can be argued that their only viable survival option in the short to medium-term is to balance the C input and output under warming via physiological adjustments, otherwise additional losses of C through increased respiration can reduce their survival and competitiveness. Physiological adjustments for the compensatory response of plant respiration could be supported by the genome size of plants (up to thousand-fold larger than many bacteria and archaea) which may contain several copies of genes that can perform similar functions under different conditions (Wendel et al., 2016), thereby providing a mechanism for physiological adjustments.

Microbes, on the other hand, generally have small genomes and low copy numbers of genes (Land et al., 2015), constraining their ability to carry out similar functions at different conditions. However, microbes have enormous advantages in terms of ecological and evolutionary adjustment to respond to climate warming mainly because of enormous diversity of microbes, the very short replication time (minute to days) and their rapid turnover rate (Schmidt et al., 2007; Singh et al., 2010). Millions of microbial species are present in small quantities of soil, but all are not active at the same time. One way that microbial communities can maintain functionality (e.g., SOM decomposition) is to have differentially active populations and dominance of microbial taxa in different seasons or under different climatic conditions. For example, active microbial populations between summer and winter seasons differ in the same soils (Matulich et al., 2015). In a global warming scenario, microbial taxa which can operate at high temperatures may become active and dominant within several generations and thus potentially change the rate of soil respiration because the overall physiology of the ecosystem has changed (Singh et al., 2010). In such a case, changes in microbial community composition could be more important in terms of effects on rates of respiration and decomposition, than physiological adjustments to climate adaptation.

The rapid turnover rate also provides an effective evolutionary mechanism for microbes that could explain the different response of microbial respiration versus plant respiration to global warming. For each round of replication, microbes can acquire genes (from other microbes, plants and animals) and lose or alter genes (e.g., mutation) in order to adjust to new climatic conditions (Schmidt et al., 2007). Such ecological and evolutionary advantages are not available for plant communities in the short to medium-term. However, it is likely that in the long-term, plant communities will also follow ecological and evolutionary mechanisms of adaptation to global warming, but longer time periods will be needed (centuries to millennia) because of the slow rate of replication and slow turnover rate in comparison to microbial communities.

higher temperatures, and a related reduction in the production of key SOM-degrading enzymes (Allison et al., 2010), then this could reduce rates of decomposition and soil microbial activity. It has been argued that one definitive test of whether soil microbes down-regulate their activity is to measure rates of respiration per unit biomass at a common temperature for warm and cool-grown populations (Bradford et al., 2008). In single species cultures of saprotrophic fungi, it was shown that biomass-specific respiration rates were indeed lower at a common temperature in populations that had been established at higher temperatures (Crowther and Bradford, 2013). This raises the potential for microbial responses to temperature to reduce the effects of warming on decomposition rates in the longer term. Therefore, the potential for a gradual down-regulation of microbial activity in warmed soils is real, and could reduce or eliminate the potential for soil C losses in response to global warming.

This issue was further explored across a range of soils sampled from the Arctic to the Amazon. Contrary to expectations, it was shown that, although there were soils in which the temperature sensitivity of decomposition was reduced by the microbial community response, overall, the response of microbial communities to warming and cooling actually increased the temperature sensitivity of decomposition (Karhu et al., 2014). This was particularly the case for soils sampled from cooler regions (Karhu et al., 2014; Hartley et al., 2008). The mechanisms underlying these responses have proven challenging to identify, but the magnitude of the respiration responses were correlated with the initial microbial community structure, while it was also demonstrated that biomass-specific respiration rates did not decline in warmer soils (Auffret et al., 2016).

One potential explanation for microbial community responses increasing the temperature sensitivity of decomposition is that the indirect effects of warming on substrate availability may be more important that the direct effects of warming on microbial physiology (Bradford, 2013). In the experiments cited earlier, microbial activity rates were compared in soils which has lost the same amount of C (Karhu et al., 2014). This was done to ensure that any down-regulation of activity was not related to substrate loss. However, if warming promotes desorption of SOM to a greater extent than adsorption (Conant et al., 2011), or makes low-quality unprotected SOM more readily decomposable (Hartley and Ineson, 2008; Conant et al., 2008a), then the effective increase in substrate availability may explain why microbial activity was higher at warmer temperatures. Previous studies have also identified apparent increases in the amount of readily decomposable organic matter when soils are incubated at higher temperatures (Townsend et al., 1997).

The explanations being proposed here can seem a little contradictory; an effective increase in substrate availability at higher temperatures results in microbial community responses enhancing the effects of warming on decomposition rates (Karhu et al., 2014), while a loss of readily decomposable C explains the reduction in respiration rates observed in long-term warming experiments (Bradford et al., 2008; Hartley et al., 2007). The key here is that changes in absolute substrate availability has been accounted in the microbial community response research (Karhu et al., 2014), and that for a given amount of C release, the response of the microbial community has enhanced the effect of a temperature change on rates of decomposition. This enhanced temperature sensitivity of decomposition likely contributes to the rapid release of readily decomposable C from warmed soils (Bradford, 2013), and the subsequent reduction in decomposition rates as the microbes are left with less readily decomposable C to breakdown—the effect of higher temperatures in terms of increasing the availability of substrates does not fully compensate for the greater rate at which these substrates are being utilized by soil microbes.

Finally, the particularly strong response observed at low temperatures also requires further attention. It could be argued that temperature changes at low temperatures may have greater effects on substrate availability than changes at higher temperatures (Kirschbaum, 1995). However, there is potentially an alternative explanation. Microbial community adaptations for surviving at low temperatures involve fundamental changes at the cellular level, with strategies that promote survival potentially reducing metabolic activity (Schimel et al., 2007). Warming such communities may allow microbes to reduce their emphasis on survival and focus more on growth and metabolic activity, and rapid increases in the abundances of functional genes involved in decomposition having been observed with in situ warming in moist acidic tundra soil (Xue et al., 2016). Therefore, empirical and modeling approaches that explicitly consider both changes in substrate availability and microbial physiology may be needed to improve understanding of why the response of microbial communities tends to increase rather than decrease the effects of a temperature change on rates of respiration and decomposition, with cold soils potentially representing a distinct situation.

In summary, overall, microbial community responses seem more likely to increase the effects of temperature changes on rates of decomposition than reduce them. Substrate availability appears likely to be the key determinant of how respiration rates will respond to warming in the longer term (Crowther et al., 2016; Bradford, 2013; Karhu et al., 2014). However, this does not mean that there will be no soils in which microbial community responses reduce the effects of a temperature change on decomposition rates, and we still lack a strong mechanistic understanding of why responses differ between soils.

8.3.2.1.3 Permafrost thaw and C release

In the context of the effects of temperature on decomposition rates, the permafrost C feedback merits special attention. Model predictions (Burke et al., 2012; Schaefer et al., 2014; Koven et al., 2015) and expert assessments (Schuur et al., 2013) have concluded that tens of billions of metric tons of C will be released from permafrost soils by the end of this century, with the rates of C release approaching those associated with land-use change and deforestation (Schuur et al., 2009). As permafrost thaws, microbes can start to breakdown previously-frozen organic matter which had effectively been locked away, out of the contemporary C cycle. Understanding the changes in microbial community structure and function that take place following thaw are critical for predicting rates of CO_2 and CH_4 release (Mackelprang et al., 2016). Perhaps unsurprisingly, microbial communities found in permafrost are highly adapted for survival at temperatures below 0°C (Jansson and Tas, 2014), with major differences being observed between the structure and function of microbial communities in active layer soils and the underlying permafrost (Hultman et al., 2015). However, following thaw, rapid changes in community composition have been observed with communities expanding the range of functions they can carry out (Coolen and Orsi, 2015), and showing a tendency to becoming more similar to those characteristic of the active layer (Mackelprang et al., 2011). This raises the potential for rapid rates of decomposition of previously-frozen organic matter, at least under conditions in which the soils remain oxygenated (Schuur et al., 2015; Schadel et al., 2016).

While rates of decomposition are likely to be much lower in anaerobic soils, microbial community changes can be particularly large when permafrost thaw results in the process of thermokarst and water-logging (Hultman et al., 2015). Under anaerobic conditions, novel methanogen phylotypes have been discovered in thaw wetlands, with these methanogens potentially playing a critical role in CH_4 release from northern peatlands (Mondav et al., 2014). For example, the abundance of *Methanoflorens stordalenmirensis* in a thawing wetland was been found to be a strong predictor of the ratio of CH_4 to CO_2 release (McCalley et al., 2014). Expert assessments have suggested that CH_4 emissions could be responsible for up to one third of the climate warming caused by the permafrost feedback (Schuur et al., 2013), so this one methanogen phylotype could play a significant role in C-cycle feedbacks to climate change. However, more recently it has been suggested that there may be less potential for the anaerobic decomposition of previously-frozen SOM to promote CH_4 release, and that the effects of permafrost thaw on CH_4 release may be controlled more by near-surface hydrology and the rate of anaerobic decomposition of recent C inputs (Cooper et al., 2017).

In summary, there have been major advances in understanding made by the growing numbers of studies carrying out detailed -omic investigations into changes that take place in microbial communities following permafrost thaw. However, arguably, integration between these studies and measurements of how rates of C and nutrient cycling have changed could be strengthened even further. Detailed comparisons with soil biogeochemical and greenhouse gas emission measurements may further help elucidate the extent to which -omic approaches can provide the understanding required for predicting future rates of CO_2 and CH_4 release.

8.3.2.2 Soil moisture

Decomposition is generally considered to have quite a broad soil moisture optimum, with rates changing relatively little at intermediate moisture levels (Falloon et al., 2011). However, due to reductions in oxygen availability at high moisture contents and reductions in rates of substrate diffusion to exoenzymes at low moisture contents, rates of decomposition can decline substantially in very wet or very dry soils. Climate change has the potential to result in changes in water-table dynamics, with deeper water tables potentially promoting greater rates of decomposition and C losses from peatland ecosystems (Davidson and Janssens, 2006). Under anaerobic conditions it has been demonstrated that reductions in the activities of a key bacterial enzyme, phenol oxidase, results in the build-up of phenolic compounds which inhibit the activities of a wide range of hydrolytic enzymes (Freeman et al., 2001, 2004). Following droughts in which water tables are lowered, the presence of oxygen can promote bacterial growth and the production of phenol oxidase, with the associated reduction in the concentration of phenolics resulting in major increases in the activities of a range of enzymes and, therefore, rates of decomposition (Fenner and Freeman, 2011). This cascading process could result in reductions in peatland C stocks if droughts become more frequent or water tables become deeper under climate change.

Given that peatlands are considered to store approximately 500 billion metric tons of C (Yu, 2012), about half the amount of C that is in the atmosphere, increased rates of decomposition in peatlands could represent an important feedback to climate change. However, as mentioned earlier, there does not seem to be clear evidence that periods of warming in the last millennium, and associated changes in precipitation patterns, have resulted in major reductions in rates of C accumulation, with the opposite in fact seeming to be the case (Charman et al., 2013). There also remains considerable observational (Smith et al., 2005; Watts et al., 2014) and modeling (Gao et al., 2013; Lawrence et al., 2015) uncertainty regarding whether permafrost thaw will increase or decrease wetland extent in different northern high-latitude regions. Overall, it is difficult to predict the extent to which organic soils will experience reductions in soil moisture, and a lowering of water tables in particular, although if these soils did become drier, the potential for considerable C losses remains.

At low moisture contents, decomposition rates decline due to slower diffusion reducing the rates of exoenzyme-catalyzed reactions (Manzoni et al., 2014). In addition, severe reductions in moisture contents can kill microbes that are not adapted to withstand droughts (Blazewicz et al., 2014). These reductions in microbial activity should reduce the amount of C released from an ecosystem during a drought, although the net effect of a drought is likely still to be a loss of C from the ecosystem because C uptake by plants has been shown to be more drought-sensitive than combined rates of plant and microbial respiration (Schwalm et al., 2010). Furthermore, when dry soils are rewetted, big pulses of CO_2 release have been observed which may result in major C losses at the end of a drought. This phenomenon has been referred to as the "Birch Effect" after it was first described in the 1950s (Birch, 1958). These pulses can release so much C that drying and rewetting cycles result in greater amounts of C release than from continuously moist soils (Fig. 8.3). However, the net effect of droughts is dependent on the frequency of rewetting events and on how dry the soils become between rewetting events (Xiang et al., 2008).

There are multiple potential sources of the CO_2 which is released during rewetting. These include: (1) the decomposition of the microbial biomass that died during the drought (Blazewicz et al., 2014); (2) the mineralization of compounds that accumulated within microbial cells during the drought to

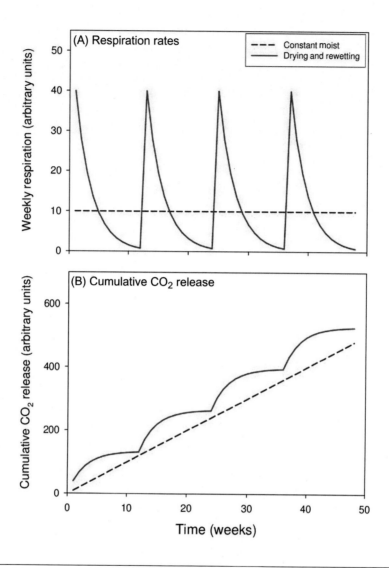

FIGURE 8.3

Representation of how the rewetting of dried soils can result in major pulses of respiration (A), which in this example result in cumulative CO_2 release being greater in the dried and rewetted soils than from soils which were maintained continuously moist (B).

aid with tolerance of the low soil moisture (Fierer and Schimel, 2003; Borken and Matzner, 2009); (3) the breakdown of organic matter that was previously protected by adsorption onto clay minerals or within aggregates, and which has been destabilized during the drying and subsequent rewetting (Xiang et al., 2008). It is essential to determine the source of the C, as there are potentially very different consequences for soil C storage. The two sources associated with microbial biomass pools suggest that the Birch Effect may have relatively little effect on long-term C storage. However, if previously-unavailable SOM is liberated by wetting and rewetting cycles, then droughts may have a major effect on the ability of mineral soils to store C. While, there remains controversy regarding the dominant source of the C released, increasing numbers of studies support the suggestion that previously-unavailable C contributes to the CO_2 pulse, and that this effect may be particularly pronounced in deeper soil layers (Schimel et al., 2011; Xiang et al., 2008). Therefore, drying and rewetting cycles may be critical in controlling soil C storage in arid and semi-arid regions and any increases in drought frequency in other regions could destabilize soil C stocks.

Soil drying and rewetting dynamics have been shown to play a fundamental role in shaping the structure of microbial communities (Evans and Wallenstein, 2014), and different taxonomic groups may play different roles in C dynamics during rewetting. For example, specific phylogenetic clusters of microbes become active at different points during rewetting and therefore may be using different substrates (Placella et al., 2012). Overall, the extent to which previously-unavailable SOM is utilized by microorganisms during soil rewetting, and the role microbial community composition plays in controlling this, requires further investigation as the implications for predicting spatial and temporal controls over soil C storage are potentially highly significant.

8.3.2.3 Priming effects

Global change may have major impacts on plant productivity, and rates of C input into soils. Counterintuitively, increased release of C into the rhizosphere may not necessarily result in greater C storage in soils; microbial activity can increase and result in greater rates of decomposition of existing SOM in a process referred to as positive priming (Kuzyakov, 2002; Fontaine et al., 2004). The overall effect on C storage in soils depends on the balance between new SOM formation from the new inputs, versus the extent to which existing organic matter is lost through positive priming (van Groenigen et al., 2014, 2015; Qiao et al., 2014). However, it also needs to be emphasized that priming effects are not universally positive, and, over the years, a number of different hypotheses have been proposed to explain why greater rates of C inputs into to soils increase, decrease, or have no effect on, the rate of the decomposition of existing SOM.

Positive priming associated with the addition of labile organic matter with high C to nutrient ratios is often considered to be associated with the stimulation of the decomposition existing SOM to meet microbial nutrient demand (Kuzyakov et al., 2000; Fontaine et al., 2003). Negative priming effects may be associated with preferential use of the added substrates. If nutrient availability is high, then microbes may shift to using the added labile substrates, rather than SOM, as their C source. Evidence for the role of nutrient availability has been demonstrated by positive priming effects being reduced when nutrients have been added in combination with the labile organic matter (Fontaine et al., 2011; Hartley et al., 2010). However, it has been demonstrated that nutrient controls over priming effects may be more complicated than previously thought, with apparent negative priming potentially being associated with microbes shifting to utilizing more recalcitrant SOM with a narrower C:N ratio (Rousk et al., 2016). The consequences of this type of N-mining response for overall C storage remain unclear, but it has been argued that overall rates of SOM decomposition could decline. Interactions between labile C inputs and microbial nutrient availability may therefore have a range of effects on decomposition and C storage, and at present it is difficult to predict the outcome in different situations.

More recently, attention has shifted to investigating which types of microorganisms are involved in priming and to identifying the sources of the primed C. There is growing evidence that fungi may play a disproportionately important role in priming effects (Fontaine et al., 2011; Rousk et al., 2016). Fungi may play a greater role in the breakdown of complex SOM than bacteria, while the production of hyphae may allow them to explore soil space more effectively giving them a greater SOM-mining capacity. In terms of the sources of the primed C, it is known that C which is 100s or 1000s of years old can be mineralized from subsoils (Fontaine et al., 2007), and it appears slow rates of decomposition in deeper soil layers may be at least partially due to the low rates of labile C input. In addition, mineral-bound (Keiluweit et al., 2015) and aggregated occluded SOM (Tian et al., 2016b) have both been shown to be potentially vulnerable to priming effects. All of these advances mean that priming research is now underpinned by a clear theoretical framework. However, the range of factors controlling the magnitude and direction of priming effects means that predicting consequences for rates of decomposition and soil C storage remains challenging.

Priming effects may be particularly important in determining whether eCO_2 results in C sequestration in soils, as well as whether positive plant growth responses can be sustained in the longer term. Increased nutrient immobilization in plant biomass, may result in productivity becoming progressively limited by the availability of key nutrients (Zaehle et al., 2015). As a result, plants have been shown to allocate greater amounts of C to the rhizosphere, increasing microbial activity and rates of decomposition, with greater rates of microbial enzyme production underpinning this response (Phillips et al., 2011, 2012; Drake et al., 2011). The greater rates of decomposition and nutrient cycling may allow positive plant growth responses to eCO_2 to be maintained, but could reduce, or eliminate, the potential for increases in soil C storage (Phillips et al., 2012; van Groenigen et al., 2014). The inverse response to that observed under eCO_2 may explain why microbial activity declines with N deposition. As N availability increases in ecosystems that were previously N limited, plants may invest less C below-ground for nutrient acquisition (Hogberg et al., 2010; Janssens et al., 2010). This reduction in C supply to the rhizosphere may reduce microbial activity and overall rates of SOM decomposition, potentially allowing SOM to accumulate.

Finally, climate change has the potential to increase plant productivity at high latitudes and altitudes, and also result in shifts in the distribution of different plant species and communities. The Arctic is greening, with greater biomass production rates and taller stature vegetation colonizing new areas (Epstein et al., 2012). The increase in C storage in plant biomass represents a negative feedback to climate change, and models have suggested that greater productivity and soil C inputs should also promote C sequestration in soils (Qian et al., 2010). However, in these ecosystems, soils stored the vast majority of ecosystem C and there is growing evidence that soil C stocks can be lower in more productive Arctic ecosystems (Hartley et al., 2012; Parker et al., 2015). These changes in plant community composition have a wide range of biological and physical impacts on these ecosystems, including impacts on snow dynamics. However, priming effects have been demonstrated as being important in controlling rates of decomposition (Hartley et al., 2012), with species in the *Betula* genus potentially acting as "ecosystem engineers" in terms of increasing rates of C and nutrient cycling (Mitchell et al., 2007, 2010). It does not appear safe to assume that gains in plant biomass at high northern latitudes will result in increases in ecosystem C storage.

In summary, plant-soil interactions associated with links between C and nutrient cycling and priming effects may be extremely important in determining how global change affects soil and ecosystem C storage.

8.3.2.4 N availability

The direct effects of N availability on rates of decomposition in soils has proven to be surprisingly difficult to determine, with seemingly contradictory findings from different ecosystems (Janssens et al., 2010; Mack et al., 2004; De Baets et al., 2016). Even in ecosystems which are considered to be N limited, N additions have been found to increase and decrease microbial activity (De Baets et al., 2016; Bragazza et al., 2012). Increases in enzyme production have been proposed to explain greater rates of SOM decomposition (Bragazza et al., 2012; Chen et al., 2017), while reductions in decomposition may be related to negative priming effects. By reducing microbial N limitation, the mining of rhizosphere SOM for N appears to decline, with microbes becoming increasingly dependent on the added N and labile SOM pools (Zang et al., 2016). Due to the relatively low abundance of labile C, microbial biomass has generally been found to decline, despite increases in CUE (Zang et al., 2016). In addition, reductions in pH (De Baets et al., 2016) and the direct inhibition of the decomposition of low-quality organic matter (Knorr et al., 2005a) may further explain why, in the majority of cases, N addition appears to reduce decomposition rates (Zang et al., 2016). Despite a few observations of increases in decomposition rates with N addition in organic soils, at present it appears that the overall effect of N deposition will probably be to reduce rates of decomposition and increase soil C storage (Janssens et al., 2010).

However, increases in soil C storage need to be placed into the context of potential increases in the emissions of other greenhouse gases. Not all ecosystems are N limited, with the productivity of tropical and subtropical ecosystems in particular being considered to be controlled more by the availability of rock-derived nutrients such as P (Vitousek, 1982). In such ecosystems, N inputs can reduce plant productivity by causing acidification and reducing the availability of other nutrients (Yavitt et al., 2011; Liu et al., 2014). Furthermore, greater rates of nitrification and denitrification will likely promote nitrous oxide (N_2O) emissions (Hall and Matson, 1999; Zheng et al., 2016), and methane (CH_4) oxidation may be inhibited by the toxicity of some nitrogenous compounds to methanotrophs as well as the potential for greater amounts of ammonium (NH_4^+) to competitively inhibit the key enzyme, methane monooxygenase (Bodelier and Steenbergh, 2014). There is growing recognition that N_2O and CH_4 emissions are extremely important in determining whether greenhouse gas fluxes to and from the terrestrial biosphere are currently reducing or increasing rates of climate change (Tian et al., 2016a), and, in terms of feedbacks to climate change, N deposition induced N_2O and CH_4 emissions may more than offset any potential gains in soil C storage (Liu and Greaver, 2009).

In summary, changes in temperature, precipitation, N deposition and eCO_2 are likely to affect C storage in soils. The balance of evidence seems to point to climate change reducing soil C storage and N deposition and eCO_2 increasing soil C storage, at least in mineral soils. However, our limited understanding of how rates of SOM stabilization and destabilization are controlled means that the relative magnitudes of these changes in soil C storage are difficult to predict, with interactions between global change drivers being particularly poorly understood. Importantly, specific microbial groups and changes in microbial physiology, especially CUE, have been identified as playing key roles in determining rates of decomposition. This further demonstrates the challenges involved in developing the mechanistic understanding required for modeling the effects of global change on soil C storage.

8.4 WHAT ARE THE PARAMETERS AND MODEL STRUCTURES USED FOR SIMULATING RESPONSES TO GLOBAL CHANGE?

8.4.1 CONVENTIONAL MODELS BASED ON MULTIPLE SOM POOLS

Many C-cycle models, including CENTURY and RothC, simulate SOM as a series of pools which differ in their intrinsic turnover rates (Jenkinson et al., 1991; Parton et al., 1987). Material is generally added to the more rapidly decomposing pools and is then gradually transferred between pools, as decomposition results in more recalcitrant organic matter being produced (Luo et al., 2016). In some models, there are also transfers back from the more recalcitrant pools to active pools as recalcitrant material is slowly broken down (Parton et al., 1987). Physical and chemical protection of SOM can be represented to some extent by making rates of transfer into the most slowly decomposing pools dependent on soil clay content. The effects of climatic factors on soil C storage can then be investigated by making rates of input and decomposition rates in each of the different pools dependent on temperature and moisture.

Models reflect the strong effects of temperature on decomposition rates using broadly exponential functions, but the specific equation used varies between different models from simple Q_{10} functions, to Arrhenius equations and functions which include thermal optima (Luo et al., 2016). There is more variation in the functions used to simulate relationships between soil moisture and decomposition rates, with the different functions resulting in very different predictions of how climate change will affect soil C storage (Falloon et al., 2011).

Models such as CENTURY and RothC have been extremely important for simulating response to global change, with, e.g., the more complex pool structure of RothC resulting in lower projected soil C losses due to global warming (Jones et al., 2005), than a model which simulated all SOM as a single homogeneous pool (Cox et al., 2000). However, as our understanding of how SOM decomposition and soil C storage has developed, it has become clear that new modeling approaches need to be developed to improve predictions of future changes in soil C storage (Wieder et al., 2015; Campbell and Paustian, 2015). This should not be seen as a criticism of the conventional SOM pool models as they reflected our understanding of SOM dynamics when they were developed and are still widely used today. However, many of the processes and responses described earlier cannot be simulated by conventional models. For example, despite the many different moisture functions, pulses of C release following the rewetting of soils generally cannot be simulated (Falloon et al., 2011; Luo et al., 2016). Furthermore, because microbial biomass is generally only considered as a pool of organic matter, microbial CUE and the production of different enzymes are not considered directly. This makes it difficult to determine the potential effects of changes in the quality and quantity of organic matter inputs on rates of SOM formation and the priming of existing SOM. Finally, the extent to which microbial community composition, and global change drivers, control rates of SOM stabilization versus destabilization needs to be determined and reflected in models (Thornley and Cannell, 2001; Conant et al., 2011; Tang and Riley, 2015). The next generation of SOM models, including those that should be incorporated into Earth systems models, will need to overcome these issues if projections of global change impacts on soil C storage are to be improved (Campbell and Paustian, 2015; Luo et al., 2016).

8.4.2 MICROBIAL MODELS

More recently, models which explicitly represent the role of soil microbial communities in C and nutrient cycling in soils have received a considerable amount of attention (Allison et al., 2010; Wieder et al., 2013). The linkage of the production of microbial biomass and SOM-degrading enzymes to SOM decomposition through Michaelis–Menten kinetics, allows the effects of environmental variables on CUE to be considered explicitly. This allows these models to simulate a much greater range of C-cycle responses to climate change (Tang and Riley, 2015; Wieder et al., 2013). Furthermore, drying and rewetting cycles can be simulated if responses are controlled by microbial biomass and enzyme dynamics. In addition, by also including links between enzyme production and decomposition, and links between C and nutrient cycling, priming effects can be simulated (Kuzyakov, 2010).

For all of these reasons, soil microbe models are attractive—it is possible to represent processes mechanistically that most soil scientists consider to be important in controlling C storage in soils. Furthermore, the incorporation of microbial physiology into SOM models has produced very different projections of changes in soil C storage than conventional models (Tang and Riley, 2015; Allison et al., 2010). To date, a major challenge has been the wide range of predictions that these models have produced and the difficulty in linking model predictions with observations (Allison et al., 2010; Wieder et al., 2013). However, because these microbial models make clear predictions regarding how decomposition rates and C storage should be affected by global change, they are able to generate testable hypotheses that empirical studies should seek to falsify.

8.5 WHAT MODEL STRUCTURES AND PARAMETERS COULD BE INCLUDED TO IMPROVE PREDICTIONS OF SOIL C STORAGE?

8.5.1 LINKS BETWEEN NUTRIENT AND CARBON CYCLING

Classical multipool SOM modeling approaches have been developed over the years to include interactions between C and nutrient cycling. This linkage has been crucial for simulating the effects of global change on C cycling in terrestrial ecosystems (Ciais et al., 2013). The new generation of microbial models have generally started out by considering the C cycle in isolation, while openly acknowledging that the next step must be to include linkages between C and nutrient cycling. Above, it is demonstrated that effects of global change on the relative availability of labile C versus nutrients, especially N, may play a critical role in determining the effects on rates of decomposition and soil C storage. The effects of temperature on microbial CUE (Sinsabaugh et al., 2016), and positive and negative priming effects induced by eCO_2 and N deposition may all be controlled by C to nutrient ratios (Phillips et al., 2011; Zang et al., 2016). Priming effects under eCO_2 will also be critical for controlling nutrient availability to plants and therefore the extent to which progressive nutrient limitation will reduce C uptake by terrestrial ecosystems this century (Drake et al., 2011). The extent to which soil nutrient availability will control responses to eCO_2 remains one of the greatest uncertainties in predicting the magnitude of C-cycle feedbacks to climate change (Ciais et al., 2013), and at present relatively few models can simulate the priming effects which may be key here. Finally, warming-induced increases in decomposition have the potential to increase nutrient availability in soils, resulting in greater plant productivity (Melillo et al., 2011) and more soil C inputs, which could offset the initial C losses (Crowther et al., 2016). This, feedback has been simulated in conventional models, but microbial models must also be able to reflect this response which has the potential to reduce or eliminate warming-induced soil C losses.

8.5.2 SOM STABILIZATION AND DESTABILIZATION

Among the most important developments in our understanding of C cycling in soils has been the recognition that the accessibility of SOM to microbes is more important than the intrinsic decomposability of the organic matter. Models are beginning to investigate ways of simulating interactions between stabilization of SOM and microbial physiology, and there have been recent developments in this area (Wieder et al., 2015; Tang and Riley, 2015). However, it is now 16 years since the potential importance of global change in influencing rates of SOM stabilization and destabilization was first proposed (Thornley and Cannell, 2001). Using a simple modeling approach it was shown that the relative temperature sensitivities of decomposition, SOM stabilization and destabilization control the effects of global warming on soil C storage. For illustrative purposes and to reiterate the importance of incorporating the kinetics of SOM stabilization and destabilization into SOM models, a simple modeling analysis is carried out below.

The model used for this simulation contains just two soil C pools: unprotected and stabilized SOM (Fig. 8.4). The rate of decomposition of the unprotected pool, as well as the rates of stabilization and destabilization can then be made dependent on temperature, allowing the model to investigate how rates of SOM stabilization and C storage may be affected by climate factors. The analysis below includes two minor differences to the model approach outlined by Thornley and Cannell (2001): (1) rather than the stabilized pool decomposing directly to release CO_2 to the atmosphere, destabilization releases C back into the unprotected pool, (2) based on more recent work, a maximum capacity for stabilizing SOM was added (ProCap), with the rate of stabilization being inversely related to how close the soil is to the stabilization capacity (Eq. 8.2) (Castellano et al., 2015). This reflects potential reductions in rates of stabilization as soil particles become saturated with organic matter (Campbell et al., 2015). Monthly rates of decomposition (Eq. 8.1) and stabilization (Eq. 8.2) were made proportional to the size of the unprotected pool (Eq. 8.3), while rates of destabilization were related to the size of the stabilized pool (all parameters are defined in Fig. 8.4 and Table 8.1).

The model parameters for the simulations presented below are in Table 8.1. The effects of a 10°C difference in temperature (ambient vs warmed) on equilibrium soil C pools were investigated under four different scenarios: (1) No protection of SOM; this effectively represents the way in which the effects of temperature on soil C storage were simulated in early Earth system models (Cox et al., 2000), albeit without the multiple pools with different intrinsic turnover rates that slow rates of C loss in responses to warming (Jones et al., 2005); (2) Decomposition, stabilization and destabilization reactions all being equally temperature sensitive ($Q_{10} = 2$); (3) Stabilization reactions being more temperature sensitive than decomposition and twice as temperature sensitive as destabilization (Thornley and Cannell, 2001); and (4) Destabilization reactions being more temperature sensitive than decomposition and twice as temperature sensitive as stabilization (Conant et al., 2011). Across all simulations the temperature sensitivity of the decomposition of unprotected SOM was maintained constant. The results in terms of soil C storage and the size of the stabilized pools are presented in Table 8.2.

Figure 8.4. Schematic showing the structure of the simple SOM model used to aid hypothesis development

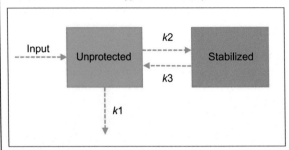

Table 8.1:Summary of the Parameters Used in the Modelling.

Parameters	Scenario 1	Scenario 2	Scenario 3	Scenario 4
Input	5	5	5	5
ProCap	100	100	100	100
k1	0.2	0.2	0.2	0.2
k2	0	0.01	0.01	0.01
k3	0	0.002	0.002	0.002
Tk1	2	2	2	2
Tk2	NA	2	3	1.5
Tk3	NA	2	1.5	3

Input *is the rate of C input per month (mg per gram soil dry weight).*
ProCap *is the maximum stabilization potential of a particular soil type (mg per gram soil dry weight).*
k1, k2, *and* **k3** *are monthly rate constants (multiplied by pool sizes as described in Eqs 8.1 — 8.3 for decomposition, stabilization and destabilization, respectively, with their values under ambient conditions being outlined in Table 8.1.*
Tk1, Tk2, Tk3 *indicate how temperature sensitive the different rate constants are (expressed as $Q_{10}s$; relative change in the reaction rate for a 10°C change in temperature).*

$$\text{Decomposition} = \text{Unprotected C} * k1 \tag{8.1}$$

$$\text{Stabilization} = \text{Unprotected C} *(k2 *(\text{ProCap} - \text{Stabilized C} /\text{ProCap})) \tag{8.2}$$

$$\text{Destabilization} = \text{Stabilized C} *k3 \tag{8.3}$$

Changes in the size of the different C pools can then be described by Eqs. (8.4) and (8.5):

$$\Delta\text{Unprotected C} = (\text{Input} + \text{Destabilization}) - (\text{Decomposition} + \text{Stabilization}) \tag{8.4}$$

$$\Delta\text{Stabilized C} = \text{Stabilization} - \text{Destabilization} \tag{8.5}$$

Ultimately the results demonstrate that it is critical that we determine the relative temperature sensitivities of processes that stabilize and destabilize SOM. Even when the temperature sensitivity of the decomposition of the unprotected pool is held constant, the reduction in soil C storage due to 10°C of warming varies from 16% to 55%, dependent on the kinetics of stabilization versus destabilization. The same argument can be applied to determining the effects on soil C storage of changes in soil moisture, and drying and rewetting cycles in particular, as well eCO_2 and N deposition. Furthermore, in terms of the fate of new C inputs under eCO_2, the consideration of a maximum capacity to stabilize SOM may result in models predicting much lower rates of C sequestration in soils than previously predicted, even before accounting for potential priming effects.

It has been argued that stabilization and destabilization process are abiotic and therefore microbes play little role in controlling them, with soil physical and chemically properties being the key (Schimel and Schaeffer, 2012). However, organic acid additions may destabilize SOM (Tian et al., 2016b; Keiluweit et al., 2015) and microbial CUE will influence the amount of organic

Table 8.2 Summary of the Key Results From the Four Different Model Scenarios Outlined in Table 8.1

Result	Scenario 1		Scenario 2		Scenario 3		Scenario 4	
	Ambient	Warmed	Ambient	Warmed	Ambient	Warmed	Ambient	Warmed
Total soil C (mg gdw^{-1})	25.0	12.5	79.2	50.3	79.2	66.6	79.2	36.0
Temp effect (% reduction in C)	50.0		36.5		15.9		54.5	
Stabilized soil C (mg gdw^{-1})	0.0	0.0	54.9	38.0	54.9	54.6	54.9	23.7

Scenario 1 involves no stabilization; Scenario 2 has equal temperature sensitivities for decomposition, stabilization and destabilization; in Scenario 3 stabilization is more temperature sensitive than decomposition and twice as temperature sensitive as destabilization; in Scenario 4 destabilization is more temperature sensitive than decomposition and twice as temperature sensitive as stabilization. The effects of a 10°C increase in temperature on equilibrium soil C stocks (the rate of C loss is not being considered here, but rather the effect on final C stocks), the percentage reduction in soil C storage caused by warming, and the amount of protected soil C, are all presented. There are clear differences between the different scenarios. In Scenarios 2–4 in which stabilization of SOM is included, despite the temperature sensitivity of SOM decomposition being held constant, differences in the relative temperature sensitivities of stabilization versus destabilization (Table 8.1) result in major differences in terms of the effects of warming on total soil C storage, and the size of the stabilized pool.

matter inputs that remain in soils and can potentially become stabilized. Ultimately, models that include both microbial dynamics and stabilization and destabilization processes may have the greatest potential for simulating the full range of global change impacts on soil C storage. Developing the understanding required to optimize model structure and provide globally-relevant parameterizations requires high levels of collaboration between modelers and empiricists. The modeling analysis presented earlier clearly does not account for links between C and nutrient cycling. Ultimately, key aspects of microbial physiology, stabilization and destabilization, and C and nutrient cycling should all be reflected within a single model. The development and testing of such a model represents a challenge that empiricists and modelers need to work together to achieve.

8.5.3 VERTICAL RESOLUTION IN SOM MODELS

Implicit within the discussion earlier has been the fact that microbial biomass and community composition, rates of new C input, nutrient availability, and total and relative amounts of unprotected versus stabilized SOM vary strongly with soil depth (von Luetzow et al., 2006). Therefore, the effects of all the different global change drivers on soil C storage may be depth-dependent and, thus, having a model that can vertically resolve C and nutrient dynamics may be crucial for predicting changes in soil C storage (Koven et al., 2013). This issue becomes even more important when vertical temperature and moisture profiles control microbial activity due to water-logging, or freezing in the case of permafrost soils. Furthermore, vertical mixing through leaching and precipitation, bioturbation or cryoturbation, and erosion and deposition, may move C and nutrients from layers where they may decompose rapidly due to greater microbial activity, or saturation of stabilization capacity, to layers where they may persist for long time periods (Harden et al., 2012; von Luetzow et al., 2008). Therefore, it is important to be able to accurately reflect the different conditions that characterize contrasting soil layers, and the rates at which C and nutrients move between these layers. Validating models against vertical C stock and radiocarbon data may be extremely valuable in determining whether new models show an enhanced ability to simulate SOM dynamics and C storage (Koven et al., 2013), although developing empirical understanding of how different global change drivers affect vertical transfers will be challenging. In summary, on top of all the other challenges for modeling, it appears that vertical resolution is likely important if models are to simulate the full range of global change impacts on soil C storage.

8.6 CURRENT CHALLENGES TO IMPROVE PREDICTIONS OF THE EFFECTS OF GLOBAL CHANGE MODULATORS ON SOIL C STORAGE, AND FUTURE PERSPECTIVES

8.6.1 MICROBIAL CUE

Throughout this chapter, the efficiency with which microbes use organic matter to produce biomass and enzymes has been emphasized as a key issue. However, CUE is not actually an easy measurement to make, especially when the effects of a global change driver like temperature need to be established. This is because microbial biomass is continuously turning over, and turnover rates are also likely to be affected by global change. For example, in ^{13}C-labeled substrate-addition experiments, CUE is often calculated based on the amount of the ^{13}C label retained in biomass divided by the amount which is lost as CO_2 (Spohn et al., 2016). The assumption, therefore, is that there has been no turnover of biomass or biomass components during the incubation period, an assumption which may not be appropriate, especially at higher temperatures. In addition, different substrates may be used in very different ways potentially telling us little about CUE of the community in general (Frey et al., 2013).

There are now a number of alternative approaches being developed which show promise for determining rates of CUE. To calculate CUE, ultimately it is necessary to quantify rates of microbial respiration versus rates of microbial biomass production. Methods that measure growth rates without adding C have been proposed as overcoming previous limitations, with the incorporation of $^{18}O_2$ from labeled water into microbial DNA having been proposed recently (Spohn et al., 2016). In addition, chemostat approaches which use continuous dilution to measure rates of biomass production may also be able quantify biomass growth rates and respiration rates, and therefore CUE (Min et al., 2016). Furthermore, the addition of low levels of compounds with positionally-labeled C atoms may allow the metabolic pathways being used by microbial communities to be identified, with associated models then calculating CUE (Dijkstra et al., 2011). Finally, there are the stoichiometric approaches mentioned earlier that use the potential activities of C, N, and P cycle enzymes to calculate CUE (Sinsabaugh et al., 2016). Given that the development of many of these methods has been relatively recent, at least in the context of their application to soil science, their relative strengths and weaknesses are still being evaluated. However, the development of a common, robust method for measuring microbial CUE would be extremely valuable for incorporating microbial physiology responses in Earth systems models.

8.6.2 ORGANIC MATTER FORMATION AND STABILIZATION

Traditional litter bag studies have provided key information about how environment conditions affect rates of decomposition. However, while understanding of how the decomposition of unprotected SOM is controlled has advanced considerably, litter bag studies do not allow for investigation of how rates of SOM formation and stabilization are controlled. When litter has been placed directly on the soil surface, it has been shown that more than half of the recorded mass loss may be associated with transfer of fragments or soluble components to the underlying soil, rather than loss as CO_2 (Rubino et al., 2010). The fate of the C transferred to the soils is then critical to determining soil C storage. The mixing of isotopically labeled litter into soils and the tracing of the C into different soil fractions (Cotrufo et al., 2015; Egli et al., 2016) and different components of the microbial community (Radajewski et al., 2000) may provide vital information on how SOM formation and stabilization are controlled. Producing highly-labeled plant residues is expensive but has great potential for advancing understanding. For example, different pathways associated with the stabilization of simple nonstructural compounds versus litter fragments were identified in a study (Cotrufo et al., 2015). The former pathway was considered to be associated with microbial processing while that latter was suggested to be more abiotic. Further study is required to determine the long-term stability of the SOM and how important different microbial communities are in the overall stabilization processes, but it is clear that detailed studies determining the fate of labeled substrates can greatly advance our understanding of how soil C storage is controlled. These types of studies have been carried out for a long time, but it is our growing appreciation of the importance of stabilization and destabilization processes that means that new studies, directed at answering key open questions, are now required. Models that can simulate links between microbial physiology and SOM stabilization are currently being developed (Tang and Riley, 2015; Wieder et al., 2015). The ideal way forward would be for these models to be used to aid hypothesis development and then for novel labeling studies to be developed to tested these hypotheses directly.

8.7 CONCLUSION

In the past decade, major advances have been made in our understanding of how soil C storage is controlled, and models can now simulate a much broader range of responses than was previously possible. Two major developments have been the recognition that it is the accessibility of SOM to microbes, rather than its chemical recalcitrance, that controls long-term storage, and the demonstration that microbial physiological responses, and in particular shifts in CUE, have the potential to strongly influence the effects of global change on soil C storage (Box 8.2). These developments, which can genuinely be referred to as paradigm shifts, have also highlighted major gaps in our empirical understanding of how soil C storage is controlled. In addition, models which incorporate the roles of microbial physiology and/or SOM stabilization dynamics have produced a wide range

---

BOX 8.2 KEY TAKE-HOME MESSAGES

1. Long-term C storage in aerobic mineral soils is now considered to be controlled more by physicochemical protection mechanisms (adsorption onto soil particles, aggregation) than by the intrinsic decomposability of the organic matter itself. Therefore, it is critical to determine how global change will affect physicochemical protection versus destabilization of soil organic matter (SOM), as well as the roles that different microbes play in these key processes. While we now have a much greater understanding of how environmental conditions control rates of decomposition, arguably our understanding of how physicochemical stabilization of SOM is controlled is much less developed.

2. Priming effects, the extent to which new C inputs affect the decomposition of existing SOM, are not simply an academic curiosity, but, rather, they have the potential to control ecosystem C storage, as well as responses to global change. The study of priming effects must bring together understanding of plant-microbe interactions, and links between C and nutrient cycling. Therefore, priming research is at the very center of efforts to improve predictions of how ecosystem and soil C storage will respond to climate change and elevated atmospheric CO_2.

3. Models have demonstrated that the incorporation of potential changes in microbial physiology, especially carbon-use efficiency (CUE), into simulations can have major effects on the predicted magnitude and direction of global change impacts on soil C storage. Thus, the development of laboratory methods than can allow for the direct measurement of microbial CUE and its response to different global change drivers may represent a really key advance. The testing of these new methods, and ideally the development of a single preferred approach, will be extremely valuable.

4. In the past 10 years, our understanding of how soil C storage is controlled has advanced rapidly with greater appreciation of the role of physicochemical protection and microbial physiology. These advances can genuinely be referred to as paradigm shifts, but they have raised as many questions as they have answered. It is really critical that modelers and empiricists now work together to answer these questions, allow the new generation of models to formulate testable hypotheses that can then be falsified with laboratory or field experiments and observations.

5. Global change is multifactor. While we must "walk before we run" and we still lack the required understanding of how single factors (e.g., temperature, soil moisture) affect some of the key processes identified earlier, it is essential that multifactorial experiments also identify interactions between global change drivers and that these can be reflected in Earth systems models.

---

of projections in terms of the effects of global change on soil C storage. Moving forward, testing the predictions of these new generation models with novel and innovative experiments will elucidate the roles of both microbial physiology and stabilization/destabilization in controlling soil C storage.

REFERENCES

Ahlstrom, A., Raupach, M.R., Schurgers, G., Smith, B., Arneth, A., Jung, M., et al., 2015. The dominant role of semi-arid ecosystems in the trend and variability of the land CO_2 sink. Science 348, 895–899.

Allison, S.D., Wallenstein, M.D., Bradford, M.A., 2010. Soil-carbon response to warming dependent on microbial physiology. Nat. Geosci. 3, 336–340.

Andrews, J.A., Schlesinger, W.H., 2001. Soil CO_2 dynamics, acidification, and chemical weathering in a temperate forest with experimental CO_2 enrichment. Global Biogeochem. Cycles 15, 149–162.

Atkin, O.K., Tjoelker, M.G., 2003. Thermal acclimation and the dynamic response of plant respiration to temperature. Trends Plant Sci. 8, 343–351.

Auffret, M.D., Karhu, K., Khachane, A., Dungait, J.A.J., Fraser, F., Hopkins, D.W., et al., 2016. The role of microbial community composition in controlling soil respiration responses to temperature. PLoS One 11, e0165448.

Birch, H.F., 1958. The effect of soil drying on humus decomposition and nitrogen availability. Plant Soil 10, 9–31.

Blazewicz, S.J., Schwartz, E., Firestone, M.K., 2014. Growth and death of bacteria and fungi underlie rainfall-induced carbon dioxide pulses from seasonally dried soil. Ecology 95, 1162–1172.

Bodelier, P.L.E., Steenbergh, A.K., 2014. Interactions between methane and the nitrogen cycle in light of climate change. Curr. Opin. Environ. Sustain. 9–10, 26–36.

Borken, W., Matzner, E., 2009. Reappraisal of drying and wetting effects on C and N mineralization and fluxes in soils. Global Change Biol. 15, 808–824.

Bradford, M.A., 2013. Thermal adaptation of decomposer communities in warming soils. Front. Microbiol. 4, 333.

Bradford, M.A., Davies, C.A., Frey, S.D., Maddox, T.R., Melillo, J.M., Mohan, J.E., et al., 2008. Thermal adaptation of soil microbial respiration to elevated temperature. Ecol. Lett. 11, 1316–1327.

Bradford, M.A., Watts, B.W., Davies, C.A., 2010. Thermal adaptation of heterotrophic soil respiration in laboratory microcosms. Global Change Biol. 16, 1576–1588.

Bragazza, L., Buttler, A., Habermacher, J., Brancaleoni, L., Gerdol, R., Fritze, H., et al., 2012. High nitrogen deposition alters the decomposition of bog plant litter and reduces carbon accumulation. Global Change Biol. 18, 1163–1172.

Burke, E.J., Hartley, I.P., Jones, C.D., 2012. Uncertainties in the global temperature change caused by carbon release from permafrost thawing. Cryosphere 6, 1063–1076.

Campbell, E.E., Paustian, K., 2015. Current developments in soil organic matter modeling and the expansion of model applications: a review. Environ. Res. Lett. 10, 123004.

Castellano, M.J., Mueller, K.E., Olk, D.C., Sawyer, J.E., Six, J., 2015. Integrating plant litter quality, soil organic matter stabilization, and the carbon saturation concept. Global Change Biol. 21, 3200–3209.

Charman, D.J., Beilman, D.W., Blaauw, M., Booth, R.K., Brewer, S., Chambers, F.M., et al., 2013. Climate-related changes in peatland carbon accumulation during the last millennium. Biogeosciences 10, 929–944.

Chen, J., Luo, Y., Li, J., Zhou, X., Cao, J., Wang, R.-W., et al., 2017. Costimulation of soil glycosidase activity and soil respiration by nitrogen addition. Global Change Biol. 23, 1328–1337.

Chen, R., Senbayram, M., Blagodatsky, S., Myachina, O., Dittert, K., Lin, X., et al., 2014. Soil C and N availability determine the priming effect: microbial N mining and stoichiometric decomposition theories. Global Change Biol. 20, 2356–2367.

Cheng, L., Booker, F.L., Tu, C., Burkey, K.O., Zhou, L., Shew, H.D., et al., 2012. Arbuscular mycorrhizal fungi increase organic carbon decomposition under elevated CO_2. Science 337, 1084–1087.

Ciais, P., Sabine, C., Bala, G., Bopp, L., Brovkin, V., Canadell, J., et al., 2013. Carbon and other biogeochemical cycles. In: Stocker, T.F., Qin, D., Plattner, G.-K., Tignor, M., Allen, S.K., Boschung, J., Nauels, A., Xia, Y., Bex, V., Midgley, P.M. (Eds.), Climate Change 2013: The Physical Science Basis. Contribution of Working Group I to the Fifth Assessment Report of the Intergovernmental Panel on Climate Change. Cambridge University Press, Cambridge and New York, NY.

Collins, M., Knutti, R., Arblaster, J., Dufresne, J.-L., Fichefet, T., Friedlingstein, P., et al., 2013. Long-term climate change: projections, commitments and irreversibility. In: Stocker, T.F., Qin, D., Plattner, G.-K., Tignor, M., Allen, S.K., Boschung, J., Nauels, A., Xia, Y., Bex, V., Midgley, P.M. (Eds.), Climate Change 2013: The Physical Science Basis. Contribution of Working Group I to the Fifth Assessment Report of the Intergovernmental Panel on Climate Change. Cambridge University Press, Cambridge and New York, NY.

Conant, R.T., Drijber, R.A., Haddix, M.L., Parton, W.J., Paul, E.A., Plante, A.F., et al., 2008a. Sensitivity of organic matter decomposition to warming varies with its quality. Global Change Biol. 14, 868–877.

Conant, R.T., Steinweg, J.M., Haddix, M.L., Paul, E.A., Plante, A.F., Six, J., 2008b. Experimental warming shows that decomposition temperature sensitivity increases with soil organic matter recalcitrance. Ecology 89, 2384–2391.

Conant, R.T., Ryan, M.G., Agren, G.I., Birge, H.E., Davidson, E.A., Eliasson, P.E., et al., 2011. Temperature and soil organic matter decomposition rates — synthesis of current knowledge and a way forward. Global Change Biol. 17, 3392–3404.

Coolen, M.J.L., Orsi, W.D., 2015. The transcriptional response of microbial communities in thawing Alaskan permafrost soils. Front. Microbiol. 6, 197.

Cooper, M.D.A., Estop-Aragones, C., Fisher, J.P., Thierry, A., Garnett, M.H., Charman, D.J., et al., 2017. Limited contribution of permafrost carbon to methane release from thawing peatlands. Nat. Clim. Change 7, 507–511.

Cotrufo, M.F., Wallenstein, M.D., Boot, C.M., Denef, K., Paul, E., 2013. The Microbial Efficiency-Matrix Stabilization (MEMS) framework integrates plant litter decomposition with soil organic matter stabilization: do labile plant inputs form stable soil organic matter? Global Change Biol. 19, 988–995.

Cotrufo, M.F., Soong, J.L., Horton, A.J., Campbell, E.E., Haddix, M.L., Wall, D.H., et al., 2015. Formation of soil organic matter via biochemical and physical pathways of litter mass loss. Nat. Geosci. 8, 776–779.

Cox, P.M., Betts, R.A., Jones, C.D., Spall, S.A., Totterdell, I.J., 2000. Acceleration of global warming due to carbon-cycle feedbacks in a coupled climate model. Nature 408, 184–187.

Crowther, T.W., Bradford, M.A., 2013. Thermal acclimation in widespread heterotrophic soil microbes. Ecol. Lett. 16, 469–477.

Crowther, T.W., Todd-Brown, K.E.O., Rowe, C.W., Wieder, W.R., Carey, J.C., Machmuller, M.B., et al., 2016. Quantifying global soil carbon losses in response to warming. Nature 540, 104–108.

Dakora, F.D., Phillips, D.A., 2002. Root exudates as mediators of mineral acquisition in low-nutrient environments. Plant Soil 245, 35–47.

Davidson, E.A., Janssens, I.A., 2006. Temperature sensitivity of soil carbon decomposition and feedbacks to climate change. Nature 440, 165–173.

De Baets, S., van de Weg, M.J., Lewis, R., Steinberg, N., Meersmans, J., Quine, T.A., et al., 2016. Investigating the controls on soil organic matter decomposition in tussock tundra soil and permafrost after fire. Soil Biol. Biochem. 99, 108–116.

Dijkstra, P., Thomas, S.C., Heinrich, P.L., Koch, G.W., Schwartz, E., Hungate, B.A., 2011. Effect of temperature on metabolic activity of intact microbial communities: evidence for altered metabolic pathway activity but not for increased maintenance respiration and reduced carbon use efficiency. Soil Biol. Biochem. 43, 2023–2031.

Drake, J.E., Gallet-Budynek, A., Hofmockel, K.S., Bernhardt, E.S., Billings, S.A., Jackson, R.B., et al., 2011. Increases in the flux of carbon belowground stimulate nitrogen uptake and sustain the long-term enhancement of forest productivity under elevated CO_2. Ecol. Lett. 14, 349–357.

Dungait, J.A.J., Hopkins, D.W., Gregory, A.S., Whitmore, A.P., 2012. Soil organic matter turnover is governed by accessibility not recalcitrance. Global Change Biol. 18, 1781–1796.

Egli, M., Hafner, S., Derungs, C., Ascher-Jenull, J., Camin, F., Sartori, G., et al., 2016. Decomposition and stabilisation of Norway spruce needle-derived material in Alpine soils using a C-13-labelling approach in the field. Biogeochemistry 131, 321–338.

Epstein, H.E., Raynolds, M.K., Walker, D.A., Bhatt, U.S., Tucker, C.J., Pinzon, J.E., 2012. Dynamics of aboveground phytomass of the circumpolar Arctic tundra during the past three decades. Environ. Res. Lett. 7, 015506.

Evans, S.E., Wallenstein, M.D., 2014. Climate change alters ecological strategies of soil bacteria. Ecol. Lett. 17, 155–164.

Falloon, P., Jones, C.D., Ades, M., Paul, K., 2011. Direct soil moisture controls of future global soil carbon changes: an important source of uncertainty. Global Biogeochem. Cycles 25, GB3010.

Fenner, N., Freeman, C., 2011. Drought-induced carbon loss in peatlands. Nat. Geosci. 4, 895–900.

Fierer, N., Schimel, J.P., 2003. A proposed mechanism for the pulse in carbon dioxide production commonly observed following the rapid rewetting of a dry soil. Soil Sci. Soc. Am. J. 67, 798–805.

Fleischer, K., Rebel, K.T., Van Der Molen, M.K., Erisman, J.W., Wassen, M.J., Van loon, E.E., et al., 2013. The contribution of nitrogen deposition to the photosynthetic capacity of forests. Global Biogeochem. Cycles 27, 187–199.

Fontaine, S., Mariotti, A., Abbadie, L., 2003. The priming effect of organic matter: a question of microbial competition? Soil Biol. Biochem. 35, 837–843.

Fontaine, S., Bardoux, G., Abbadie, L., Mariotti, A., 2004. Carbon input to soil may decrease soil carbon content. Ecol. Lett. 7, 314–320.

Fontaine, S., Barot, S., Barre, P., Bdioui, N., Mary, B., Rumpel, C., 2007. Stability of organic carbon in deep soil layers controlled by fresh carbon supply. Nature 450, 277–280.

Fontaine, S., Henault, C., Aamor, A., Bdioui, N., Bloor, J.M.G., Maire, V., et al., 2011. Fungi mediate long term sequestration of carbon and nitrogen in soil through their priming effect. Soil Biol. Biochem. 43, 86–96.

Fowler, D., Coyle, M., Skiba, U., Sutton, M.A., Cape, J.N., Reis, S., et al., 2013. The global nitrogen cycle in the twenty-first century. Philos. Trans. R. Soc. B-Biol. Sci. 368, 20130164.

Freeman, C., Ostle, N., Kang, H., 2001. An enzymic 'latch' on a global carbon store – a shortage of oxygen locks up carbon in peatlands by restraining a single enzyme. Nature 409, 149.

Freeman, C., Ostle, N.J., Fenner, N., Kang, H., 2004. A regulatory role for phenol oxidase during decomposition in peatlands. Soil Biol. Biochem. 36, 1663–1667.

Frey, S.D., Lee, J., Melillo, J.M., Six, J., 2013. The temperature response of soil microbial efficiency and its feedback to climate. Nat. Clim. Change 3, 395–398.

Galloway, J.N., Dentener, F.J., Capone, D.G., Boyer, E.W., Howarth, R.W., Seitzinger, S.P., et al., 2004. Nitrogen cycles: past, present, and future. Biogeochemistry 70, 153–226.

Gao, X., Schlosser, C.A., Sokolov, A., Anthony, K.W., Zhuang, Q., Kicklighter, D., 2013. Permafrost degradation and methane: low risk of biogeochemical climate-warming feedback. Environ. Res. Lett. 8, 035014.

Giardina, C.P., Ryan, M.G., 2000. Evidence that decomposition rates of organic carbon in mineral soil do not vary with temperature. Nature 404, 858–861.

van Groenigen, K.J., Qi, X., Osenberg, C.W., Luo, Y., Hungate, B.A., 2014. Faster decomposition under increased atmospheric CO_2 limits soil carbon storage. Science 344, 508–509.

van Groenigen, K.J., Xia, J., Osenberg, C.W., Luo, Y., Hungate, B.A., 2015. Application of a two-pool model to soil carbon dynamics under elevated CO_2. Global Change Biol. 21, 4293–4297.

Hall, S.J., Matson, P.A., 1999. Nitrogen oxide emissions after nitrogen additions in tropical forests. Nature 400, 152–155.

Harden, J.W., Koven, C.D., Ping, C.L., Hugelius, G., Mcguire, A.D., Camill, P., et al., 2012. Field information links permafrost carbon to physical vulnerabilities of thawing. Geophys. Res. Lett. 39, L15704.

Hartley, I.P., Ineson, P., 2008. Substrate quality and the temperature sensitivity of soil organic matter decomposition. Soil Biol. Biochem. 40, 1567–1574.

Hartley, I.P., Heinemeyer, A., Ineson, P., 2007. Effects of three years of soil warming and shading on the rate of soil respiration: substrate availability and not thermal acclimation mediates observed response. Global Change Biol. 13, 1761–1770.

Hartley, I.P., Hopkins, D.W., Garnett, M.H., Sommerkorn, M., Wookey, P.A., 2008. Soil microbial respiration in arctic soil does not acclimate to temperature. Ecol. Lett. 11, 1092–1100.

Hartley, I.P., Hopkins, D.W., Sommerkorn, M., Wookey, P.A., 2010. The response of organic matter mineralisation to nutrient and substrate additions in sub-arctic soils. Soil Biol. Biochem. 42, 92–100.

Hartley, I.P., Garnett, M.H., Sommerkorn, M., Hopkins, D.W., Fletcher, B.J., Sloan, V.L., et al., 2012. A potential loss of carbon associated with greater plant growth in the European Arctic. Nat. Clim. Change 2, 875–879.

Hartmann, D.L., Klein Tank, A.M.G., Rusticucci, M., Alexander, L.V., Brönnimann, S., Charabi, Y., et al., 2013. Observations: atmosphere and surface. In: Stocker, T.F., Qin, D., Plattner, G.-K., Tignor, M., Allen, S.K.,

Boschung, J., Nauels, A., Xia, Y., Bex, V., Midgley, P.M. (Eds.), Climate Change 2013: The Physical Science Basis. Contribution of Working Group I to the Fifth Assessment Report of the Intergovernmental Panel on Climate Change. Cambridge University Press, Cambridge and New York, NY.

He, Y., Trumbore, S.E., Torn, M.S., Harden, J.W., Vaughn, L.J.S., Allison, S.D., et al., 2016. Radiocarbon constraints imply reduced carbon uptake by soils during the 21st century. Science 353, 1419−1424.

Hogberg, M.N., Briones, M.J.I., Keel, S.G., Metcalfe, D.B., Campbell, C., Midwood, A.J., et al., 2010. Quantification of effects of season and nitrogen supply on tree below-ground carbon transfer to ectomycorrhizal fungi and other soil organisms in a boreal pine forest. New. Phytol. 187, 485−493.

Hopkins, F.M., Torn, M.S., Trumbore, S.E., 2012. Warming accelerates decomposition of decades-old carbon in forest soils. Proc. Natl. Acad. Sci. U.S.A. 109, E1753−E1761.

Hugelius, G., Strauss, J., Zubrzycki, S., Harden, J.W., Schuur, E.A.G., Ping, C.L., et al., 2014. Estimated stocks of circumpolar permafrost carbon with quantified uncertainty ranges and identified data gaps. Biogeosciences 11, 6573−6593.

Hultman, J., Waldrop, M.P., Mackelprang, R., David, M.M., Mcfarland, J., Blazewicz, S.J., et al., 2015. Multi-omics of permafrost, active layer and thermokarst bog soil microbiomes. Nature 521, 208−212.

Janssens, I.A., Dieleman, W., Luyssaert, S., Subke, J.A., Reichstein, M., Ceulemans, R., et al., 2010. Reduction of forest soil respiration in response to nitrogen deposition. Nat. Geosci. 3, 315−322.

Jansson, J.K., Tas, N., 2014. The microbial ecology of permafrost. Nat. Rev. Microbiol. 12, 414−425.

Jenkinson, D.S., Adams, D.E., Wild, A., 1991. Model estimates of CO_2 emissions from soil in response to global warming. Nature 351, 304−306.

Jones, C., Mcconnell, C., Coleman, K., Cox, P., Falloon, P., Jenkinson, D., et al., 2005. Global climate change and soil carbon stocks; predictions from two contrasting models for the turnover of organic carbon in soil. Global Change Biol. 11, 154−166.

Kaiser, M., Kleber, M., Berhe, A.A., 2015. How air-drying and rewetting modify soil organic matter characteristics: an assessment to improve data interpretation and inference. Soil Biol. Biochem. 80, 324−340.

Kallenbach, C.M., Frey, S.D., Grandy, A.S., 2016. Direct evidence for microbial-derived soil organic matter formation and its ecophysiological controls. Nat. Commun. 7, 13630.

Karhu, K., Auffret, M.D., Dungait, J.A.J., Hopkins, D.W., Prosser, J.I., Singh, B.K., et al., 2014. Temperature sensitivity of soil respiration rates enhanced by microbial community response. Nature 513, 81−84.

Keiluweit, M., Bougoure, J.J., Nico, P.S., Pett-Ridge, J., Weber, P.K., Kleber, M., 2015. Mineral protection of soil carbon counteracted by root exudates. Nat. Clim. Change 5, 588−595.

Kirschbaum, M.U.F., 1995. The temperature-dependence of soil organic-matter decomposition, and the effect of global warming on soil Organic-C storage. Soil Biol. Biochem. 27, 753−760.

Kirschbaum, M.U.F., 2004. Soil respiration under prolonged soil warming: are rate reductions caused by acclimation or substrate loss? Global Change Biol. 10, 1870−1877.

Kirschbaum, M.U.F., 2006. The temperature dependence of organic-matter decomposition − still a topic of debate. Soil Biol. Biochem. 38, 2510−2518.

Knorr, M., Frey, S.D., Curtis, P.S., 2005a. Nitrogen additions and litter decomposition: a meta-analysis. Ecology 86, 3252−3257.

Knorr, W., Prentice, I.C., House, J.I., Holland, E.A., 2005b. Long-term sensitivity of soil carbon turnover to warming. Nature 433, 298−301.

Koven, C.D., Riley, W.J., Subin, Z.M., Tang, J.Y., Torn, M.S., Collins, W.D., et al., 2013. The effect of vertically resolved soil biogeochemistry and alternate soil C and N models on C dynamics of CLM4. Biogeosciences 10, 7109−7131.

Koven, C.D., Schuur, E.A.G., Schaedel, C., Bohn, T.J., Burke, E.J., Chen, G., et al., 2015. A simplified, data-constrained approach to estimate the permafrost carbon-climate feedback. Philos. Trans. R. Soc. A Math. Phys. Eng. Sci. 373, 20140423.

Kuzyakov, Y., 2002. Review: Factors affecting rhizosphere priming effects. J. Plant Nutr. Soil Sci. 165, 382–396.

Kuzyakov, Y., 2010. Priming effects: interactions between living and dead organic matter. Soil Biol. Biochem. 42, 1363–1371.

Kuzyakov, Y., Friedel, J.K., Stahr, K., 2000. Review of mechanisms and quantification of priming effects. Soil Biol. Biochem. 32, 1485–1498.

Land, M., Hauser, L., Jun, S.R., Nookaew, I., Leuze, M.R., Ahn, T.H., et al., 2015. Insights from 20 years of bacterial genome sequencing. Funct. Integr. Genomics 15, 141–161.

Larmola, T., Bubier, J.L., Kobyljanec, C., Basiliko, N., Juutinen, S., Humphreys, E., et al., 2013. Vegetation feedbacks of nutrient addition lead to a weaker carbon sink in an ombrotrophic bog. Global Change Biol. 19, 3729–3739.

Lawrence, D.M., Koven, C.D., Swenson, S.C., Riley, W.J., Slater, A.G., 2015. Permafrost thaw and resulting soil moisture changes regulate projected high-latitude CO_2 and CH_4 emissions. Environ. Res. Lett. 10, 11.

Leakey, A.D.B., Ainsworth, E.A., Bernacchi, C.J., Rogers, A., Long, S.P., Ort, D.R., 2009. Elevated CO_2 effects on plant carbon, nitrogen, and water relations: six important lessons from FACE. J. Exp. Bot. 60, 2859–2876.

Liski, J., Ilvesniemi, H., Makela, A., Westman, C.J., 1999. CO_2 emissions from soil in response to climatic warming are overestimated – the decomposition of old soil organic matter is tolerant of temperature. Ambio 28, 171–174.

Liu, J.X., Zhang, D.Q., Huang, W.J., Zhou, G.Y., Li, Y.L., Liu, S.Z., 2014. Quantify the loss of major ions induced by CO_2 enrichment and nitrogen addition in subtropical model forest ecosystems. J. Geophys. Res.-Biogeosci. 119, 676–686.

Liu, L.L., Greaver, T.L., 2009. A review of nitrogen enrichment effects on three biogenic GHGs: the CO_2 sink may be largely offset by stimulated N_2O and CH_4 emission. Ecol. Lett. 12, 1103–1117.

Liu, W.X., Zhang, Z., Wan, S.Q., 2009. Predominant role of water in regulating soil and microbial respiration and their responses to climate change in a semiarid grassland. Global Change Biol. 15, 184–195.

von Luetzow, M., Kogel-Knabner, I., Ekschmitt, K., Matzner, E., Guggenberger, G., Marschner, B., et al., 2006. Stabilization of organic matter in temperate soils: mechanisms and their relevance under different soil conditions – a review. Eur. J. Soil Sci. 57, 426–445.

von Luetzow, M., Kogel-Knabner, I., Ludwig, B., Matzner, E., Flessa, H., Ekschmitt, K., et al., 2008. Stabilization mechanisms of organic matter in four temperate soils: development and application of a conceptual model. J. Plant Nutr. Soil Sci. 171, 111–124.

Luo, Y.Q., Ahlstrom, A., Allison, S.D., Batjes, N.H., Brovkin, V., Carvalhais, N., et al., 2016. Toward more realistic projections of soil carbon dynamics by Earth system models. Global Biogeochem. Cycles 30, 40–56.

Mack, M.C., Schuur, E.A.G., Bret-Harte, M.S., Shaver, G.R., Chapin, F.S., 2004. Ecosystem carbon storage in arctic tundra reduced by long-term nutrient fertilization. Nature 431, 440–443.

Mackelprang, R., Waldrop, M.P., Deangelis, K.M., David, M.M., Chavarria, K.L., Blazewicz, S.J., et al., 2011. Metagenomic analysis of a permafrost microbial community reveals a rapid response to thaw. Nature 480, 368–371.

Mackelprang, R., Saleska, S.R., Jacobsen, C.S., Jansson, J.K., Tas, N., 2016. Permafrost meta-omics and climate change. In: Jeanloz, R., Freeman, K.H. (Eds.) Annual Review of Earth and Planetary Sciences, Vol 44.

Manzoni, S., Schaeffer, S.M., Katul, G., Porporato, A., Schimel, J.P., 2014. A theoretical analysis of microbial eco-physiological and diffusion limitations to carbon cycling in drying soils. Soil Biol. Biochem. 73, 69–83.

Matulich, K.L., Weihe, C., Allison, S.D., Amend, A.S., Berlemont, R., Goulden, M.L., et al., 2015. Temporal variation overshadows the response of leaf litter microbial communities to simulated global change. ISME J. 9, 2477–2489.

McCalley, C.K., Woodcroft, B.J., Hodgkins, S.B., Wehr, R.A., Kim, E.H., Mondav, R., et al., 2014. Methane dynamics regulated by microbial community response to permafrost thaw. Nature 514, 478−481.

Melillo, J.M., Steudler, P.A., Aber, J.D., Newkirk, K., Lux, H., Bowles, F.P., et al., 2002. Soil warming and carbon-cycle feedbacks to the climate system. Science 298, 2173−2176.

Melillo, J.M., Butler, S., Johnson, J., Mohan, J., Steudler, P., Lux, H., et al., 2011. Soil warming, carbon-nitrogen interactions, and forest carbon budgets. Proc. Natl. Acad. Sci. U.S.A. 108, 9508−9512.

Midgley, M.G., Phillips, R.P., 2016. Resource stoichiometry and the biogeochemical consequences of nitrogen deposition in a mixed deciduous forest. Ecology 97, 3369−3378.

Min, K., Lehmeier, C.A., Ballantyne, F., Billings, S.A., 2016. Carbon availability modifies temperature responses of heterotrophic microbial respiration, carbon uptake affinity, and stable carbon isotope discrimination. Front. Microbiol. 7, 2083.

Mitchell, R.J., Campbell, C.D., Chapman, S.J., Osler, G.H.R., Vanbergen, A.J., Ross, L.C., et al., 2007. The cascading effects of birch on heather moorland: a test for the top-down control of an ecosystem engineer. J. Ecol. 95, 540−554.

Mitchell, R.J., Campbell, C.D., Chapman, S.J., Cameron, C.M., 2010. The ecological engineering impact of a single tree species on the soil microbial community. J. Ecol. 98, 50−61.

Mondav, R., Woodcroft, B.J., Kim, E.H., McCalley, C.K., Hodgkins, S.B., Crill, P.M., et al., 2014. Discovery of a novel methanogen prevalent in thawing permafrost. Nat. Commun. 5, 3212.

Nazaries, L., Tottey, W., Robinson, L., Khachane, A., Abu Al-Soud, W., Sorensen, S., et al., 2015. Shifts in the microbial community structure explain the response of soil respiration to land-use change but not to climate warming. Soil Biol. Biochem. 89, 123−134.

Nie, M., Lu, M., Bell, J., Raut, S., Pendall, E., 2013a. Altered root traits due to elevated CO_2: a meta-analysis. Global Ecol. Biogeogr. 22, 1095−1105.

Nie, M., Pendall, E., Bell, C., Gasch, C.K., Raut, S., Tamang, S., et al., 2013b. Positive climate feedbacks of soil microbial communities in a semi-arid grassland. Ecol. Lett. 16, 234−241.

NOAA National Centers for Environmental Information, 2017. State of the climate: global climate report for annual 2016. < https://www.ncdc.noaa.gov/sotc/global/201613 > (accessed 01.12.17).

Parker, T.C., Subke, J.-A., Wookey, P.A., 2015. Rapid carbon turnover beneath shrub and tree vegetation is associated with low soil carbon stocks at a subarctic treeline. Global Change Biol. 21, 2070−2081.

Parton, W.J., Schimel, D.S., Cole, C.V., Ojima, D.S., 1987. Analysis of factors controlling soil organic-matter levels in great-plains grasslands. Soil Sci. Soc. Am. J. 51, 1173−1179.

Phillips, R.P., Finzi, A.C., Bernhardt, E.S., 2011. Enhanced root exudation induces microbial feedbacks to N cycling in a pine forest under long-term CO_2 fumigation. Ecol. Lett. 14, 187−194.

Phillips, R.P., Meier, I.C., Bernhardt, E.S., Grandy, A.S., Wickings, K., Finzi, A.C., 2012. Roots and fungi accelerate carbon and nitrogen cycling in forests exposed to elevated CO_2. Ecol. Lett. 15, 1042−1049.

Placella, S.A., Brodie, E.L., Firestone, M.K., 2012. Rainfall-induced carbon dioxide pulses result from sequential resuscitation of phylogenetically clustered microbial groups. Proc. Natl. Acad. Sci. U.S.A. 109, 10931−10936.

Poulter, B., Frank, D., Ciais, P., Myneni, R.B., Andela, N., Bi, J., et al., 2014. Contribution of semi-arid ecosystems to interannual variability of the global carbon cycle. Nature 509, 600−603.

Qian, H., Joseph, R., Zeng, N., 2010. Enhanced terrestrial carbon uptake in the Northern High Latitudes in the 21st century from the Coupled Carbon Cycle Climate Model Intercomparison Project model projections. Global Change Biol. 16, 641−656.

Qiao, N., Schaefer, D., Blagodatskaya, E., Zou, X., Xu, X., Kuzyakov, Y., 2014. Labile carbon retention compensates for CO_2 released by priming in forest soils. Global Change Biol. 20, 1943−1954.

Radajewski, S., Ineson, P., Parekh, N.R., Murrell, J.C., 2000. Stable-isotope probing as a tool in microbial ecology. Nature 403, 646−649.

Riggs, C.E., Hobbie, S.E., Bach, E.M., Hofmockel, K.S., Kazanski, C.E., 2015. Nitrogen addition changes grassland soil organic matter decomposition. Biogeochemistry 125, 203−219.

Rousk, K., Michelsen, A., Rousk, J., 2016. Microbial control of soil organic matter mineralization responses to labile carbon in subarctic climate change treatments. Global Change Biol. 22, 4150−4161.

Rubino, M., Dungait, J.A.J., Evershed, R.P., Bertolini, T., De Angelis, P., D'Onofrio, A., et al., 2010. Carbon input belowground is the major C flux contributing to leaf litter mass loss: evidences from a C-13 labelled-leaf litter experiment. Soil Biol. Biochem. 42, 1009−1016.

Rustad, L.E., Campbell, J.L., Marion, G.M., Norby, R.J., Mitchell, M.J., Hartley, A.E., et al., 2001. A meta-analysis of the response of soil respiration, net nitrogen mineralization, and aboveground plant growth to experimental ecosystem warming. Oecologia 126, 543−562.

Schadel, C., Schuur, E.A.G., Bracho, R., Elberling, B., Knoblauch, C., Lee, H., et al., 2014. Circumpolar assessment of permafrost C quality and its vulnerability over time using long-term incubation data. Global Change Biol. 20, 641−652.

Schadel, C., Bader, M.K.F., Schuur, E.A.G., Biasi, C., Bracho, R., Capek, P., et al., 2016. Potential carbon emissions dominated by carbon dioxide from thawed permafrost soils. Nat. Clim. Change 6, 950−953.

Schaefer, K., Lantuit, H., Romanovsky, V.E., Schuur, E.A.G., Witt, R., 2014. The impact of the permafrost carbon feedback on global climate. Environ. Res. Lett. 9, 085003.

Schimel, J., Balser, T.C., Wallenstein, M., 2007. Microbial stress-response physiology and its implications for ecosystem function. Ecology 88, 1386−1394.

Schimel, J.P., Schaeffer, S.M., 2012. Microbial control over carbon cycling in soil. Front. Microbiol. 3, 348.

Schimel, J.P., Wetterstedt, J.A.M., Holden, P.A., Trumbore, S.E., 2011. Drying/rewetting cycles mobilize old C from deep soils from a California annual grassland. Soil Biol. Biochem. 43, 1101−1103.

Schmidt, M.W.I., Torn, M.S., Abiven, S., Dittmar, T., Guggenberger, G., Janssens, I.A., et al., 2011. Persistence of soil organic matter as an ecosystem property. Nature 478, 49−56.

Schmidt, S.K., Costello, E.K., Nemergut, D.R., Cleveland, C.C., Reed, S.C., Weintraub, M.N., et al., 2007. Biogeochemical consequences of rapid microbial turnover and seasonal succession in soil. Ecology 88, 1379−1385.

Schuur, E.A.G., Vogel, J.G., Crummer, K.G., Lee, H., Sickman, J.O., Osterkamp, T.E., 2009. The effect of permafrost thaw on old carbon release and net carbon exchange from tundra. Nature 459, 556−559.

Schuur, E.A.G., Abbott, B.W., Bowden, W.B., Brovkin, V., Camill, P., Canadell, J.G., et al., 2013. Expert assessment of vulnerability of permafrost carbon to climate change. Clim. Change. 119, 359−374.

Schuur, E.A.G., Mcguire, A.D., Schadel, C., Grosse, G., Harden, J.W., Hayes, D.J., et al., 2015. Climate change and the permafrost carbon feedback. Nature 520, 171−179.

Schwalm, C.R., Williams, C.A., Schaefer, K., Arneth, A., Bonal, D., Buchmann, N., et al., 2010. Assimilation exceeds respiration sensitivity to drought: a FLUXNET synthesis. Global Change Biol. 16, 657−670.

Singh, B.K., Millard, P., Whiteley, A.S., Murrell, J.C., 2004. Unravelling rhizosphere-microbial interactions: opportunities and limitations. Trends Microbiol. 12, 386−393.

Singh, B.K., Bardgett, R.D., Smith, P., Reay, D.S., 2010. Microorganisms and climate change: terrestrial feedbacks and mitigation options. Nat. Rev. Microbiol. 8, 779−790.

Sinsabaugh, R.L., Turner, B.L., Talbot, J.M., Waring, B.G., Powers, J.S., Kuske, C.R., et al., 2016. Stoichiometry of microbial carbon use efficiency in soils. Ecol. Monogr. 86, 172−189.

Smith, L.C., Sheng, Y., Macdonald, G.M., Hinzman, L.D., 2005. Disappearing Arctic lakes. Science 308, 1429.

Spohn, M., Klaus, K., Wanek, W., Richter, A., 2016. Microbial carbon use efficiency and biomass turnover times depending on soil depth − implications for carbon cycling. Soil Biol. Biochem. 96, 74−81.

Tang, J.Y., Riley, W.J., 2015. Weaker soil carbon-climate feedbacks resulting from microbial and abiotic interactions. Nat. Clim. Change 5, 56−60.

Terrer, C., Vicca, S., Hungate, B.A., Phillips, R.P., Prentice, I.C., 2016. Mycorrhizal association as a primary control of the CO_2 fertilization effect. Science 353, 72−74.

Thornley, J.H.M., Cannell, M.G.R., 2001. Soil carbon storage response to temperature: an hypothesis. Ann. Bot. 87, 591−598.

Tian, H., Lu, C., Ciais, P., Michalak, A.M., Canadell, J.G., Saikawa, E., et al., 2016a. The terrestrial biosphere as a net source of greenhouse gases to the atmosphere. Nature 531, 225−228.

Tian, J., Pausch, J., Yu, G.R., Blagodatskaya, E., Kuzyakov, Y., 2016b. Aggregate size and glucose level affect priming sources: a three-source-partitioning study. Soil Biol. Biochem. 97, 199−210.

Townsend, A.R., Vitousek, P.M., Desmarais, D.J., Tharpe, A., 1997. Soil carbon pool structure and temperature sensitivity inferred using CO_2 and (13)CO_2 incubation fluxes from five Hawaiian soils. Biogeochemistry 38, 1−17.

Treat, C.C., Natali, S.M., Ernakovich, J., Iversen, C.M., Lupascu, M., Mcguire, A.D., et al., 2015. A pan-Arctic synthesis of CH_4 and CO_2 production from anoxic soil incubations. Global Change Biol. 21, 2787−2803.

Van Oijen, M., Schapendonk, A., Hoglind, M., 2010. On the relative magnitudes of photosynthesis, respiration, growth and carbon storage in vegetation. Ann. Bot. 105, 793−797.

Vicca, S., Fivez, L., Kockelbergh, F., Van Pelt, D., Segers, J.J.R., Meire, P., et al., 2009. No signs of thermal acclimation of heterotrophic respiration from peat soils exposed to different water levels. Soil Biol. Biochem. 41, 2014−2016.

Vitousek, P., 1982. Nutrient cycling and nutrient use efficiency. Am. Nat. 119, 553−572.

Watts, J.D., Kimball, J.S., Bartsch, A., Mcdonald, K.C., 2014. Surface water inundation in the boreal-Arctic: potential impacts on regional methane emissions. Environ. Res. Lett. 9, 075001.

Wendel, J.F., Jackson, S.A., Meyers, B.C., Wing, R.A., 2016. Evolution of plant genome architecture. Genome Biol. 17, 37.

Wieder, W.R., Bonan, G.B., Allison, S.D., 2013. Global soil carbon projections are improved by modelling microbial processes. Nat. Clim. Change 3, 909−912.

Wieder, W.R., Grandy, A.S., Kallenbach, C.M., Taylor, P.G., Bonan, G.B., 2015. Representing life in the Earth system with soil microbial functional traits in the MIMICS model. Geosci. Model Dev. 8, 1789−1808.

Wiesmeier, M., Hubner, R., Sporlein, P., Geuss, U., Hangen, E., et al., 2014. Carbon sequestration potential of soils in southeast Germany derived from stable soil organic carbon saturation. Global Change Biol. 20, 653−665.

Wilson, G.W.T., Rice, C.W., Rillig, M.C., Springer, A., Hartnett, D.C., 2009. Soil aggregation and carbon sequestration are tightly correlated with the abundance of arbuscular mycorrhizal fungi: results from long-term field experiments. Ecol. Lett. 12, 452−461.

Xiang, S.R., Doyle, A., Holden, P.A., Schimel, J.P., 2008. Drying and rewetting effects on C and N mineralization and microbial activity in surface and subsurface California grassland soils. Soil Biol. Biochem. 40, 2281−2289.

Xue, K., Yuan, M.M., Shi, Z.J., Qin, Y.J., Deng, Y., Cheng, L., et al., 2016. Tundra soil carbon is vulnerable to rapid microbial decomposition under climate warming. Nat. Clim. Change 6, 595−600.

Yavitt, J.B., Harms, K.E., Garcia, M.N., Mirabello, M.J., Wright, S.J., 2011. Soil fertility and fine root dynamics in response to 4 years of nutrient (N, P, K) fertilization in a lowland tropical moist forest, Panama. Austral Ecol. 36, 433−445.

Yu, Z.C., 2012. Northern peatland carbon stocks and dynamics: a review. Biogeosciences 9, 4071−4085.

Zaehle, S., Medlyn, B.E., De kauwe, M.G., Walker, A.P., Dietze, M.C., Hickler, T., et al., 2014. Evaluation of 11 terrestrial carbon-nitrogen cycle models against observations from two temperate Free-Air CO_2 Enrichment studies. New Phytol. 202, 803−822.

Zaehle, S., Jones, C.D., Houlton, B., Lamarque, J.F., Robertson, E., 2015. Nitrogen availability reduces CMIP5 projections of twenty-first-century land carbon uptake. J. Clim. 28, 2494–2511.

Zang, H.D., Wang, J.Y., Kuzyakov, Y., 2016. N fertilization decreases soil organic matter decomposition in the rhizosphere. Appl. Soil Ecol. 108, 47–53.

Zheng, M.H., Zhang, T., Liu, L., Zhu, W.X., Zhang, W., Mo, J.M., 2016. Effects of nitrogen and phosphorus additions on nitrous oxide emission in a nitrogen-rich and two nitrogen-limited tropical forests. Biogeosciences 13, 3503–3517.

CHAPTER

PROJECTING SOIL C UNDER FUTURE CLIMATE AND LAND-USE SCENARIOS (MODELING)

9

Marta Dondini[1], Mohamed Abdalla[1], Fitri K. Aini[2], Fabrizio Albanito[1], Marvin R. Beckert[1], Khadiza Begum[1], Alison Brand[1], Kun Cheng[3], Louis-Pierre Comeau[4], Edward O. Jones[1], Jennifer A. Farmer[1], Diana M.S. Feliciano[1], Nuala Fitton[1], Astley Hastings[1], Dagmar N. Henner[1], Matthias Kuhnert[1], Dali R. Nayak[1], Joseph Oyesikublakemore[1], Laura Phillips[5], Mark I.A. Richards[1], Vianney Tumwesige[6], William F.A. van Dijk[7], Sylvia H. Vetter[1], Kevin Coleman[8], Joanne Smith[1] and Pete Smith[1]

[1]University of Aberdeen, Aberdeen, United Kingdom [2]CIFOR—Center for International Forestry Research Bogor, Indonesia [3]Nanjing Agricultural University, China [4]Fredericton Research and Development Centre, Fredericton, Canada [5]World Food Programme, Dhaka, Bangladesh [6]African Centre for Clean Air, Kampala, Uganda [7]Province of Noord-Holland, Haarlem, The Netherlands [8]Rothamsted Research, Harpenden, United Kingdom

9.1 INTRODUCTION

Soils globally represent the most significant long-term organic carbon (C) store in terrestrial ecosystems, containing 4.5 times as much C as all living biomass and 3.1 times as much as the atmosphere (McClean et al., 2015). Therefore, soil organic carbon (SOC) dynamics have become increasingly important in many research and policy areas (Manlay et al., 2007), ranging from small-scale projects to preserve or improve soil health, to large-scale climate change mitigation strategies (Lal 2004; Powlson et al., 2011). The soil system is heterogeneous and complex and direct SOC measurements alone do not easily support these types of efforts. Simulation models, however, provide the capacity for numeric evaluation of SOC after changes in land uses at different time and spatial scales. This has led to an expanding use of soil models specifically to predict SOC dynamics in order to apply policies or to make decisions on land use and management (Campbell and Paustian, 2015).

Different types of models have been developed in an attempt to quantify C in soil, including empirical and process-based multicompartment models. These models have varying levels of complexity and their utility will depend on the datasets available to drive them (Dondini et al., 2009). In empirical modeling, there is no attempt to model the processes that result in changes in soil C—the model is a mathematical formula that has been fitted to reproduce the available data and can then be used to predict other values within similar environmental conditions (Lawson and Tabor, 2001). By contrast, process-based models have been developed from an understanding of how soil C is affected by soil properties, land management, and weather fluctuations. These models

Soil Carbon Storage. DOI: https://doi.org/10.1016/B978-0-12-812766-7.00009-3

have varying levels of complexity and the choice of model depends on the data available to drive the simulation as well as the conditions used to develop and test the model.

The objective of this work is to describe the structure and development of models that have been widely used at international level to assess the impact of land-use and climate change on SOC stocks. We also aim to describe the versatility of model applications and their importance to disentangle local and global socioeconomic-environmental issues by reporting practical applications of such models from field to global scale.

9.2 EMPIRICAL MODELS

Empirical models seek to parameterize a hypothesized relationship between variables, typically known as the dependent and independent variables. The structure of the model is determined by the statistical relationships observed within experimental data, where the hypothesis statement is translated into a simple mathematical representation. The goal in this case is prediction of the value of the dependent variable, not an explanation of the nature of the relationship between the variables (Hillier et al., 2016).

9.2.1 GREENHOUSE GAS EMISSIONS CALCULATORS

The simplest empirical model is a linear one—this is used, e.g., in the emission factor methods of the Intergovernmental Panel on Climate Change (IPCC) Guidelines for National Greenhouse Gas (GHG) Inventories (IPCC, 2006). From this simple approach, several tools have been developed that integrate a number of such empirical equations into a complete model for C assessment; one example of this is the Cool Farm Tool developed by Hillier et al. (2011).

The Cool Farm Tool is a GHG emissions calculator which allows users to estimate annual GHG emissions associated with the production of crops or livestock products, following the emissions from production to the farm gate (Hillier et al., 2011). It comprises a generic set of empirical models that are used to estimate full farm-gate product emissions. The model has several submodels breaking down the overall emission by GHG emitted and farm management practices. The GHG emissions from the production and distribution of a range of fertilizer types was taken from the Ecoinvent database (Ecoinvent Centre, 2007). For nitrous oxide and nitric oxide emissions related to fertilizer application, the multivariate empirical model of Bouwman et al. (2002)—which is based on a global dataset of over 800 sites—was used. Soil C stock changes were estimated using the IPCC Tier 1 method (IPCC, 2006). After changes in management practice related to tillage or soil C inputs, soil C stocks change by an amount determined in Ogle et al. (2005) for a period of 20 years. The effect of manure and compost addition on soil C stocks are derived from those of Smith et al. (1997) in which relationships were established using medium/long-term data from EU15 countries. A simplified model was developed from ASABE technical standards (ASABE, 2006a,b) for fuel use as a function of machinery operation for tilling, drilling, seeding, and harvest operations for differing soil types and crop yields.

The mitigation option tool, developed for the Climate Change, Agriculture and Food Security program of CGIAR, is another example of a tool used to estimate GHG from baseline management options in agriculture. The mitigation option tool accommodates a wide range of users, experts to nonexperts, depending on objectives and issues such as time constraints and information available. It requires little input data and has the unique characteristic of suggesting management options that have the potential to further increase C sequestration in soils without risking crop yields. By providing a quick assessment of the C sequestration from current management practices, and of the practices that can increase the potential for soil C sequestration, these tools are extremely useful to inform policy-makers in the design of more effective policies to support the implementation of sustainable agricultural practices.

9.2.2 MODELS OF CHANGES IN SOIL CARBON

An example of an empirical model used to determine soil C stocks is the "C response function" (CRF) concept. The CRFs are representations of the average annual change in soil C following changes in land management, and they can also be used to show the cumulative change in soil C over time. The CRF curves are developed by using published reviews and analytical data, each describing a number of long-term, paired field experiments that quantify changes in soil C in response to changes in land use and management. The development of each CRF curve is based on analysis of one or more datasets, each describing a number of long-term, paired field experiments. The difference in soil C between the control and experimental plot for each field experiment in the data set is averaged across all experiments to estimate the mean change in soil C associated with a specific change in management. The CRF curves are developed by choosing a regression algorithm that best represents the estimated trend in soil C change over time, while ensuring that the sum of annual changes in soil C is equal to the previously estimated cumulative change in soil C (McClean et al., 2015; van der Weerden et al., 2012). In order to provide an estimate of the uncertainty surrounding mean changes in soil C, the 95% confidence intervals are given for each CRF curve. Standard error and sample size are also often given so that other confidence intervals can be calculated.

9.3 PROCESS-BASED MODELS

Process-based models focus on the processes mediating the movement and transformations of matter or energy. Each soil organic matter (SOM) pool within a model is characterized by its position in the model structure and its decay rate. Decay rates are usually expressed by first-order rate kinetics (Paustian, 1994) with respect to the concentration (Conc) of the pool

$$\frac{d\text{Conc}}{dt} = -k\text{Conc}$$

where t is the time and k is the decay constant.

Here we give a description of the most common models based on the complexity of the process description and the types of nutrients modeled.

9.3.1 SIMPLE MODELS THAT INCLUDE CARBON ONLY

The simplest approach used to model SOM turnover is to describe the SOC as pools with different turnover rates—these models predict SOC only and require minimal data inputs, including soil properties, meteorological data, and land-use type, to initialize the simulations. The advantage of this approach is that the models can predict soil C sequestration under a wide range of ecosystems (e.g., from natural forest to managed arable land) and at different scales (from site to regional). Because of their simplicity and minimal input data requirements, these models are easily understood and used by nonexpert users. However, because these models have been developed to describe only SOC in the soil, the impacts of nutrients on SOM turnover are not taken into account.

RothC is an example of a simple process-based model that includes C only. It simulates the turnover of organic C in nonwaterlogged topsoil (Coleman and Jenkinson, 1996) using a monthly time step to calculate total SOC. The model has been widely tested and used at the plot, field, regional and global scales, using data from long-term field experiments throughout the world. The data required to run the model are: monthly rainfall and evaporation or potential evapotranspiration (mm), monthly air temperature (°C), clay content (%), an estimate of the decomposability of the incoming plant material, monthly soil cover (whether the soil is bare or vegetated), monthly input of plant residues (t C ha^{-1}) and monthly input of farmyard manure (t C ha^{-1}) if any. The model performs two types of simulations: "Direct" that uses the known input of organic C to the soil to calculate the SOC; and "inverse" that evaluates the input of organic C required to maintain the stock of SOC.

RothC uses a pool type approach, describing SOC as pools of inert organic matter, humus, microbial biomass, resistant plant material, and decomposable plant material (Fig. 9.1).

During the decomposition process, material is exchanged between the SOC pools according to first-order rate equations. These equations are characterized by a specific rate constant for each pool, and are modified according to rate modifiers which are dependent on the temperature, moisture, and crop cover of the soil. The decomposition process results in gaseous losses of carbon dioxide (CO_2). In Fig. 9.1 we report the original RothC structure (Coleman and Jenkinson, 1996), but other RothC model structures can been found in several publications, such as Li Liu et al., 2009.

9.3.2 SIMPLE MODELS THAT INCLUDE CARBON AND NITROGEN

The ECOSSE model (Estimate Carbon in Organic Soils—Sequestration and Emissions) is an example of a simple model that can be used for both C and nitrogen (N) simulation (Smith et al., 2010). It was developed by combining and adapting RothC (Coleman and Jenkinson, 1996) and a mineral soil model (SUNDIAL, Bradbury et al., 1993) to allow organic soils in Scotland to be simulated, which were previously not well represented in models (Smith et al., 2007). Since its inception, it has been modified for use internationally (Bell et al., 2012) and evaluated using measurements in both organic and mineral soils.

ECOSSE uses a pool-based approach with C and N transferred between pools. As in RothC, the soil pools used are described as biomass (active), humus (stabilized) and inert organic matter, and plant litter is described as decomposable and resistant plant material. The base rate of exchange

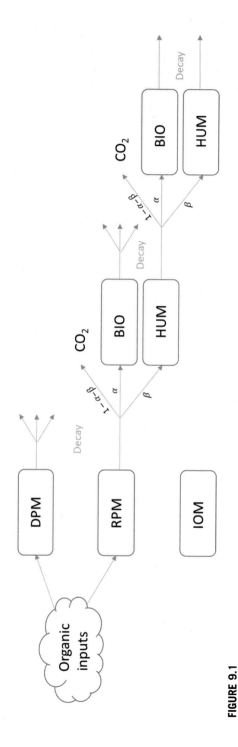

FIGURE 9.1

Structure of the RothC carbon sequestration model. Key: DPM is Decomposable Plant Material; RPM is Resistant Plant Material; BIO is Microbial Biomass; HUM is Humified Organic Matter; and IOM is Inert Organic Matter, and α, β and $(1 - \alpha - \beta)$ are the proportions of BIO, HUM and CO_2 produced on aerobic decomposition.

Adapted from Bradbury, N.J., Whitmore, A.P., Hart, P.B.S., Jenkinson, D.S., 1993. Modelling the fate of nitrogen in crop and soil in the years following application of 15N-labelled fertilizer to winter wheat. J. Agric. Sci. 121, 363–379 and Coleman, K., Jenkinson, D.S., 1996. RothC-26.3 – a model the turnover of carbon in soil. In: Powlson, D.S.,Smith, P., Smith, J.U. (Eds.). Evaluation of Soil Organic Matter Models Using Existing Long-TermDatasets. NATO ASI Series I. Springer, Berlin, pp. 237–246.

between the pools is specific to the pools in question and is then adjusted according to rate modifiers that describe the impact of environmental factors on the processes—these include pH, moisture and temperature. Soil texture is used to determine the efficiency of the decomposition (i.e., the amount of CO_2 lost on decomposition). Under aerobic conditions, the decomposition process results in gaseous losses of CO_2; under anaerobic conditions losses as methane dominate. Nitrogen released from decomposing SOM as ammonium or added to the soil may be nitrified to nitrate. Carbon and N may be lost from the soil by the processes of leaching, denitrification, volatilization or plant uptake—or C and N may be returned to the soil by plant inputs, inorganic fertilizers, atmospheric deposition or organic amendments.

9.3.3 MODELS THAT INCLUDE COMPLEX DESCRIPTIONS OF CARBON AND NITROGEN DYNAMICS

More complex models have been developed using the pool concept described earlier, with extra complexity to provide scope for the model to be applied at ecosystem level. These models couple descriptions of decomposition and denitrification processes, as influenced by the soil environment, to predict C and N turnover. Often, such models are used to examine the impacts of management and climate change in agriculture at site and regional scale. These types of models are highly amenable, allowing the user to describe the effect of various management and climate scenarios on a wide range of ecosystems. The user has full control of a large number of parameters, which need to be accurately determined to allow a successful simulation.

The DeNitrification DeComposition (DNDC) model is an example of a model that includes detailed descriptions of the processes of C and N dynamics. It was first described by Li et al. (1992). The first versions (1.0−7.0) of DNDC consisted of three main submodels for simulating nitrous oxide and N emissions; (1) soil-climate/thermal-hydraulic flux submodel, (2) decomposition submodel, and (3) denitrification submodel. During the following two decades many additions were made to the early version of DNDC. Li (2000) reorganized the model into two components incorporating six submodels (Fig. 9.2) and this new structure formed the basis of many DNDC-based models. Component 1 links ecological drivers to soil environmental variables and consists of the soil climate, crop growth, and decomposition submodels. Component 2 links soil environmental factors to trace gases and consists of the already known denitrification submodel and two additional submodels for nitrification and fermentation.

The DNDC model can be run on a site-specific or regional basis. For most input variables, default values are set, but many can and should be changed by the user in order to adequately describe the particular situation. Some input variables are mandatory and need to be set with individual values. These are location (latitude), weather data (daily mean air temperature and precipitation as minimum), soil bulk density, pH, and SOC at the surface (0−10 cm). The mandatory input variables together with land use, crop type, soil texture, and management practices will be sufficient to run the model. Among the most important output values for DNDC are daily reports on weather, soil climate, and soil C to N ratio in the pools, C and N fluxes, water balance, crop yields, and field management for the modeled site for each simulated year.

Over the past 20 years, many versions of DNDC have been developed and published, both for regional application (e.g., UK-DNDC) and for specific uses (e.g., Crop-DNDC, Wetland-DNDC,

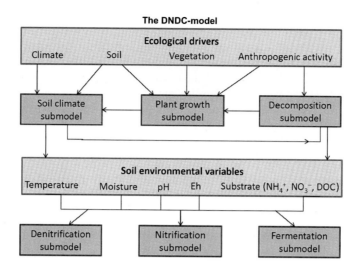

FIGURE 9.2

Structure of the two-component DNDC model with six submodels: Soil climate, crop growth, decomposition, denitrification, nitrification and fermentation.

Adapted from Li, C.S., 2000. Modeling trace gas emissions from agricultural ecosystems. Nutr. Cycling Agroecosyst. 58, 259–276.

Forest-DNDC). In some cases, DNDC has been coupled with market management models to include economic impacts of policy (e.g., DNDC-Europe). Due to the default values that are provided, DNDC is relatively easy to use and can easily be used by inexperienced modelers. The model is freely available.

9.3.4 MODELS THAT INCLUDE DESCRIPTIONS OF THE DYNAMICS OF A RANGE OF NUTRIENTS

Quantifying nutrient availability is crucial to understanding the interaction between plant and soil processes—these mechanisms relate to litter quantity and quality, and so are important drivers for SOM accumulation. The prediction of nutrient cycling aims to quantify the availability in time and space of nutrient elements in soil and to assess likely effects on plant growth and on nutrient fluxes, which can affect water and air quality. Quantifying nutrient availability requires an understanding of the rates of nutrient input, transformation and loss from the soil. The most appropriate approach to modeling nutrient interactions may vary with the ecosystem and with the data available to run the model.

DAYCENT is an example of a C model that includes simulation of the dynamics of a range of nutrients. It was developed by a team at the Natural Resource Ecology Laboratory at Colorado State University in Fort Collins (Parton et al., 1998). It is the daily time-step version of the 1994 monthly CENTURY model (Parton, 1998), also developed by the Natural Resource Ecology Laboratory at Colorado State University. The DAYCENT model is a terrestrial ecosystem model that simulates C and N cycles for forest, arable and grassland ecosystems. There is also an option to consider the phosphorous and sulfur cycles, if needed. Fluxes from the atmosphere to plant and soil are considered in simple approaches as atmospheric CO_2 concentration and N deposition. Submodels are included that describe plant productivity, phenology, decomposition of dead plant material and SOC, soil water and temperature dynamics, and GHG fluxes—these are described in detail by Del Grosso et al. (2001). Required input variables are physical soil properties (e.g., soil texture, field capacity, wilting point, bulk density, and pH), climate data, and management information. The management information provided depends on the land use simulated—for grassland it includes grazing, for forests it includes thinning and fire (forest), while for cropland it includes tillage, fertilizer inputs, irrigation and sowing, and harvest dates. DAYCENT is a one-dimensional model developed for site simulations, but it can also be applied on a regional scale.

9.3.5 MICROBIAL MECHANISMS AND SOIL PROCESS-BASED MODELS

A key similarity across all of the process-based models discussed earlier is the representation of organic matter decomposition as a first-order process. First-order models assume that the activity of decomposers only depends on temperature, pH, clay content, and moisture. This assumption implies that the microbial biomass and composition are not directly represented in the models, but only indirectly via the outcome of temperature and moisture effects on the rate of decomposition (Pagel et al., 2016). One limitation of this approach is that the effects of the changes in microbial community composition due to new conditions are not directly represented in the models. Recent evidence from empirical studies suggests microbial communities may shift in composition, adapt physiologically, or evolve in response to environmental changes, such as warming, N addition, and altered precipitation (Allison and Martiny, 2008; Hawkes et al., 2011). Furthermore, management techniques, such as plowing or no-till, and organic amendments, such as manure or straw incorporation, change the composition of the soil biota ecosystem, and hence the SOM decomposition rate.

van Groenigen et al. (2011) attempted to compare direct measurements of soil C to predictions made by RothC and a cohort model. They reported on soil C sequestration beneath a 9-year-old tillage and straw management experiment in an Irish winter wheat field, to estimate the decomposition rate of crop residue under different tillage management practices. Correlation between modeled and observed SOC were achieved by varying the size and decay rate of each pool and for each treatment, therefore not developing a mathematical function to describe the effects of different management practices on the soil biota ecosystem and processes. However, insufficient experimental evidence has been provided from various environments to enable robust process-based modeling of these affects. Salinity also effects the soil biota and, again, SOC and input decomposition rates have to be modified in models such as RothC to implicitly model the effect, although the actual soil biota processes are not explicitly modeled. Despite the drawbacks in describing soil decomposition by first-order process, all of the models used to assess SOC stocks in the most recent IPCC assessment (IPCC, 2014), use the same first-order assumption. Including models which can represent microbial mechanism in soils would increase the diversity of model predictions. This would help to prevent the biases which can arise from averaging the predictions of an ensemble of models that all make the same first-order assumptions (Knutti et al., 2008).

One of the main challenges in including microbial mechanisms in process-based models is to define which of these mechanisms should be scaled up from plot to regional level. One approach would be to use plot data to inform the models, which could then be modified by new mechanistic equations for including microbial processes before validating the model developed using independent data. However, this approach could lead to at least two sources of error on the simulated values at both the spatial and temporal scales. Many large-scale models operate with a spatial resolution that could potentially include high levels of microbial diversity and heterogeneity. Also, soil models at a large spatial scale are generally used to simulate soil processes over time (decades). It is unclear if plot-scale measurements, which are meant to describe microbial responses on a short-term basis, could be applied to a higher temporal scale without loss of accuracy in the model predictions (Todd-Brown et al., 2012). In the future, the increased use of new technologies, such as remote sensing and precision farming, will help in reducing the granularity of our knowledge of the spatial variability of soil, soil water, plant yields and GHG emissions. The application of remote sensing will improve the accuracy and resolution of land use maps to less than 10 m resolution (current land use maps are available at 100 m × 100 m resolution)—these new maps could be then used for models' parameterization. Precision farming, and the associated sensors that enable 1 m × 1 m resolution detail of field soil and crop condition, will allow maps of crop yield to be made. This information can be used with new informatics technology, which will enable these large spatial datasets to be used to drive high spatial and temporal resolution models.

Another approach to better represent soil C cycling processes in current models would be to quantify functional trait in microbial communities and to link these traits to key factors controlling the soil decomposition and degradation processes. There is a body of research, particularly in India investigating the impact of soil biota on fertility and the use of different biological inoculates to increase crop yields (e.g., Pandya and Saraf, 2010a,b), and hence organic input and SOC. This will lead to a better understanding of the function of different taxa of soil biota. Consequently, a few models have been proposed to explore possible microbial roles in SOC dynamics (Wieder et al., 2015), but these models need rigorous evaluation with observations before they can be incorporated into large-scale soil process-based models (Luo et al., 2016).

9.4 EXAMPLES OF MODEL APPLICATION FOR PREDICTING SOIL ORGANIC CARBON CHANGES

Soil models are useful tools to estimate the effect of "disturbance" events on soil C dynamics—disturbances such as climate change, land management, land cover, and land-use change have been widely represented in models, while soil erosion and extreme events have been found difficult to model and are not directly used in soil process-based models (Box 9.1). Here we present a selection of studies where soil models have been applied from field to global scale to predict SOC changes under different vegetation types.

9.4.1 SIMULATION OF CARBON SEQUESTRATION AT FIELD PLOT SCALE

9.4.1.1 Impact of land-use change from grassland to woodland at Glensaugh

The Glensaugh Research Station in rural Aberdeenshire in Scotland is an experimental site where conversion from grassland to woodland was undertaken almost 30 years ago. The site was set up to investigate the impact of afforestation of pasture on animal output (Sibbald et al., 2001). Three tree

---

BOX 9.1 IMPACT OF SOIL EROSION AND EXTREME EVENTS ON SOIL ORGANIC CARBON (SOC)

This text box shows relevant aspects of SOC modeling, which are not yet well represented in SOC model approaches. Two of these aspects are the impact of soil erosion and the impact of extreme events on SOC. Extreme event is a general term and there are several definitions available to define an event as extreme. Here we refer to extreme events as "an episode or occurrence in which a statistically rare or unusual climatic period alters ecosystem structure and/or functions well outside the bounds of what is considered typical or normal variability" (Reichstein et al., 2013). In the context of soil C, these are mainly extreme climate and weather events.

Soil erosion results from extreme precipitation and storm events, and includes both wind and water erosion. Here we focus on the erosion by water, which affects a larger area (751 versus 296 Mha land affected by water and wind erosion, respectively) and erodes more sediment compared to wind erosion (Lal, 2003). The scientific debate about the impact of soil erosion on the SOC is controversial; while some studies come to the conclusion that erosion causes C losses, others show that it enhances soil C accumulation (Doetterl et al., 2016). Despite its high relevance for global C dynamics, the impact of soil erosion on the global C budget is not yet quantified (Lal, 2003; Müller-Nedebock and Chaplot, 2015) and it is rarely considered in biogeochemical models. EPIC (Williams, 1990) and CENTURY (Lugato et al., 2016) are biogeochemical models that contains an erosion routine, the RUSLE model (Renard, 1997), a revised version of the universal soil loss equation (USLE; Wischmeier and Smith, 1978). The USLE model, and its modifications, simulates sediment detachment using empirical approaches based on relative simple factors such as precipitation, soil properties, slope and tillage. The disadvantage of this approach is that sediment deposition is not simulated.

Extreme events are not explicitly considered in SOC model approaches. Thresholds in the models consider limitations or impacts affected by soil water content, soil temperature or nutrient concentration in the soil without considering these explicitly as extreme events. Therefore, some direct impacts (e.g., drought might reduce respiration rates) can be simulated, whereas indirect impacts (e.g., a lag effect of respiration as the soil microbial community might be affected by a drought) won't be considered in the model approach (Frank et al., 2015). The limitations in modeling extreme events include a lack of observations describing large-scale impacts and a lack of standardization of experimental designs. Moreover, several processes may be too sensitive or too detailed to be implemented within a model—e.g., microorganisms are responsible for C sequestration, but the specific communities or activity are not directly considered in the models.

As extreme events and soil erosion are hardly considered in SOC models, more experimental data are needed to understand their impacts on SOC and to calibrate and validate soil process-based models. A standardized experimental and observational framework would be beneficial so that the collection of comparable modeling-friendly datasets may be realized.

---

species, namely scots pine, hybrid larch, and sycamore were planted at 400 trees ha^{-1} silvopastural configuration, which allows for animal grazing between the rows of trees. The same species were also planted at 2500 trees ha^{-1} in farm woodland plots that have received no thinning since the site was established. Both approaches integrate trees into farmland, either spatially segregated in farm woodland or integrated as silvopasture. The site was sampled for total soil C and labile, stabilized and inert C fractions in 2012 (Beckert et al., 2016). In both silvopasture and farm woodland, SOC was found to be greater compared to the pasture treatment. While woodland and silvopasture plots had similar levels of total SOC, silvopasture showed levels of stabilized C comparable to pasture.

The RothC model was used to investigate how C stocks will develop in the different land-use systems at the Glensaugh site, assuming that land management remains constant. The RothC model was first run from the year of tree planting (1988) to the year of sampling (2012), assuming equilibrium at each site. Comparison with measured fractions showed that this assumption only holds true for the pasture site, which had seen no change in management. To investigate how C stocks will develop up to the year 2040 taking actual C quality into account, the model was initialized with measured fractions to replace equilibrium pools. Initializations with fractionation data resulted in the prediction of an increase in C stocks at all wooded sites, particularly in the silvopastoral systems, which showed evidence of combined pasture/forest C stabilization mechanisms. The initialization revealed a slightly increased accumulation rate after 2020 compared to 2012−20 before it levels off in ca. 2030, indicating that initial increase in respiration is negated when the systems reach a more mature age. The results at site level agree with the results of large-scale modeling (Section 9.4.3.1), showing that afforestation of grassland soils could have a positive impact on SOC in the long term.

9.4.1.2 Impact of Climate Change on Grassland and Arable Systems in Ireland

Grasslands represent an effective option for C sequestration in soils. However, predictions of increase in SOC are associated with a great uncertainty (Freibauer et al., 2004; Vleeshouwers and Verhagen, 2002). Croplands have less SOC than grassland (Cole et al., 1993) as a result of several factors including soil disturbance, less return of plant residues to the soil, less below-ground biomass, and no grazing (Franzluebbers et al., 2000). Here we present a study where measured and simulated net ecosystem exchange (NEE) values from a managed grassland and a spring barley field, in Ireland, were compared with simulated NEE to validate the latest version (9.5) of the DNDC (www.dndc.sr.unh.edu; Li et al., 1992) model and to estimate present and future NEE and SOC (Abdalla et al., 2013). The averages measured NEE for the grassland during the experimental period (2003−06) was calculated as $-212 \, \text{g C m}^{-2}$. The DNDC model predicted seasonal trends of NEE effectively for 2003 and 2004 but overestimated carbon losses in 2006 (Fig. 9.3A).

The root mean square error (RMSE) values were small and ranged from 0.20 to $0.22 \, \text{g C m}^{-2}$ with an overall RMSE of $0.21 \, \text{g C m}^{-2}$. The relative deviation (RD) between the measured and simulated NEE values was also small ($+30\%$) except in the year 2006 when it was $+45\%$. The average annual values of NEE, GPP and Reco, over the measurement period (2003−07) were -189, 906 and $715 \, \text{g C m}^{-2}$, respectively. The DNDC model effectively predicted the seasonal trend of NEE at the spring barley field (Fig. 9.3B). The RMSE values from the comparison between daily simulated and measured NEE are small, ranging from 0.09 to $0.16 \, \text{g C m}^{-2}$ indicating a good fit between the model and simulated values. The RD values between the measured and predicted NEE values ranged from -13% to $+100\%$, with the highest RDs in 2004 ($+100\%$) and 2005 ($+92\%$). These poor RD were mainly due to the DNDC overestimation of NEE peaks during the growing seasons.

In future simulations to 2060, SOC at the grassland site was predicted to decrease by $2\%-3\%$ by the year 2060 for all climate scenarios. At the arable site, the SOC was also predicted to decrease, but only by $1\%-2\%$. This indicates that the soil C systems for the two ecosystems are not in equilibrium. The cropland was historically under grassland prior to 1990 and, therefore, continues to lose C. The grassland had been tilled and reseeded with perennial ryegrass in 2001 and, therefore, will take time to reach a new equilibrium after the tillage disturbance. In both the arable and grassland case, water stress would affect crop yields (Hastings et al., 2010) and thereby, the amount of carbon input. The model effectively predicted seasonal and annual changes in NEE at both sites, and responded appropriately to changes in air temperature, timing of precipitation events, and management, which have a strong influence on the seasonal NEE. These results suggest that the DNDC model is a valid tool for predicting the consequences of climate change on NEE and SOC from arable and grassland ecosystem.

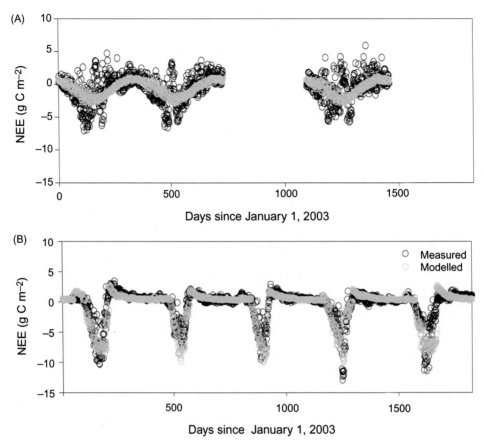

FIGURE 9.3

Measured (*dark circle*) and simulated (*light circle*) NEE for (A) the grassland and (B) arable fields during the experimental period (grassland experimental period: 2003–06; arable experimental period: 2003–07).

Adapted from Abdalla, M., Saunders, M., Hastings, A., Williams, M., Smith, P., Osborne, B., et al., 2013. Simulating the impacts of land use in Northwest Europe on Net Ecosystem Exchange (NEE): the role of arable ecosystems, grasslands and forest plantations in climate change mitigation. Sci. Total Environ. 465, 325–336.

9.4.1.3 Impact of rice management in Bangladesh

In Bangladesh, rice occupied 70% of all agricultural land in 2016, accounting for 7% of the world's total harvested area (FAOSTAT, 2016). Due to different physiological characteristics, such as the need of continuous flooding of water to provide the best growth environment, rice can sequester more C relative to upland crops and offers substantial mitigation potential (Smith et al., 2008). The DAYCENT model was used to simulate SOC sequestration potential under different N management and mitigation options applied at two rice sites in Bangladesh. In this study, all model parameters, except for the plant growth, were set to default values based on previous literature (Cheng et al., 2014). Values of the plant growth parameter, were adjusted to 3.50 for rice while for wheat it was set to 2.00, and was fixed for all treatments. Annualized C stock changes were calculated as the difference of the SOC stock of the mitigation scenario and the SOC of the baseline scenario normalized by time period. The management treatments at the sites included application of N as mineral N, organic manure alone and in combination with N applications (Karim et al., 1995; Egashira et al., 2003, 2005). There was a significant agreement between measured and simulated SOC at both sites under single nutrient management practices (Fig. 9.4A and B). A systematic

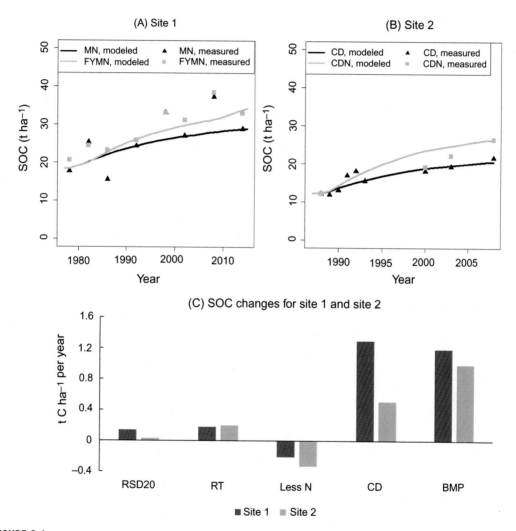

FIGURE 9.4

Simulated (*line*) and measured (*points*) SOC values at 20 cm depth over 20 years under different treatment for the period of 1978−2015 and 1988−2008 for site 1 (A) and site 2 (B) respectively. (C) Indicates modeled annualized SOC stock changes under different mitigation scenarios of two test sites for the period of 1988−2008. (*MN*, mineral N; *FYMN*, farmyard manure + mineral N; *CD*, cowdung; *CDN*, cowdung + mineral N; *RSD20*, 20% residue return; *RT*, reduced tillage; *BMP*, best management practice, RSD20 + RT + less N + CD).

underestimation of SOC was observed at Site 1 (combination of manure and N treatments), which could be attributed to a reduction of plant inputs and suggesting that less N application through manure was limiting plant production.

   Mitigation options considered including reduced tillage (sowing with less disturbance to the top-soil in place of tractor plowing), a reduction in residue removal, replacement of mineral fertilizer by manure, combined application of fertilizer and manure, and an integrated scenario of inorganic fertilizer, manure addition, less residue removal, and reduced tillage. All tested mitigation options increased SOC in comparison to the standard procedures, except for the scenario with lower N application, which shows a slight decrease in SOC contents (Fig. 9.4C). The integrated scenario, which combines mineral N and manure applications with reduced tillage and increased residue incorporation, appears to be the best management practice for both sites. Despite the limited availability of long-term field data for tropical rice cropland, the results suggest that the DAYCENT model could be a powerful tool for exploring mitigation potentials of rice in Bangladesh.

9.4.2 SIMULATING CARBON SEQUESTRATION AT FARM SCALE

Whole farm modeling attempts to simulate not only C sequestration, but also to determine the impact of C sequestration on crop and animal production, water use, fuel availability, labor, and finances so that the feedback of these factors on the potential for C sequestration can be accounted for. Whole farm modeling is particularly important in low input, close-to-subsistence farming, where the potential for external inputs to the farm from inorganic fertilizers and organic resources is minimal. Such systems are often also severely limited in organic resources, with important competing uses for the organic resources that are available, such as for household energy provision, animal feed, and building. In such situations, it becomes important to model, not only the impact of the different types of organic amendment on potential C sequestration, but also to estimate the amount of material that is left over and can be added to the soil. Whole farm modeling of C sequestration attempts to account for these competing uses, and works through the impact of using resources in different ways on the quality and quantity of C inputs to the soil (Fig. 9.5). One

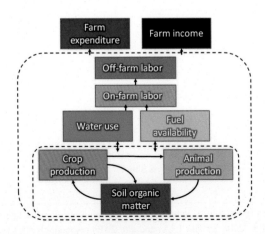

FIGURE 9.5

Whole farm modeling, accounting for the feedback between soil organic matter on crop and animal production, water use, fuel availability, labor, and finances.

example of this is seen in Hawassa, Ethiopia, where soils are often highly depleted in SOM, and so C sequestration is important, not only for the environment, but also to improve soil fertility and hence productivity.

Whole farm modeling of C sequestration starts with some form of accounting—what goes where and how is it used? The nature of this depends on the input variables available to the user— when working with data provided by subsistence farmers the number of animals that must be fed is usually known, but the amount of home-produced crop fed to each animal may not be known. In this case, a simple model or look-up table of feed requirements can be used (e.g., Herrero et al., 2013). Similarly, the farmer knows what crops are grown, but the yield may not be measured as it is mainly consumed within the household. Therefore, a simple crop model is needed to estimate yield and the impact of different management decisions on crop production (e.g., Leith, 1972; Reid, 2002; Zaks et al., 2007).

Having accounted for the different uses of organic resources, a SOM model is then used to determine the impact of adding differently treated organic wastes to the soil. This was simulated by Smith et al. (2014) using a variant of RothC (Coleman and Jenkinson, 1996), showing more rapid C sequestration per unit of starting material if the organic wastes are added as compost or biochar, rather than applying it fresh or as bioslurry (Fig. 9.6). After application of organic materials stops (after 20 years in this example), the C content of the soil returns to the starting position within 100 years for the fresh residue, compost, and bioslurry amended soils. However, if the biochar contains a high proportion of inert organic material (currently an area of uncertainty), then the C sequestered by biochar application remains in the soil. Long-term experiments on impact of biochar on SOC

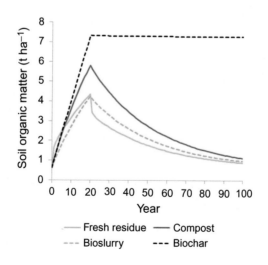

FIGURE 9.6

Rate of carbon sequestration for application continued over 20 years of differently treated organic residues derived from 1 t ha^{-1} per year of carbon in fresh residue.

Adapted from Smith, J., Abegaz, A., Matthews, R., Subedi, M., Orskov, R., Tumwesige, V., et al., 2014. What is the potential for biogas digesters to improve soil carbon sequestration in Sub-Saharan Africa? Biomass Bioenergy 70, 73–86.

BOX 9.2 MODELING IMPACT OF BIOCHAR APPLICATION ON SOIL ORGANIC CARBON (SOC)

Biochar is a more stabilized form of C obtained from thermal decomposition of raw biomass. Because of its high recalcitrant nature and slow turnover rate, biochar has been identified as one of the promising option to mitigate climate change. However, modeling biochar is still in its infancy and only few models have been recently developed, or modified, to account for the effects of biochar on SOC. For example, Woolf and Lehmann (2012), and Smith et al. (2014), modified the turnover rates of the labile organic C (LOC) pool in the RothC model to simulate impact of biochar on SOC sequestration. Priming effects of biochar on LOC was also included in the model by altering the decomposition rate coefficients of the resistant plant material (RPM) and decomposable plant material (DPM). Positive priming effect—i.e., the increase in mineralization of LOC—was modeled by increasing RPM and DPM decomposition rate coefficients by an amount proportional to the concentration of biochar C in the soil. Negative priming effect—i.e., an increase in the fraction of LOC transferred to the stable organo-soil-mineral fraction—was modeled as an increase in the fraction of DPM and RPM that is transferred to the humus pool (HUM) rather than mineralized to CO_2.

Lychuk et al. (2015) modified the Environmental policy Integrated Climate (EPIC) model by developing a set of new algorithms to determine the impact of biochar amendment on SOC sequestration, as well as other soil and crop parameters (e.g., CEC, pH, bulk density, and corn yield). In the EPIC model, SOC is split into three compartments— i.e., microbial biomass, slow humus and passive humus. To account for biochar applications, the total biochar C is allocated to the three pools as follows: 60% to the slow humus pool, 38% to the passive humus pool, and only 2% to the metabolic pool. Recently, Archontoulis et al. (2016) developed a biochar submodel within the Agricultural Production Systems sIMulator (APSIM) model. The APSIM model divided the SOC into three pools—i.e., microbial biomass pool, humic pool and inert pool—but the fresh organic matter is accounted as a separate pool, which is also divided in three subpools. Archontoulis et al. (2016) introduced an additional biochar C pool to the model, which represents both labile and recalcitrant components and varies according to the type of biochar; a new double exponential decay function has been also introduced to calculate the biochar decomposition rate. Priming effects of biochar and the impact of biochar on N mineralization, soil CEC, soil pH, ammonium adsorption and desorption, soil water and bulk density have also been included in the biochar submodel.

Despite the late developments in modeling biochar at field scale, more long-term field trials are required to better understand the relationship between soil C sequestration and biochar applications and to consequently develop, calibrate and validate soil models.

---

dynamics and soil fertility are still limited and there are very few simulation studies on biochar and its effect on agricultural soil. Moreover, only few models have been developed to account for the effects of biochar on SOC, as discussed in Box 9.2.

The real value of the whole farm model is to then use these simulations to try out different options. For example, if organic wastes are composted rather than applying them as fresh farmyard manure, how will this affect C sequestration? Identifying these positive feedbacks will provide important information for better management of subsistence farms. Similarly, identifying negative feedbacks will highlight practices that result in a reduction in the overall productivity of the farm, so helping to reduce soil degradation.

9.4.3 REGIONAL SCALE

9.4.3.1 Potential for carbon sequestration with land-use change

Currently the Scottish Government has committed to increase the amount of forest by approximately 100,000 ha per year as part of a national strategy of reducing GHG emissions by 42% by

2020 and 80% by 2050. Several models (e.g., RothC, Century) have been used to study C sequestration due to land-use change. This section describes the application of the ECOSSE model (Smith et al., 2010) to analyze the long-term change in soil C stocks with afforestation of nonforest soils, aiming to identify regions that would provide most C benefit if reforested.

To achieve this, high resolution (1 ha grid) land use data from the Integrated Administrative and Control System was used to identify the dominant land use; cropland, grassland, forestry, and seminatural land. Masks of productive agricultural land and current forest were applied to the land use database and this was then combined with the Scottish Soils Knowledge and Information Base (SSKIB) and long-term climate input data from the UK Metrological Office. Each land-use change to forestry was assumed to take place in this decade (2010's). Suitability masks of 12 different forest compositions were applied and soil C was simulated only for areas where land-use change was deemed suitable.

Fig. 9.7 details the change in soil C after land-use conversion from crop, grass and seminatural land to native conifer forest, which is the forest type with the greatest extent of suitability in Scotland. Values outline the average annual loss in soil C for the first 20 years after planting. Across Scotland, conversion from arable and grassland to forest typically resulted in an increase in soil C where in some cases, after conversion, C accumulated up to 0.69 t C ha^{-1} per year on mineral soils. By contrast, land-use change to seminatural soils, which typically were defined as occurring on peaty soils, lead to an emission of soil C at a rate of up to 5 t C ha^{-1} per year in the most extreme cases. While changing to forest tends to enhance C sequestration in arable and grassland soils, mass conversion may not be economically viable or sustainable as removal of productive land can increase Scotland's reliability on food or cereal imports. While unmanaged seminatural land may be an obvious alternative, in some cases the management involved in converting these soils into a forest may lead to long-term losses in soil C, despite any increases in plant C inputs. These results suggest that while, theoretically, conversion to forest maybe a long-term approach to enhancing C removals, to implement such a mitigation strategy, especially in Scotland, detailed analysis on the impacts on soil C losses in different areas should be undertaken. A similar approach was used by Pogson et al. (2016) and Richards et al. (2016) to investigate the impact on SOC of land-use change across the United Kingdom. Pogson et al. (2016) developed the ELUM Software Package, which is based on the ECOSSE model, to spatially predict the net soil GHG balance of land-use change to grow energy crops in the United Kingdom up to 2050. The results of the model application demonstrated that wood and perennial grass production on arable land sequestered SOC, on grassland it was neutral and on forest it emitted CO_2.

9.4.3.2 Carbon losses from tropical peatlands undergoing land-use change to oil palm

Tropical peatlands are hugely underresearched compared to their temperate counterparts, with approaches to sampling and interpretation of peat properties still evolving to more "tropically" appropriate methods (Farmer et al., 2011). As such, there are considerable data limitations when it comes to modeling scenarios of climate and land-use change on tropical peats. Some process-based models, such as RothC and ECOSSE could potentially be used to model C dynamics in tropical peats (Farmer et al., 2011), and are currently undergoing modification to be made more applicable in scenarios where the soil is accumulating C (i.e., an intact peatland scenario) before undergoing land-use change. The HPMTrop (Kurinato et al., 2015) is the first process-based model to simulate

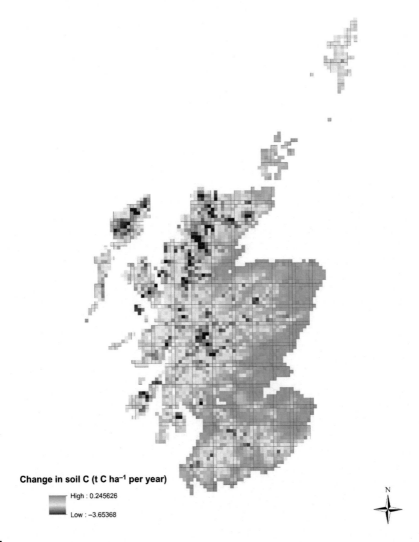

Change in soil C (t C ha⁻¹ per year)

High : 0.245626

Low : −3.65368

N

FIGURE 9.7

Change in soil C (t C ha^{-1} per year) after conversion from grass, crop or seminatural land to Forestry. Values represent the average annual change in soil C for the first 20 years after conversion.

long-term (decadal to millennial) C accumulation dynamics in tropical peat ecosystems. It has been applied to simulate peat accumulation in Indonesian peat swamp forests and to study the impact of land-use change of these areas to oil palm plantations (Kurinato et al., 2015). The modeled average peat accumulation rates and the mean annual C losses due to conversion to oil palm were comparable to literature values—however the limited published values restricted model evaluation (Dommain et al., 2011).

Hooijer et al. (2012) measured and then modeled subsidence rates in oil palm plantations on Sumatran peatlands and an empirical model, the Tropical Peatland Plantation-Carbon Assessment Tool (TROPP-CAT), was developed from this data to provide a user-friendly tool to predict soil C and CO_2 emissions from drained tropical peat soils (Farmer et al., 2014). The model uses simple input values to determine the rate of subsidence, of which the oxidizing proportion results in CO_2 emissions. Although based on a number of assumptions, evaluation across sites of various ages showed simulations of net CO_2 fluxes from the soil to be within 6% of measured CO_2 emissions and within the range of measurement error.

In tropical peat soils, positive correlation has been observed between mean water table depth and net C loss, heterotrophic emissions and total emissions (Carlson et al., 2015) which is also observed in Northern peat soils (Abdalla et al., 2016). This relationship can be used to make predictions on emissions under future drainage scenarios. However, several studies have found discrepancies between empirical model outputs and experimental data (e.g., Allison et al., 2010; Davidson et al., 2012; Wieder et al., 2013), likely to be due to the omission of key factors, such as direct microbial control of soil C dynamics and brief soil respiration increase due to warming. To partially remedy these discrepancies, annual rhythm oscillation models have been suggested (Comeau, 2016). The novelty and advantage of a rhythm oscillation method over the traditional empirical approaches is that it automatically provides the annual flux amplitude and the peak emission time. In addition, the oscillation curves are not biased due to possible delay in microbial activity response to temperature change and other environmental variables that affect soil C dynamics. As tropical peatland research continues to develop with more datasets becoming available, an enhanced understanding of the dynamics of tropical peat formation and soil properties and characteristics will make for improved modeling of the impacts of land-use change on these soils.

9.4.4 GLOBAL SCALE

9.4.4.1 The impact of growing bioenergy crops on carbon stocks

Quantitative and qualitative global datasets on the environmental effects of land use and land-use change are still scarce, making climate mitigation analysis difficult. In addition, there is still a lack of information on where, at what rates, and what type of land cover is affected by land-use change. In that respect, highly productive food croplands are unlikely to be used for bioenergy, but in many regions of the world a proportion of cropland is being abandoned, particularly marginal croplands, and some of this land is now being used for bioenergy. Albanito et al. (2016) used a number of harmonized geographically explicit datasets and process-based biogeochemical models to assess the global climate change mitigation potential of cropland when converted to bioenergy production (C_4 grass, short rotation coppice woody crops as willow and poplar) or reforested. This study, in particular, identified areas where cropland is so productive that it may never be converted, and assess the potential of the remaining cropland to mitigate climate change by identifying which alternative land use provides the best climate benefit: C_4 grass bioenergy crops, coppiced woody energy crops, or allowing forest regrowth to create a C sink.

The average cropland C loss resulting from land-use change was calculated as the difference in C between annual bioenergy crop yields and cropland yields aggregated over 20 years. The global forest C stocks scenario was developed using the IPCC 2006 Tier-1 method for estimating vegetation C stocks. The potential distribution and forest vegetation C stocks were obtained using the LPJmL-DGVM v3.1 model simulations. In the comparison with cropland, the C sequestration in forests was calculated by applying the factors representing percentage of final biomass C stock accumulated after 20 years (F_{20}). F_{20} was estimated by integrating, over a 100-year timescale, the IPCC default dry matter biomass annual increments in aboveground biomass in naturally regenerated forest classified below and above 20 years of age (IPCC-GPG-LULUCF, 2006). Total SOC change in reforested cropland was assumed to be equal to 53% of the initial SOC occurring in cropland (Guo and Gifford, 2002) adjusted by the percentage of biomass stock accumulated after 20 years.

Across 1.11 billion hectares of global agricultural land, Albanito et al. (2016) reported that approximately 420.1 Mha would be more suitable for food crop production and therefore excluded from conversion to bioenergy crops or reforestation. Over a 20 year rotation horizon, 597.7 Mha of croplands could potentially be converted to bioenergy crops or forest, sequestering approximately 13.8 Pg C in soil (Fig. 9.8). An area of 384.9 Mha has annual extractable C of C_4 bioenergy crops that is equal to or lower than cropland, but nevertheless sequesters approximately 10.3 Pg C in soil.

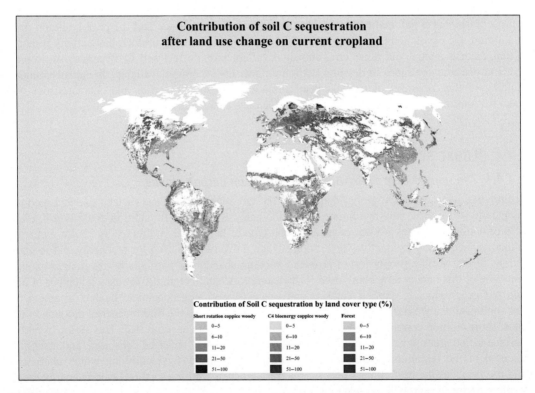

FIGURE 9.8

Potential contribution (%) of soil C sequestration to the total C savings occurring from the conversion of rainfed and irrigated high-input croplands to C_4 bioenergy crops, short rotation coppice wood land and forests.

In Asia (continental and insular) the replacements of croplands with C_4 bioenergy crops have the potential to sequester 3.6 Pg C in soil across 66.1 Mha of cropland. On approximately 26.3 Mha of cropland, short rotation of woody crops has greater or equal C mitigation potential to C_4 bioenergy crops and forest, giving a potential sequestration in soil of 0.8 Pg C (Fig. 9.7). Finally, approximately 186.5 Mha reforestation of cropland would be the best climate mitigation option, saving a total of ~8.4 Pg C in biomass and ~2.7 Pg C in the soil (Fig. 9.7). It is important to note, however, that this study does not present these projections as a scenario of land-use change where bioenergy crops or forests should replace cropland, which will depend on many other factors, not least of which is the need to produce food—rather it is to show where there could be a climate benefit if this land were to be converted.

9.5 POLITICAL ASPECTS AND CONCLUDING REMARKS

In 2015, the world defined and committed itself to striving toward the UN Sustainable Development Goals (UN SDG) (UNDP, 2015), in which the historic Paris Climate Agreement (PCA) was signed under the UN Framework Convention on Climate Change (UNFCCC, 2015), and was also the UN International Year of Soils (UN, 2015).

The agreement of the UN SDG and the PCA could not have set up a better legacy for the UN International Year of Soils, since soils are recognized as being critical to the delivery of both. A number of the UN SDG are underpinned by healthy soil C stocks, including the following Sustainable Development Goals (SDGs), among them: SDG 1—no poverty—in developing countries, a large proportion of the population rely on the land for their livelihoods, and productive land relies on healthy soils (Smith et al., 2013), SDG 2—zero hunger—soils underpin the production of safe and nutritious food (Keesstra et al., 2016), SDG 13—climate action—soil C sequestration offers climate mitigation (Smith, 2016) and makes ecosystems more resilient to future climate change (Smith et al., 2016a), and SDG 15—life on land—healthy ecosystems are founded on healthy soils (Smith et al., 2015).

By linking international, national and local policies, and action frameworks to the PCA, governments can develop more comprehensive and robust approaches to climate change, food security, soil protection, sustainable land management, water management, and energy generation (Chan et al., 2015; Casado-Asensio et al., 2016). However, there is often a difference in objectives between practitioners at various levels and policy makers, particularly in the agricultural sector, with respect to priorities for resource and land management (Casado-Asensio et al., 2016; Bodansky et al., 2014). This disconnect requires robust institutional support to encourage inclusivity in decision making, increase the dissemination of policies, offer financial assistance and access to markets and provide insurance for climate risks. These actions will require collaborative action from both the public and the private sector. In this context it is crucial to explore the relationship between farmers' attitudes and their farming practices, as well as informing decision makers regarding the social impacts of their decisions. This aspect is discussed in more detail in Box 9.3.

Given the role of soils, and soil C, in delivering the UN SDGs and the PCA, the accurate modeling of soil C stocks has never been more significant. There is a pressing need to develop, test and challenge our soil C models to meet the challenges facing humanity in the 21st century.

BOX 9.3 TRANSLATING SCIENTIFIC SOIL CARBON MODELS TO THE FARMING COMMUNITY

Scientific models predicting the effects of farming practice and land-use change on C emissions and sequestration provide a very valuable tool that can guide policy-makers, industry, and individual farmers to make changes for a more sustainable agricultural sector. Greenhouse gas (GHG) calculator tools such as the Cool Farm Tool, C-Plan and CCAFS-Mitigation option tool are being used as a platform to translate scientific models to the daily farming practice (Hillier et al., 2011; Whittaker et al., 2013). These tools aim to encourage farmers to change their behavior by raising awareness of the negative outcomes of their farming practice on GHG emissions and help them to take informed decisions on alternatives. This approach has for a long time been a popular strategy in promoting proenvironmental behavior in various contexts (Stern, 2011). Although it has been proven to be effective in increasing people's knowledge, it has minimal effects in changing actual behavior (Abrahamse et al., 2005; Gardner and Stern, 2002; Stern, 2011). To effectively motivate farmers to take up mitigation measures, it is recommended that information provision from GHG calculators be combined with other psychological interventions. To effectively create a bespoke intervention aiming at a specific psychological factor, it is recommended to first assess which factors underlie the willingness of farmers to take up mitigation measures. Psychological models, such as the Theory of Planned Behaviour (Ajzen, 1991), can provide a good starting point to assess the significance of a number of factors such as attitude toward proenvironmental measures, social pressure, group pressure, or self-identity (Van Dijk et al., 2015, 2016). For example, if the model indicates that peer pressure is related to the motivation of farmers to take up mitigation measures, benchmarking would be an effective intervention. This can be done by organizing plural workshops in which farmers collectively run a GHG calculator for their farms and receive information on how their outcomes compare to their peers. Benchmarking has been proven to be effective at increasing farmers' intentions and uptake of proenvironmental measures (Lokhorst et al., 2010). However, combining different interventions can further increase the uptake of measures. For example, combining benchmarking with public commitment making, in which farmers commit themselves in front of fellow participants of the workshop to certain measures, has been demonstrated to even further increase the willingness and uptake of these measures (Lokhorst et al., 2010). In conclusion, GHG calculator tools are very valuable tools to translate scientific carbon models to the farming community by providing information on how to decrease GHG emissions, but to successfully establish a change in the daily practice it is recommended to combine these tools with other psychological interventions and communication strategies.

Whatever type of models are used to meet future challenges, it is important that they continue to be tested using appropriate data, and that they are used in regions and for land uses where they have been developed and validated. As new uses of land are developed, models should continue to be validated and modified if necessary, so that they are still appropriate. In addition, in many situations the type of model used, will be dependent on the input data available. Models such as DAYCENT, ECOSSE, and the Cool Farm Tool are ideal for assessing soil C sequestration under future climate and land use, but if insufficient data is available, then less data-intensive models (e.g., RothC, statistical techniques) should be used.

It is also important that the best data available are readily accessible, whether this is decomposition pot experiments, long-term experiments, soil maps, or satellite data. The development of the technologies of remote sensing and precision farming will provide high resolution data and advances in informatics will enable their use in developing higher resolution and more detailed process-based models. It is extremely important that experimentalists/data curators are involved in the modeling process, as modelers need to know if analytical methods have changed over time or between different counties, what quality control has been used on the data, and how missing data has been addressed.

BOX 9.4 TAKE HOME MESSAGE

- Soil models are essential tools to understand the effects of land and climate change, from field to global scale.
- Soil models are crucial tools to up-scale and interpolate point/site/field information to larger scales in a quantitative way.
- In order to provide meaningful and useful soil C predictions, uncertainties in model outputs should always be quantified.
- Whatever type of models are used to meet future challenges, it is important that they continue to be tested using appropriate data.
- As new uses of land are developed, models should continue to be validated and modified if necessary, so that they remain appropriate.
- It is extremely important that experimentalists/data curators are involved in the modeling process, as modelers need to know if analytical methods have changed over time, what quality control has been used on the data and how missing data has been addressed.
- Calibrated and validated models can be used by experimentalists to provide information on data acquisition and to develop new research hypothesis.
- Greenhouse gas (GHG) calculator tools are very valuable tools to translate scientific carbon models to the farming community by providing information on how to decrease GHG emissions, but to successfully establish a change in the daily practice it is recommended to combine these tools with other psychological interventions and communication strategies.
- By linking international, national and local policies, and action frameworks to the Paris Climate Agreement, governments can develop more comprehensive and robust approaches to climate change, food security, soil protection, sustainable land management, water management, and energy generation.
- There is often a difference in objectives between practitioners at various levels and policy makers with respect to priorities for resource and land management. This disconnect requires robust institutional support to encourage inclusivity in decision-making.

With good quality data and timely modifications, soil C models will be able to help meet the challenges of the future (Box 9.4).

REFERENCES

Abdalla, M., Saunders, M., Hastings, A., Williams, M., Smith, P., Osborne, B., et al., 2013. Simulating the impacts of land use in Northwest Europe on Net Ecosystem Exchange (NEE): the role of arable ecosystems, grasslands and forest plantations in climate change mitigation. Sci. Total Environ. 465, 325–336.

Abdalla, M., Hastings, A., Truu, J., Espenberg, M., Mander, Ü., Smith, P., 2016. Emissions of methane from northern peatlands: a review of management impacts and implications for future management options. Ecol. Evol. 6, 7080–7102.

Abrahamse, W., Steg, L., Vlek, C., Rothengatter, T., 2005. A review of intervention studies aimed at household energy conservation. J. Environ. Psychol. 25, 273–291.

Ajzen, I., 1991. The theory of planned behavior. Organ. Behav. Human Decision Process. 50, 179–211.

Albanito, F., Beringer, T., Corstanje, R., Poulter, B., Stephenson, A., Zawadzka, J., et al., 2016. Carbon implications of converting cropland to bioenergy crops or forest for climate mitigation: a global assessment. GCB-Bioenergy 8, 81–95.

Allison, S.D., Martiny, J.B.H., 2008. Resistance, resilience, and redundancy in microbial communities. Proc. Natl. Acad. Sci. U.S.A. 105 (Suppl. 1), 11512–11519.

Allison, S.D., Wallenstein, M.D., Bradford, M.A., 2010. Soil-carbon response to warming dependent on microbial physiology. Nat. Geosci. 3, 336–340.

Archontoulis, S.V., Huber, I., Miguez, F.E., Thorburn, P.J., Rogovska, N., Laird, D.A., 2016. A model for mechanistic and system assessments of biochar effects on soils and crops and trade-offs. GCB-Bioenergy 8, 1028–1045.

ASABE, 2006a. Agricultural Machinery Management Data. American Society of Agricultural and Biological Engineers Standard ASAE EP496.3. ASABE, St Joseph, MI, pp. 385–390.

ASABE, 2006b. Agricultural Machinery Management Data. American Society of Agricultural and Biological Engineers Standard ASAE EP496.3. ASABE, St Joseph, MI, USA, pp. 391–398.

Beckert, M.R., Smith, P., Lilly, A., Chapman, S.J., 2016. Soil and tree biomass carbon sequestration potential of silvopastoral and woodland-pasture systems in North East Scotland. Agroforestry Syst. 90, 371–383.

Bell, M., Jones, E., Smith, J., Smith, P., Yeluripati, J., Augustin, J., et al., 2012. Simulation of soil nitrogen, nitrous oxide emissions and mitigation scenarios at 3 European cropland sites using the ECOSSE model. Nutr. Cycling Agroecosyst. 92, 161–181.

Bodansky, D., Hoedl, S.A., Metcalf, G.E., Stavins, R.N., 2014. Facilitating Linkage of Heterogeneous Regional, National, and Sub-National Climate Policies Through a Future International Agreement. Harvard Project on Climate Agreements, Cambridge, MA.

Bouwman, A.F., Boumans, L.J.M., Batjes, N.H., 2002. Modeling global annual N_2O and NO emissions from fertilized fields. Global Biogeochem. Cycles 16, 1080.

Bradbury, N.J., Whitmore, A.P., Hart, P.B.S., Jenkinson, D.S., 1993. Modelling the fate of nitrogen in crop and soil in the years following application of ^{15}N-labelled fertilizer to winter wheat. J. Agric. Sci. 121, 363–379.

Campbell, E., Paustian, K., 2015. Current developments in soil organic matter modeling and the expansion of model applications: a review. Environ. Res. Lett. 10, 123004.

Carlson, K.M., Goodman, L.K., May-Tobin, C.C., 2015. Modeling relationships between water table depth and peat soil carbon loss in Southeast Asian plantations. Environ. Res. Lett. 10, 074006.

Casado-Asensio, J., Drutschinin, A., Corfee-Morlot, J., Campillo, G., 2016. Mainstreaming Adaptation in National Development Planning. OECD Development Co-operation Working Papers, No. 29, OECD Publishing, Paris.

Chan, S., Asselt, H., Hale, T., Abbott, K.W., Beisheim, M., Hoffmann, M., et al., 2015. Reinvigorating international climate policy: a comprehensive framework for effective nonstate action. Global Policy 6, 466–473.

Cheng, K., Ogle, S.M., Parton, W.J., Pan, G., 2014. Simulating greenhouse gas mitigation potentials for hinese croplands using the DAYCENT ecosystem model. Global Change Biol. 20, 948–962.

Cole, C.V., Flach, K., Lee, J., Sauerbeck, D., Stewart, B., 1993. Agricultural sources and sinks of carbon. Water Air Soil Pollut. 70, 111–122.

Coleman, K., Jenkinson, D.S., 1996. RothC-26.3 – a model the turnover of carbon in soil. In: Powlson, D.S., Smith, P., Smith, J.U. (Eds.), Evaluation of Soil Organic Matter Models Using Existing Long-Term Datasets. NATO ASI Series I. Springer, Berlin, pp. 237–246.

Comeau, L.P., 2016. Carbon Dioxide Fluxes and Soil Organic Matter Characteristics on an Intact Peat Swamp Forest, a Drained and Logged Forest on Peat, and a Peatland Oil Palm Plantation in Jambi, Sumatra, Indonesia (Ph.D. thesis). University of Aberdeen.

Davidson, E.A., Samanta, S., Caramori, S.S., Savage, K., 2012. The Dual Arrhenius and Michaelis–Menten kinetics model for decomposition of soil organic matter at hourly to seasonal time scales. Global Change Biol. 18, 371–384.

Del Grosso, S.J., Parton, W.J., Mosier, A.R., Hartman, M.D., Brenner, J., Ojima, D.S., et al., 2001. Simulated interaction of carbon dynamics and nitrogen trace gas fluxes using the DAYCENT model. In: Schaffer, M., Ma, L., Hansen, S. (Eds.), Modeling Carbon and Nitrogen Dynamics for Soil Management. CRC Press, Boca Raton, FL, pp. 303–332.

van Dijk, W.F.A., Lokhorst, A.M., Berendse, F., de Snoo, G.R., 2015. Collective agri-environment schemes: How can regional environmental cooperatives enhance farmers' intentions for agri-environment schemes? Land Use Pol. 42, 759−766.

van Dijk, W.F.A., Lokhorst, A.M., Berendse, F., de Snoo, G.R., 2016. Factors underlying farmers' intentions to perform unsubsidised agri-environmental measures. Land Use Pol. 59, 207−216.

Doetterl, S., Berhe, A.A., Nadeu, E., Wang, Z., Sommer, M., Fiener, P., 2016. Erosion, deposition and soil carbon: a review of process-level controls, experimental tools and models to address C cycling in dynamic landscapes. Earth-Sci. Rev. 154, 102−122.

Dommain, R., Couwenberg, J., Joosten, H., 2011. Development and carbon sequestration of tropical peat domes in south-east Asia: links to post-glacial sea-level changes and Holocene climate variability. Quat. Sci. Rev. 30, 999−1010.

Dondini, M., Hastings, A., Saiz, G., Jones, M.B., Smith, P., 2009. The Potential of *Miscanthus* to sequester carbon in soils: comparing field measurements in Carlow, Ireland to model predictions. GCB-Bioenergy 1, 413−425.

Ecoinvent Centre, 2007. Ecoinvent datav2.0. Ecoinvent reports No. 1e25, Swiss Centre for Life Cycle Inventories, Dübendorf, Switzerland.

Egashira, K., Han, J., Karim, A., 2003. Evaluation of long-term application of organic residues on accumulation of organic matter and improvement of soil chemical properties in a clay terrace soil of Bangladesh. J. Fac. Agric., Kyushu Univ. 48, 227−236.

Egashira, K., Han, J., Satake, N., Nagayama, T., Mian, M.J.A., Moslehuddin, A.Z.M., 2005. Field experiment on long-term application of chemical fertilizers and farmyard manure in floodplain soil of Bangladesh. J. Fac. Agric., Kyushu Univ. 50, 861−870.

FAOSTAT, 2016. Food and agriculture organization of the United Nations. <http://faostat3.fao.org/browse/D/FS/E> (accessed 05.10.16.).

Farmer, J., Matthews, R., Smith, J.U., Smith, P., Singh, B.K., 2011. Assessing existing peatland models for their applicability for modelling greenhouse gas emissions from tropical peat soils. Curr. Opin. Environ. Sustain. 3, 339−349.

Farmer, J., Matthews, R., Smith, P., Smith, J.U., 2014. The Tropical Peatland Plantation-Carbon Assessment Tool: estimating CO_2 emissions from tropical peat soils under plantations. Mitigation Adapt. Strategies Global Change 19, 863−885.

Frank, D., Reichstein, M., Bahn, M., Thonicke, K., Frank, D., Mahecha, M.D., et al., 2015. Effects of climate extremes on the terrestrial carbon cycle: concepts, processes and potential future impacts. Global Change Biol. 21, 2861−2880.

Franzluebbers, A.J., Stuedemann, J.A., Schomberg, H.H., Wilkinson, S.R., 2000. Soil organic C and N pools under long-term pasture management in the Southern Piedmont USA. Soil Biol. Biochem. 32, 469−478.

Freibauer, A., Rounsevell, M.D.A., Smith, P., Verhagen, J., 2004. Carbon sequestration in the agricultural soils of Europe. Geoderma 122, 1−23.

Gardner, G.T., Stern, P.C., 2002. Environmental Problems and Human Behavior, second ed. Pearson Custom Publishing, Boston, MA.

Guo, L.B., Gifford, R.M., 2002. Soil carbon stocks and land use change: a meta analysis. Global Change Biol., 8, 345−360.

van Groenigen, K.J., Hastings, A., Forristal, D., Roth, B., Jones, M., Smith, P., 2011. Soil C storage as affected by tillage and straw management: an assessment using field measurements and model predictions. Agric. Ecosyst. Environ. 140, 218−225.

Hastings, A.F., Wattenbach, M., Eugster, W., Li, C., Buchmann, N., Smith, P., 2010. Uncertainty propagation in soil greenhouse gas emission models: an experiment using the DNDC model and at the Oensingen cropland site. Agric. Ecosyst. Environ. 136, 97−110.

Hawkes, C.V., Kivlin, S.N., Rocca, J.D., Huguet, V., Thomsen, M.A., Suttle, K.B., 2011. Fungal community responses to precipitation. Global Change Biol. 17, 1637−1645.

Herrero, M., Havlík, P., Valin, H., Notenbaert, A., Rufino, M.C., Thornton, P.K., et al., 2013. Biomass use, production, feed efficiencies and greenhouse gas emissions from global livestock systems. PNAS 110, 20888−20893.

Hillier, J., Walter, C., Malin, D., Garcia-Suarez, T., Mila-i-Canals, L., Smith, P., 2011. A farm-focused calculator for emissions from crop and livestock production. Environ. Model. Softw. 9, 1070−1078.

Hillier, J., Abdalla, M., Bellarby, J., Albanito, F., Datta, A., Dondini, M., et al., 2016. Mathematical modeling of greenhouse gas emissions from agriculture for different end users. In: Del Grosso, S., Ahuja, L., Parton, W. (Eds.), Synthesis and Modeling of Greenhouse Gas Emissions and Carbon Storage in Agricultural and Forest Systems to Guide Mitigation and Adaptation. Advances in Agricultural Systems Modeling 6. ASA, CSSA, and SSSA, Madison, WI, pp. 197−228.

Hooijer, A., Page, S., Jauhiainen, J., Lee, W.A., Lu, X.X., Idris, A., et al., 2012. Subsidence and carbon loss in drained tropical peatlands. Biogeosciences 9, 1053−1071.

IPCC, 2006. Agriculture, forestry and other land use. In: Eggleston, H.S., Buendia, L., Miwa, K., Ngara, T., Tanabe, K. (Eds.), IPCC Guidelines for National Greenhouse Gas Inventories. National Greenhouse Gas Inventories Programme, vol. 4. IGES, Japan.

IPCC, 2014. Climate Change 2014−Impacts, Adaptation and Vulnerability: Regional Aspects. Cambridge University Press, Cambridge.

Karim, A., Egashira, K., Yamada, Y., Haider, J., Nahar, K., 1995. Long-term application of organic residues to improve soil properties and to increase crop yield in terrace soil of Bangladesh. J. Fac. Agric., Kyushu Univ. 39, 149−165.

Keesstra, S.D., Bouma, J., Wallinga, J., Tittonell, P., Smith, P., Cerdà, A., et al., 2016. The significance of soils and soil science towards realization of the UN sustainable development goals. Soil 2, 111−128.

Knutti, R., Allen, M.R., Friedlingstein, P., Gregory, J.M., Hegerl, G.C., Meehl, G.A., et al., 2008. Review of uncertainties in global temperature projections over the twenty-first century. J. Clim. 21, 2651−2663.

Kurinato, S., Warren, M., Talbot, J., Kauffman, B., Murdiyarso, D., Frolking, S., 2015. Carbon accumulation of tropical peatlands over millennia: a modeling approach. Global Change Biol. 21, 431−444.

Lawson, D.A., Tabor, J.H., 2001. Stopping distances: an excellent example of empirical modelling. Teach. Math. Appl. 20, 66−74.

Lal, R., 2003. Soil erosion and the global carbon budge. Environ. Int. 29, 437−450.

Lal, R., 2004. Soil carbon sequestration impacts on global climate change and food security. Science 304, 1623−1627.

Leith, H., 1972. Modelling the primary productivity of the world. Nature and Resources, UNESCO VIII 2, 5−10.

Li, C., Frolking, S., Frolking, T.A., 1992. A model of nitrous oxide evolution from soil driven by rainfall events: 1. Model structure and sensitivity. J. Geophys. Res. 97 (D9), 9759−9776.

Li, C.S., 2000. Modeling trace gas emissions from agricultural ecosystems. Nutr. Cycling Agroecosyst. 58, 259−276.

Li Liu, D., Chan, K.Y., Conyers, M.K., 2009. Simulation of soil organic carbon under different tillage and stubble management practices using the Rothamsted carbon model. Soil Tillage Res. 104, 65−73.

Lokhorst, A.M., van Dijk, J., Staats, H., van Dijk, E., de Snoo, G., 2010. Using tailored information and public commitment to improve the environmental quality of farm lands: an example from the Netherlands. Hum. Ecol. 38, 113−122.

Lugato, E., Paustian, K., Panagod, P., Jones, A., Borelli, P., 2016. Quantifying the erosion effect on current carbon budget of European agricultural soils at high spatial resolution. Global Change Biol. 22, 1976−1984.

Luo, Y., Ahlström, A., Allison, S.D., Batjes, N.H., Brovkin, V., Carvalhais, N., et al., 2016. Toward more realistic projections of soil carbon dynamics by Earth system models. Global Biogeochem. Cycles 30, 40−56.

Lychuk, T.E., Izaurralde, R.C., Hill, R.L., McGill, W.B., Williams, J.R., 2015. Biochar as a global change adaptation: predicting biochar impacts on crop productivity and soil quality for a tropical soil with the Environmental Policy Integrated Climate (EPIC) model. Mitigation Adapt. Strategies Global Change 20, 1437−1458.

Manlay, R.J., Feller, C., Swift, M.J., 2007. Historical evolution of soil organic matter concepts and their relationships with the fertility and sustainability of cropping systems. Agric. Ecosyst. Environ. 119, 217−233.

McClean, G.J., Rowe, R.L., Heal, K.V., Cross, A., Bending, G.D., Sohi, S.P., 2015. An empirical model approach for assessing soil organic carbon stock changes following biomass crop establishment in Britain. Biomass Bioenergy 83, 141−151.

Müller-Nedebock, D., Chaplot, V., 2015. Soil carbon losses by sheet erosion: a potentially critical contribution to the global carbon cycle. Earth Surf. Process. Landforms 40, 1803−1813.

Ogle, S.M., Breidt, F.J., Paustian, K., 2005. Agricultural management impacts on soil organic carbon storage under moist and dry climatic conditions of temperature and tropical regions. Biogeochemistry 72, 87−121.

Pagel, H., Poll, C., Ingwersen, J., Kandeler, E., Streck, T., 2016. Modeling coupled pesticide degradation and organic matter turnover: from gene abundance to process rates. Soil Biol. Biochem. 103, 349−364.

Pandya, U., Saraf, M., 2010a. Application of fungi as a biocontrol agent and their biofertilizer potential in agriculture. J. Adv. Dev. Res. 1, 90−99.

Pandya, U., Saraf, M., 2010b. Role of single fungal isolates and consortia as plant growth promoters under saline conditions. Res. J. Biotechnol. 5, 5−9.

Parton, W.J., 1998. The CENTURY model. In: Powlson, D.S., Smith, P., Smith, J.H. (Eds.), Evaluation of Soil Organic Matter Models. Springer Berlin Heidelberg, pp. 283−291.

Parton, W.J., Hartman, M.D., Ojima, D.S., Schimel, D.S., 1998. DAYCENT and its land surface submodel: description and testing. Global Planet. Change 19, 35−48.

Paustian, K., 1994. Modelling soil biology and biochemical processes for sustainable agriculture research. In: Pankhurst, C.E., Doube, B.M., Gupta, V.V.S.R., Grace, P.R. (Eds.), Soil Biota: Management in Sustainable Farming Systems. CSIRO, Australia, pp. 182−193.

Pogson, M., Richards, M., Dondini, M., Jones, E.O., Hastings, A., Smith, P., 2016. ELUM: a spatial modelling tool to predict soil greenhouse gas changes from land conversion to bioenergy in the UK. Environ. Model. Softw. 84, 458−466.

Powlson, D.S., Gregory, P.J., Whalley, W.R., Quinton, J.N., Hopkins, D.W., Whitmore, A.P., et al., 2011. Soil management in relation to sustainable agriculture and ecosystem services. Food Policy 36, S72−S87.

Reichstein, M., Bahn, M., Ciais, P., Frank, D., Mahecha, M.D., Seneviratne, S.I., et al., 2013. Climate extremes and the carbon cycle. Nature 500, 287−295.

Reid, J.B., 2002. Yield response to nutrient supply across a wide range of conditions 1. Model derivation. Field Crops Res. 77, 161−171.

Renard, K.G., 1997. Predicting Soil Erosion by Water: A Guide to Conservation Planning with the Revised Universal Loss Soil Equation (RUSLE). USDA-ARS, Washington, DC.

Richards, M., Pogson, M., Dondini, M., Jones, E.O., Hastings, A., Henner, D.N., et al., 2016. High-resolution spatial modelling of greenhouse gas emissions from land-use change to energy crops in the United Kingdom. GCB-Bioenergy. Available from: https://doi.org/10.1111/gcbb.12360.

Sibbald, A.R., Eason, W.R., Mcadam, J.H., Hislop, A.M., 2001. The establishment phase of a silvopastoral national network experiment in the UK. Agroforestry Syst. 53, 39−53.

Smith, J., Abegaz, A., Matthews, R., Subedi, M., Orskov, R., Tumwesige, V., et al., 2014. What is the potential for biogas digesters to improve soil carbon sequestration in Sub-Saharan Africa? Biomass Bioenergy 70, 73−86.

Smith, J.U., Gottschalk, P., Bellarby, J., Chapman, S., Lilly, A., Towers, W., et al., 2010. Estimating changes in national soil carbon stocks using ECOSSE — a new model that includes upland organic soils. Part I. Model description and uncertainty in national scale simulations of Scotland. Clim. Res. 45, 179—192.

Smith, P., 2016. Soil carbon sequestration and biochar as negative emission technologies. Global Change Biol. 22, 1315—1324. Available from: https://doi.org/10.1111/gcb.13178.

Smith, P., Powlson, D.S., Glendining, M.J., Smith, J.U., 1997. Potential for carbon sequestration in European soils: preliminary estimates for five scenarios using results from long-term experiments. Global Change Biol. 3, 67—79.

Smith, P., Smith, J., Flynn, H., Killham, K., Rangel-Castro, I., Foereid, B., et al., 2007. ECOSSE: Estimating Carbon in Organic Soils-Sequestration and Emissions: Final Report.

Smith, P., Martino, D., Cai, Z., Gwary, D., Janzen, H., Kumar, P., et al., 2008. Greenhouse gas mitigation in agriculture. Phil. Trans. R. Soc. Lond. Series B, Biol. Sci. 363 (1492), 789—813.

Smith, P., Ashmore, M., Black, H., Burgess, P.J., Evans, C., Quine, T., et al., 2013. The role of ecosystems and their management in regulating climate, and soil, water and air quality. J. Appl. Ecol. 50, 812—829.

Smith, P., Cotrufo, M.F., Rumpel, C., Paustian, K., Kuikman, P.J., Elliott, J.A., et al., 2015. Biogeochemical cycles and biodiversity as key drivers of ecosystem services provided by soils. Soil 1, 665—685.

Smith, P., House, J.I., Bustamante, M., Sobocká, J., Harper, R., Pan, G., et al., 2016a. Global change pressures on soils from land use and management. Global Change Biol. 22, 1008—1028.

Stern, P.C., 2011. Contributions of psychology to limiting climate change. Am. Psychol. 66, 303—314.

Todd-Brown, K.E.O., Hopkins, F.M., Kivlin, S.N., Talbot, J.M., Allison, S.D., 2012. A framework for representing microbial decomposition in coupled climate models. Biogeochemistry 109, 19—33.

UNFCCC, 2015. Framework convention on climate change. <http://unfccc.int/paris_agreement/items/9485. php> (accessed 28.11.16.).

United Nations, 2015. International year of soils. <http://www.fao.org/soils-2015/en/> (accessed 28.11.16.).

United Nations Development Programme, 2015. Sustainable development goals. <http://www.undp.org/content/undp/en/home/sustainable-development-goals.html> (accessed 28.11.16.).

Vleeshouwers, L.M., Verhagen, A., 2002. Carbon emissions and sequestration by agricultural land use: a model study for Europe. Global Change Biol. 8, 519—530.

van der Weerden, T.J., Kelliher, F.M., de Klein, C.A.M., 2012. Influence of pore size distribution and soil water content on nitrous oxide emissions. Soil Res. 50, 125—135.

Whittaker, C., McManus, M.C., Smith, P., 2013. A comparison of carbon accounting tools for arable crops in the United Kingdom. Environ. Model. Softw. 46, 228—239.

Wieder, W.R., Bonan, G.B., Allison, S.D., 2013. Global soil carbon projections are improved by modelling microbial processes. Nat. Clim. Change 3, 909—912.

Wieder, W.R., Allison, S.D., Davidson, E.A., Georgiou, K., Hararuk, O., He, Y., et al., 2015. Explicitly representing soil microbial processes in Earth system models. Global Biogeochem. Cycles 29, 1782—1800.

Williams, J.R., 1990. The erosion productivity impact calculator (EPIC) model: a case history. Phil. Trans. R. Soc. Lond. 329, 421—428.

Wischmeier, W.H., Smith, D.D., 1978. Predicting Rainfall Erosion Losses — A Guide Planning. U.S. Department of Agriculture, Agriculture Handbook No. 537.

Woolf, D., Lehmann, J., 2012. Modelling the long-term response to positive and negative priming of soil organic carbon by black carbon. Biogeochemistry 111, 83—95.

Zaks, D.P.M., Ramankutty, N., Barford, C.C., Foley, J.A., 2007. From Miami to Madison: investigating the relationship between climate and terrestrial net primary production. Global Biogeochem. Cycles 21, GB3004.

FURTHER READING

Abdalla, M., Jones, M., Yeluripati, J., Smith, P., Burke, J., Williams, M., 2010. Testing DayCent and DNDC model simulations of N_2O fluxes and assessing the impacts of climate change on the gas flux and biomass production from a humid pasture. Atmos. Environ. 44, 2961–2970.

Herrero, M., Henderson, B., Havlík, P., Thornton, P.K., Conant, R.T., Smith, P., et al., 2016. Greenhouse gas mitigation potentials in the livestock sector. Nat. Clim. Change 6, 452–461.

Hollis, J.M., 2008. Unpublished internal report prepared by Cranfield University for the Defra PS2225 project.

Hooijer, A., Page, S., Canadell, J.G., Silvius, M., Kwadijk, J., Wösten, J.H.M., et al., 2010. Current and future CO_2 emissions from drained peatlands in Southeast Asia. Biogeosciences 7, 1505–1514.

Nijbroek, R.P., Andelman, S.J., 2016. Regional suitability for agricultural intensification: a spatial analysis of the Southern Agricultural Growth Corridor of Tanzania. Int. J. Agric. Sustain. 14 (2), 231–247.

Smith, P., Davis, S.J., Creutzig, F., Fuss, S., Minx, J., Gabrielle, B., et al., 2016b. Biophysical and economic limits to negative CO_2 emissions. Nat. Clim. Change 6, 42–50.

Tilman, D., Clark, M., 2015. Food, agriculture & the environment: can we feed the world & save the earth? Daedalus 144, 8–23.

Watson, J.E., Iwamura, T., Butt, N., 2013. Mapping vulnerability and conservation adaptation strategies under climate change. Nat. Clim. Change 3, 989–994.

Index

Printed in the United States
By Bookmasters